Handbook of Fiber Optics

Handbook of Fiber Optics
Theory and Applications

Chai Yeh

Department of Electrical Engineering and Computer Science
The University of Michigan
Ann Arbor, Michigan

Academic Press, Inc.
Harcourt Brace Jovanovich, Publishers
San Diego   New York   Boston
London   Sydney   Tokyo   Toronto

Copyright © 1990 by Academic Press, Inc.
All Rights Reserved.
No part of this publication may be reproduced or transmitted in any form or by any means, electronic or mechanical, including photocopy, recording, or any information storage and retrieval system, without permission in writing from the publisher.

Academic Press, Inc.
San Diego, California 92101

United Kingdom Edition published by
Academic Press Limited
24–28 Oval Road, London NW1 7DX

Library of Congress Cataloging-in-Publication Data

Yeh, Chai.
    Handbook of fiber optics : theory and applications / by Chai Yeh.
        p.   cm.
    "February 1989."
    Includes bibliographical references.
    ISBN   0-12-770455-8   (alk. paper)
    1. Fiber optics. 2. Telecommunication systems. 3. Optical communications.   I. Title.
    TA1800.Y44   1990
    621.36'92--dc20   89-17501
                                                              CIP

Printed in the United States of America
90 91 92 93   9 8 7 6 5 4 3 2 1

To the memory of my parents
Yunquing Yeh and Ayehwa Ho

Contents

3. Wave Propagation in Lightguides

4. Auxiliary Components for Optical Fiber Systems

5. Optical Fiber Measurements

PART TWO   Semiconductor Light Sources and Detectors

6. Semiconductor Light Sources: Light-Emitting Diodes and Injection Lasers

Foreword

Fiber optics is optical technology's nova. It existed for many years as a novel and respectable branch of optics, with such products as fiber optics probes and face plates. One would have expected it to remain in this condition. Then, the unexpected happened, with Corning's development of fibers with extremely low attenuation, on the order of decibels per kilometer. This development opened the door to a new and vastly enlarged domain—telecommunications. Wideband communication signals could now be transported on light beams conducted by optical fibers instead of on an electrical transmission line. The potential annual dollar volume jumped from millions to billions. With this development, back in the 1960s, came exuberant predictions about the impact. To some they sounded exorbitant; certain wildly optimistic claims about technological developments had been heard many times in the past, only to be forgotten. But fiber optics was different. Within a short time, some of the most astonishing predictions began to materialize right on schedule. Indeed, optical fibers shortly began to supplement underground cables in telephone systems, and military aircraft became wired with optical fibers.

Such on-time performance is rare in technology. Even in those instances where true revolution is impending, there is usually an unanticipated time lag.

Fiber optics and lasers have a close affinity; each has enhanced the importance of the other. The long-distance transmission capability of the fiber would be of little practical importance without the laser to provide highly concentrated and spectrally narrow light. Thus, the laser makes fiber optics more useful, and at the same time fiber optics offers another major application for the laser.

Lasers of course would be enormously important even without fiber optics since they nowadays have so many applications. But it was not always so. As with fiber optics, the invention of the laser brought immediately the realization that it was surely destined to have far reaching impact. But unlike low attenuation optical fibers, the utilization was slow to arrive. For many years the laser was a solution in search of an application. Everyone knew that great things would come, but it was hard to fathom just what they would be. Holography became a solution looking for an application. Another major anticipated laser application was communication, which was then understood to involve free space propagation, an idea which has only limited success. We now realize that communication is indeed one of the major applications destined for the laser, but with the laser beams propagating in fibers, not free space. It required fiber optics to bring this early proposed laser application to fruition.

Low attenuation fiber optics is thus alone among the major recently emerging optical technologies to fulfill with reasonable promptness its early promise. In partnership with the laser, the fiber has seen the early applications expand and a plethora of new ones develop.

Professor Yeh has spent a long, productive career working in the areas of microwave semiconductor devices and electromagnetic wave theory. His vast experience eminently qualifies him as an author in this important area of optical fibers and he has come up with a volume that will become a widely useful reference.

Emmett N. Leith
Department of Electrical Engineering
and Computer Science
University of Michigan

Preface

The science of optical fibers is a fascinating field. Not too long ago, very few people would believe that one could transmit telephone conversations over a single glass fiber that outperformed a pair of copper wires or a coaxial cable in transmission properties. Now, people are anticipating that a significant changeover from the existing wires and coaxial cables to optical fibers in telecommunications and other applications will occur within this decade. Indeed, the art of optical fiber technology is developing rapidly and is beginning to mature and reach a fully commercial stage. The time is ripe to consider training our young scientists and engineers to develop this expanding field.

This book is intended to provide state-of-the-art information on many aspects of fiber optics. It contains four parts. After an introductory remark (Chapter 1) that reviews the advantages of optical fibers in telecommunication applications, the historical development, and future trends, we begin Part One, "The Optical Fibers," consisting of four chapters (Chapters 2–5). In Chapter 2, optical fibers are classified and characterized. Fabrication of silica fibers is described in detail. In Chapter 3, the transmission properties, including the loss mechanism, and the dispersion properties of optical fibers are discussed. Auxiliary components for optical fiber systems such as connectors, splices, switches, and couplers are discussed in Chapter 4. Chapter 5 describes the measurements of some important fiber parameters. Part Two (Chapters 6 and 7) deals with semiconductor light sources and detectors for optical fiber communication systems. Chapter 6 covers light-emitting diodes (LEDs) and injection lasers. Chapter 7 discusses

PIN diodes (P-type intrinsic N-type junction diode), avalanche photodiodes (APD), and photoconductor diodes (PCD). Part Three (Chapters 8 and 9) is devoted to optical fiber communication systems.
Chapter 8 concerns the design and implementation of optical fiber
communication systems. Chapter 9 highlights recent activities in optical fiber communication system development around the world.
This includes the newly developed local-area networks (LANs), metropolitan-area networks (MANs), and integrated-service-digital networks (ISDNs). The last part of the book (Chapters 10–13) is devoted
to optical fiber sensors and other applications. Chapter 10 introduces
the physical phenomena useful in optical fiber sensors. It is followed
by a description of some examples of pure-fiber and remote-fiber sensors in Chapter 11. Chapter 12 describes other applications of fiber
optics in industry and medicine. The book concludes with a short
introduction to the integrated optics useful in fiber optics in general
and optical fiber telecommunication applications in particular. Tech-
nically, the advantages of optical fibers in telecommunication and
data transmission applications over the existing copper-wire pairs,
coaxial lines, and waveguides are overwhelmingly recognized. These
advantages include vastly increased bandwidth, low line losses, increased channel capacity, and immunity to electromagnetic interference. Economically, it is equally convincing that the lighter weight
and smaller size of optical fibers offer even lower installation and operating costs over existing wired connections. We are witnessing a
changeover of the long-haul transmission system from coaxial (submarine) cables to optical fiber cables in every part of the world in
rapid succession. With the development of LANs, MANs, and ISDNs,
we are anticipating that complete metropolitan telephone and data
transmission services will be replaced by optical fiber systems in the
near future.

The presentation of this book is geared to practicing engineers and
scientists as well as technicians with some college-level training in
fundamental science and mathematics and who need just a little re-
freshing to stimulate their interest in engaging this task of new development. In mentioning a device or a component, the objective of using this device or component is first identified. A brief description of
the working principles is given, more or less as physical interpretations, sometimes with the help of a few simple equations if necessary.

Although optical fiber technology originally emerged as a major in-
novation in telecommunications, its applications are developing and
expanding rapidly into practically every branch of science and engineering. Even in medical science, the use of optical fiber sensing
shows great success. Control engineering is excited by the fact that
optical fiber sensors offer many advantages over existing devices in
dealing with hostile environments. Optical fiber sensors offer light
weight, small size, simple implementation, and low cost. The future
of fiber optics is very bright indeed.

But no derivations of equations are included. Then a guideline is given for developing these devices or components to achieve desired objectives. The results are displayed in either graphs or tabulated form for easy identification and quick comparison. Examples of calculations are given as often as deemed necessary. Ample references are cited in each chapter for readers who are interested in further investigation.

I hope this book will achieve the following objectives:

1. To provide state-of-the-art information in the field of fiber optics
2. To cover a broader field than other existing books of a similar nature (optical fiber sensor is a new addition)
3. To stimulate better understanding of the new technology
4. To make use of up-to-date information effectively

Acknowledgments

I wish to thank Professor Emmett N. Leith at the University of Michigan and Professor Yin Yeh at the University of California, Davis, for their valuable comments and suggestions during the development of the book. Thanks are due to Miss Deborah J. Yeh for her meticulous editing of the manuscript. I also wish to express my appreciation for the encouragement and cooperation I received from my colleagues and staff of the department of Electrical Engineering and Computer Science at the University of Michigan during the preparation of the book. Finally, I thank my wife, Ida Shuyen Chiang, for her patience and continued encouragement.

Chai Yeh

1

Introduction

General Introduction

One of the most interesting developments in recent years in the field of telecommunication and data transmission systems is the use of optical fibers to carry information in a way similar to that employing radiowaves and microwaves. The impetus responsible for this development is partly technical and partly economical. Technologically, significant advances have been achieved in the past decade to prove that lightwave transmission over optical fibers is far superior in performance than that which can be obtained over wires and microwave links. Typically, optical fiber can support a wider transmission bandwidth (0.1–1000 GHz), has lower transmission loss per unit length (0.15–5 dB/km), and is impervious to electromagnetic interference. Economically, the increasing cost and demand for high data rate or large bandwidth per transmission channel and the lack of available space in congested conducts in every metropolitan area for telephone and data transmission provide an atmosphere to favor a new and less costly system. The optical fiber system fulfills both demands. A typical cabled fiber weighs about 3 kg/km, costs under $500/km or less, and yields a longer repeater spacing up to 100 km or more at a data rate of at least several hundred megabits per second. The economical advantage of an optical fiber system over wired and coaxial links becomes overwhelming.

A typical optical fiber communication system employs the elementary setup shown in Fig. 1.1. The inputs to the system may consist of data, video, and telephone conversations, or any combination of these signals.

1

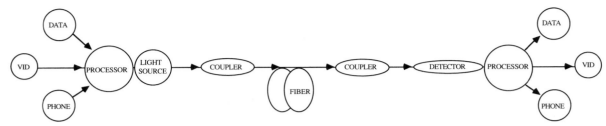

Figure 1.1 A typical optical fiber telecommunication system.

They are usually coded and multiplexed corresponding to a spectrum including the infrared and visible wavelength. Even 0.1 percent of the carrier frequency can provide a bandwidth greater than 1000 GHz. Within this bandwidth, the entire broadcase spectrum, including television, telephone conversations, and many data links, can be multiplexed in the processor. The modulated lightwave is then launched into the optical fiber through a coupler. A length of optical fiber serves as a transmission line that guides the message-carrying lightwave to the receiving terminal. The transmission of the lightwave along the fiber suffers transmission loss and dispersion as in any other transmission medium. When the lightwave finally reaches the receiving end, it is coupled into a photodetector to convert its information back into electrical signal, which is then amplified and reshaped, if necessary, and directed to the appropriate channels through another processor for local reception.

This description represents the initial phase of the optical fiber applications to telecommunications. As the technology matures, the overwhelming advantage of using optical fibers over the wired pairs and coaxial cables is recognized. Wider applications of fibers are being developed for large-scale, high-data-rate communication systems. These systems include the local-area networks (LANs), metropolitan-area networks (MANs), and integrated-service digital networks (ISDNs).

Still another new noncommunication application of optical fiber developed immediately. This is the use of optical fibers as sensors. The practical area of application includes the detection of energies of many forms (chemical, electrical, mechanical, magnetic, and thermal energies) either with or without the use of transducers, and with sensitivity and versatility superior to those of conventional design. The ability of optical fiber sensors to operate under most hostile environmental conditions and the potential for very low cost, lightweight, small integrated sensor units are all valuable assets of optical fiber sensors.

Image-transferring systems represent another rapidly growing application of the fiber optics. Using flexible fiber bundle and rigid faceplate, simple and very inexpensive copying machine and display systems can be built.

Historical Developments

Historically, the idea of using lightwaves for signaling was tested by Alexander Graham Bell as early as 1880. In fact, he was granted a patent entitled the "photophone." In his original invention, sunlight was modulated by means of a comb-like grid attached to a flexible diaphragm. Voice caused the grid to move with respect to a similar fixed grid. This caused the light to vary in intensity. At the receiving end, a selenium detector was used to translate these variations into electric current to recreate the speech through a telephone receiver. The invention did not yield a commercial success for two reasons: (1) the medium for lightwave transmission was air, which is very lossy; and (2) sunlight is an incoherent light source that was very inefficient for this application. But the potential use of lightwaves for communication remained attractive in many minds.

During World War II, as most developments in radio communication were shifted into radar works, lightwave research was temporarily suspended. Efforts were diverted to developing microwave and millimeter wave for military applications. In the 30–300-GHz range of the spectrum, a circular hollow metallic waveguide was found to have the property that for a particular mode, the TE-01 mode, the attenuation of the transmitted electromagnetic radiation in the guide decreases as the frequency increases. A metallic waveguide, 5 cm in diameter, was actually built for a 50-GHz wave, and the reported loss between repeaters (15–30 km) was about 2 dB/km. Unfortunately, the cost was far too expensive for commercial applications. Besides, when compared to the use of lightwaves, the available bandwidth of the millimeter wave is far too narrow for future expansion. However, the earlier fiber optic systems were conceived by people in the millimeter-wave business. For a time, most researchers viewed a fiber as essentially a lightwave pipe in which electromagnetic energy is propagating by multireflections from the walls of the pipe.

After the invention of the laser (1960), lightwave research resumed vigorously. Researchers on laser light transmission soon found out that on a rainy or foggy day, the transmission may suffer a 300-dB/km loss in open atmosphere. This lead to the suggestion to enclose the light beam in a tube. However, a light beam tends to spread as it travels down the tube. Lens-train has to be used to confine the beam by focusing it axially. Gas-filled pipes were also suggested and built to improve beam confinement. It was also found that to minimize the beam spread, lenses must be placed confocally. This means a large number of lenses must be built into the pipe. The mounting cost of construction, the increased absorption loss of the lenses, and the increased alignment difficulty made this design impractical, and it was soon abandoned.

In the meantime, electromagnetic theory of microwave dielectric rods was developed. Extending the theory to include glass fibers of smaller diameter followed. Bundles of glass fibers were fabricated and experimented. In earlier fibers, a loss over 1000 dB/km was reported. It was estimated that for glass fibers to be economical for communication applications, a loss below 20 dB/km would be necessary.

Eventually, two important developments exerted major influence on the development of optical fibers. The first was the discovery of low-loss glass fiber for supporting the lightwave transmission. Kao, Hockman, and co-workers in Britain (1966) [1] proposed that silica-based glass fibers could be made to transmit lightwaves with losses less than 20 dB/km. Corning Glass Works succeeded in fabricating such low-loss fibers in 1970. The second was the invention of continuous wave (CW), room-temperature diode lasers (1970) that provided coherent light sources so desperately needed to make possible radio-like telecommunications on optical fibers. From there on, intensive research aimed at fundamental understanding of the loss mechanisms and the dispersion properties of glass fibers pressed forward. Today, laboratory samples of fibers having a loss of 0.15 dB/km at 1.55 μm have been reported.

The Optical Fibers

Optical fibers are dielectric waveguides. They are designed to guide lightwaves along their length. For easy fabrication and implementation, silica preform is usually drawn into fibers of circular cross section. In general, optical fibers consist of a cylindrical core with a refractive index n_1, surrounded by a cladding with an index n_2, with $n_1 > n_2$. If both n_1 and n_2 are uniform across the cross sections, the fiber is known as a *step-index fiber* (SI). If n_1 varies with the core radius, it is a *graded-index fiber* (GI). Also, if the core diameter is small, say, within 10 μm, the fiber is a single mode (SMF). If the core diameter is greater than 10 μm, it is usually a multimode fiber (MMF). A more detailed classification of the fibers will be given in Chapter 2.

The guiding properties of optical fibers have been studied extensively either by the electromagnetic theory [2] or by the geometric optical theory [3] or both [4]. We shall adopt these results freely without derivation. It suffices here to say that lightwaves can be confined to the core of the fiber if $n_1 > n_2$ and the light source is entering the fiber and at an angle close to its axis.

Just like other transmission mediums, optical fiber suffers power loss and signal dispersion when signal is propagating along the fiber length. It is the extremely low loss of the fiber and the possibility of zero dispersion at certain wavelengths that promote us to adopt the fiber for replacing the existing wired or piped signal transmission system in the immediate future.

A planar lightguide has also been developed for integrated optics, which consists of a narrow strip of optical fiber guide material diffused into an appropriate substrate.

Present Optical Fiber Communication Systems

Commercial optical fiber communication systems operating at wavelengths from 0.8 to 0.9 μm have been in existence since 1977 [5]. Hundreds of MMF links for interoffice trunking, subscriber's loop, and cable TV using MMF have been installed in the United States, Japan, the United Kingdom, and France since 1980. Most of these systems use MMF with transmission losses of 4−6 dB/km, including installation losses. Typical bandwidths are several hundred megahertz-kilometers. They generally use (AlGaAs/GaAs) light-emitting diodes (LEDs) and laser diodes (LDs) as light sources and use silicon PIN (p−i−n diode) and avalanche photodiodes (APDs) as detectors. The operating speed of these systems is from 10 to 100 Mbit/s with repeater spacings of 5−10 km. These systems are generally referred to as *first-generation systems*.

Second-generation systems use wavelengths around 1.3 μm to take advantage of lower fiber loss and zero dispersion. The application of these systems can be divided into two categories: (1) small- to moderate-capacity systems, using MMF with LEDs or multimode lasers as light sources and (2) large-capacity long-haul systems using SMF with conventional lasers as light sources. In the first category, the distance involved is short, such that one does not have to consider repeater spacings. Also, as the systems operate in the wavelength region of minimum dispersion, the spectral properties required for light sources can be relaxed. LEDs can be used advantageously because they are relatively cheap and have long life. Germanium (Ge) PIN photodiode and APDs can be used as detectors. The operating and installation cost can be low; thus, second-generation systems could become the prevalent systems in lightwave communications. In the second category, repeater spacing and system reliability become important considerations. Single-mode fiber and more sophisticated laser light sources can be advantageously utilized. The United States, the United Kingdom, and Japan all have systems in this wavelength range in commercial operation.

Future Outlook of Optical Fiber Technologies

For a telecommunication system using optical fiber technology, the trend is to use longer wavelengths around 1.55 μm to take full advantage of the ultralow fiber loss at this wavelength. This is the *third-generation system,* or the future optical fiber system of communications. Single-mode fiber is used for wideband or high-speed applications. The problem of spectral broadening becomes very acute.

The future of the third-generation system will depend on the development of stable, low-spectral-width lasers with adequately long life and the availability of low-dispersion SMF operating within this wavelength range.

A great amount of research work is being carried out in all parts of the world to promote optical fiber technology. In general, these activities can be grouped in several categories.

Fiber Materials

Silica fiber is the essential building block of optical fiber communication systems. Earlier researchers have concentrated on improving the purity of silica fiber and have achieved fiber losses as low as 0.14 dB/km at a wavelength of 1.55 μm. A dispersion minimum at 1.3 μm has been located and can be shifted to longer wavelengths by adjusting the design and doping. While research continues on purifying silica fibers, work has begun on new materials that will be capable of even lower losses and wavelength beyond 2 μm. Intrinsic attenuation below 0.1 dB/km between 2 and 3 μm and a minimum dispersion at 1.7 μm has been reported for a Sb_2O_3-doped GeO_2 fiber. Other materials under investigation include fluoride glasses and other heavy-metal halides that may have theoretical minima as low as 3×10^{-4} dB/km occur at 2–12 μm, and calcogenide glasses prepared by the zone-refining technique estimated to be below 10^{-2} dB/km between 3 and 35 μm have been reported. The door is wide open for imaginative researchers.

Light Sources and Detectors

To take advantage of low fiber loss and zero dispersion at longer wavelengths, the trend in optical fiber communication systems has been steadily shifting toward longer wavelengths [6]. Development of adequate light sources and detectors must keep pace. Highly stable, single-frequency (spectral width < 1 Å) low threshold current, CW, room temperature, and long-lived lasers become the goals of the search [7]. A GaAsSb/AlGaAsSb laser has been reported in the 1.7-μm range. New laser materials that emit light at a wavelength ranging from 7 to 15 μm will be needed for new fibers. Detectors with high sensitivity and low noise are badly needed. Most of the commercially available detectors for this wavelength range are manufactured using Ge as base materials. Compound semiconductor materials such as InGaAsP/InP are currently under intensive investigation.

The interest in the organic-to-inorganic semiconductor heterojunctions is on the rise. These thin-films can be used as building blocks for varieties of fiber optic devices suitable for high-bandwidth, high-speed applications [8].

Coherent detection is another addition to the detection technique in recent years. With narrow spectral-width lasers and improved laser stability, true radiowave-like coherent detection becomes practical.

An improvement of 10 dB in detection sensitivity of the original system with coherent detection can easily be achieved.

Nonlinear Effects

To date, our experiences with optical fibers have been limited primarily to low-power-level light sources such that only linear effects on fibers are realized. As the demand for higher power, long-haul transmission increases, however, nonlinear effects on fiber must be considered. It is found that stimulated Brillouin scattering may initiate cross-talk in fibers at a power level above 4 mW at 1.3 μm, and 2.5 mW at 1.5 μm. Intensity-dependent refractive index may lead to significant effect on dispersion over long distances. Further investigations into these effects are important.

Components

Although fiber loss has been reduced considerably, down to 0.14 dB/km for SMF, the losses in connector, splices, and other components are still very high. Typically, a single connector can have losses in excess of 0.5 dB, which limits system performance. Optical coupling between optical fibers and source detectors still leave room for improvement. Fiber splice techniques, connectors, and couplers are discussed in Chapter 4 [9], where loss mechanism and design considerations are illustrated. Switches and modulators are another area where integrated optics may provide important advancement (see Chapter 13) [10].

Integrated-Optics Fiber Devices and Circuits and Integrated Optics

The operating wavelength of an optical fiber device is on the order of micrometers (μm), the same order of magnitude as integrated circuit linewidth—device size. A merging of the optical fiber and integrated circuit technologies therefore would seem natural. In fact, two lines of development under that suggestion are already under way: (1) the integration of optical fiber devices with electronic devices and (2) the integrated optics.

Semiconductor substrates for lightwave devices, such as InP and GaAs, possess large carrier mobility, are also widely used as substrates for integrated circuits and electronic devices. The possibility of integrating monolithically high-speed and low-noise electronic devices with lightwave devices to obtain high-speed integrated optoelectronic devices is quite promising. Already, a laser and field-effect transistor (FET) driver integrated with an InP substrate has been reported. Integration of PIN photodiode and a FET for high-speed and low-noise detection, as well as monolithic integration of a detector and an LED, have also been explored.

Optical fiber systems work at optical frequencies. In practice, signals, in either optical or electrical form, are usually processed in the form of electrical variations, although optical transformations are necessary before transmission onto the fiber. This process is modulation. At the receiving end, optical information is again converted back to electrical signals through detection and processing. Sometimes, it may take an additional conversion for optical presentation. Would it be more direct if some of these processes could be carried out as optical processing instead of using the double conversions? Such possibilities are also under intensive investigation. Filters, couplers, and switches are candidates for these purposes. A family of devices is being developed that allows signals to be injected into or extracted from fiber links optically, as discussed in greater detail in Chapter 13.

Multiplexing

One of the simplest ways to increase the transmission capacity of an optical fiber transmission system is by multiplexing. Multiplexing is the simultaneous transmission of a number of independent signals along one communication channel of high capacity. Multiplexing can be carried out in two ways: by wavelength-division multiplexing (WDM) and by time-division multiplexing (TDM). Sometimes a frequency-division multiplexing (FDM) scheme, which is equivalent to WDM, is also used. In a WDM scheme, each signal will modulate its own optical carrier of one wavelength. Several carriers whose wavelengths are spaced out within the bandwidth of the channel are multiplex-coupled to the optical fiber for transmission. At the receiving end, the optical wave is separated in wavelengths by interference or diffraction grating type filters to recover individual signals. A WDM or FDM scheme is appropriate for analog signals. In a TDM scheme, pulse-code-modulated signals from several sources are allowed to transmit only during a specified time interval in sequence within each pulse period. The portion of time slot not used by an individual short pulse is now occupied by other pulses carrying useful information. A synchronizing pulse is sent along with each time interval such that only the designated receiver at that particular time interval will be operating to accept the information. Details of operation of these schemes is discussed in Chapter 8.

Coherent Optics

Coherent optical communication is a relatively new addition to the optical fiber communication system, although a similar scheme has been in use for broadcasting, satellite, and microwave radio transmission for over several decades (since 1955). Coherent system requires a highly coherent high-frequency source. Ordinary semiconductor lasers can hardly be considered as coherent light sources. Recently, improved spectral purity and frequency-stabilized complex lasers become available. This makes the application of coherent technology to optical fiber communication possible and profitable. With a narrow-linewidth laser source, it is possible to directly amplitude-, phase-, or frequency-modulate the optical carrier. With the advances in integrated optics technology, it is possible to build devices that will perform modulation processes such as the amplitude-shift keying (ASK), phase-shift keying (PSK), and frequency-shift keying (FSK). At the receiver, a heterodyne or homodyne technique can be applied to result significant improvement in receiver sensitivity, increasing in repeater spacing and allow frequency and WDM. Chapter 8 devotes some space to discuss these subjects.

System Development

In the initial stage of the fiber optic system development, the goal was to secure point-to-point communications at a best and a most economical way: low cost, low fiber losses, least number of repeater stations, minimum dispersion, and highest speed of transmission. Many long-haul optical fiber cables have been built or are in the process of building all around the world to realize the fruits of research in recent years.

Lately attention has rapidly been shifted from long-haul services to local services. There are LANs to deal with interoffice trunk and data transmission, MANs, and ISDNs to integrate all kinds of services, including voice, video, data transmission, and even iterative community service into the local subscriber loop. These system developments are discussed in Chapter 8.

Optical Fiber Sensing Devices

A separate branch of the developing technology using optics fibers is the optical fiber sensors. We have learned that optical fibers have many advantages over metallic wires. Besides the advantages of low loss and low cost, they are impervious to many environmental changes such as the electromagnetic interferences and electrical noises. However, environment does have influence on the operating characteristics of optical fibers. For example, temperature, electric or magnetic field, and even the radius of curvature of fiber bending may affect optical fiber performance. We therefore should exercise extreme caution to prevent or alleviate these effects. On the other hand, we can also take advantage of the sensitivity of optical fibers to these environmental effects by building sensors to gauge these effects. In fact, with little care and ingenuity, many types of sensors can be built to sense various physical perturbances such as acoustic, magnetic, temperature, and rotations. Indeed, with this dielectric construction, it can be used in environments involving high voltages, high temperatures, electrical noise, chemical corrosion, or other stressing conditions.

Two types of sensors are under development: pure optical sensors and remote telemetry sensors. Pure-fiber sensors depend on environmentally induced changes to light as it travels through a fiber. One of two things happens to light in response to an external effect: the light can (1) leak from the core into the cladding, resulting eventually in its absorption, or (2) can be forced to travel a different distance than the light going through a separate optical path. The resulting phase difference in the two lightwaves is detected by combining them and observing their interference pattern. In remote telemetric sensors, optical fibers are used merely as a means to transmit the sensed information for remote control. Both types of sensors are under development, as discussed in Part Four of this book. Other fiber optic applications, including industrial and medical applications, are discussed in Chapter 12.

References

1. K. C. Kao and G. A. Hockman, Dielectric fibre surface waveguide for optical frequencies. *Proc. IEEE* **113,** 1151–1158 (1966).
2. E. Snitzer, Cylindrical dielectric waveguide modes. *J. Opt. Soc. Am.* **51,** 491–498 (1961).
3. D. Marcuse, *Light Transmission Optics.* Van Nostrand-Reinhold, New York, 1972.
4. D. Marcuse, *Theory of Dielectric Optical Waveguide.* Academic Press, New York, 1974.
5. S. E. Miller, E. A. J. Marcatili, and T. Li, *Proc. IEEE* **61** (12), 1704–1726 (1973); S. E. Miller, T. Li, and E. A. J. Marcatili, *ibid.* **61** (12), 1726–1751 (1973).
6. Y. Suometsu, Long-wavelength optical fiber communication. *Proc. IEEE* **71** (6), 692–721 (1983).
7. C. D. Chaffee, *The Rewiring of America: The Fiber Optics Revolution.* Academic Press, San Diego, California, 1988.
8. S. R. Forest. Organic-on-inorganic semiconductor heterojunctions: building blocks for the next generation of optoelectronics devices. *IEEE Circuits and Devices Magazine,* **5,** (3), 33–37 (1989).
9. D. Botez and G. J. Herskowitz, Components for optical communications systems: A review. *Proc. IEEE* **68** (6), 689–731 (1980).
10. R. C. Alferness, Integrated optics and optoelectronics. *Proc. IEEE* **75** (11), 142–157 (1987).

The Optical Fiber

Part One consists of four chapters, Chapters 2–5 inclusive. In Chapter 2, optical fibers are classified and fiber parameters are defined. This is followed by fiber fabrication techniques. Optical cable design considerations are also mentioned. In Chapter 3, a brief theoretical discussion of both the planar and cylindrical lightguides is given. A general description of the fiber loss mechanism and dispersion effect follows. Fiber optic components, including the splicers, connectors, couplers, and switches are discussed in Chapter 4. Measurement techniques are covered in Chapter 5.

2

The Optical Fibers

Introduction

In this chapter, parameters pertinent to the characterization of optical fibers will be introduced. Materials for fiber fabrication are selected, and the effects of doping on these materials are discussed. Methods of fabrication are introduced. Fiber drawing, coating, and jacketing techniques are described. A short description of the fiber cabling technique is included at the end of this chapter.

Classification of Optical Fibers and Fiber Parameters

All optical fibers consist of a core having an index of refraction slightly higher than that of a surrounding cladding. The border between the core and the cladding may be sharp or graduated. Generally, optical fibers are of two types: the single-mode fiber (SMF) and the multimode fiber (MMF). A SMF has a smaller core diameter and can support only one mode of propagation, albeit in two mutually orthogonal polarizations. The refractive index profile of its core is usually a step-index type. In a MMF, the core diameter is usually much larger than that of a SMF. The core index profile of a MMF can be either a step-index or a graded-index type. In a graded-index MMF, the core index profile is designed to vary with the radius to meet the requirement for a minimum modal dispersion effect. A plastic lossy coating and a thick jacket usually surround the cladding to prevent damage and to increase the strength of the fiber.

13

Fiber Parameters

To describe optical fibers more specifically, let us define a few parameters [1]. Let n_1 and n_2 be the refractive indices of the core and cladding, respectively. A new parameter Δ can be defined such that

$$\Delta = \frac{n_1^2 - n_2^2}{2n_1^2} \cong \frac{n_1 - n_2}{n_1} \tag{2-1}$$

where Δ is a measure of the relative difference between the core refractive index and its cladding value. For an optical fiber to guide lightwave effectively, $\Delta \ll 1$. Typically, $\Delta = 0.002$ for a SMF and $\Delta = 0.02$ for a MMF.

A second parameter V [1] is defined as

$$V = \frac{2\pi a}{\lambda} \left(n_1^2 - n_2^2 \right)^{1/2} = \frac{2\pi a}{\lambda} n_1 \sqrt{2\Delta} \tag{2-2}$$

where $2a$ is the diameter of the core, λ is the free-space wavelength of the light source, and V is the normalized frequency of the fiber.

If one borrows a term frequently used in geometric optics, the numerical aperture NA, one finds that $NA = n_1 \sqrt{2\Delta}$ [2] so that

$$V = \frac{2\pi a}{\lambda} NA \tag{2-3}$$

Equations (2-2) and (2-3) define the key parameter V of the fiber with which one can define the types of fiber core scientifically.

Fiber Classification and Design Considerations

A detailed classification of the most commonly used optical fibers are listed in Table 2.1. The fibers are arranged in two general categories, SMF and MMF. Under the SMF division, there are subdivisions as (a) simple single-mode fiber, (b) w-fiber, (c) quadrupled clad, and (d) triangle fiber. The polarization-controlled SMFs are also listed under this category as (e) elliptical core fiber and (f) the bow-tie fiber. Under the multimode fibers, we list (g) the step-index type and (h) the graded-index type. The parabolic index fiber is a special case of (h) where $g = 2$. Under each type, the fiber cross-sectional view, the picture of its refractive-index profile, n-profile expressions (except the polarization control types), typical core diameter, cladding diameter, the numerical aperture NA, and some typical applications are listed for easy reference. The purpose of these designs is explained in the text whenever the particular fiber is introduced.

With the parameters V and Δ of a fiber defined as in the preceding section, "Fiber Parameters," one is able to classify fibers more precisely. For $V < 2.405$, the fiber can support only one mode and is

Table 2.1 Classification of Optical Fibers

	SMF				Polarization control		MMF	
	(a) Simple SI	(b) w-Fiber	(c) Quadrupole clad	(d) Triangle	(e) Elliptical core	(f) Bow-tie fiber	SI	GI
Cross-sectional view								
Radii	b, a	a, b, c	a, b, c, d, e	a, b			a, b	a, b
n-Profile	n_1, n_2	n_1, n_2	n_1, n_2, n_3, n_4, n_5	n_1, n_2	—	—	n_1, n_2	n_1, n_2, $n(r)$
Profile expression	$n = n_1$ $0 < r < a$ $n = n_2$ $a < r < b$ $n_1 < n_2$	$n = n_1$ $0 < r < a$ $n = n_2$ $a < r < b$ $n = n_3$ $b < r < c$ $n_1 > n_3 > n_2$	$n = n_1$ $0 < r < a$ $n = n_2$ $a < r < b$ $n = n_3$ $b < r < c$ $n = n_4$ $c < r < d$ $n = n_5$ $d < r < e$ $n_1 > n_3 > n_5 > n_4 > n_2$	$n = n_1\left[1 - 2\Delta\frac{r}{a}\right]$ $0 < r < a$ $\Delta = (n_1 - n_2)/n_1$ $n = n_2$ $a < r < b$ $n_1 > n_2$	Core is elliptical in shape but uniform	Two additional stripes of birefringence material partially surrounding the core	$n = n_1$ $a < r < b$ $n = n_2$ $0 < r < a$ $n_1 > n_2$	$n = n(r)$ $= n_1\left[1 - 2\Delta\left(\frac{r}{a}\right)^g\right]^{1/2}$ $0 < r < a$ $g > 1$ $\Delta = (n_1 - n_2)/n_1$ $n = n_2$ $= n_1\sqrt{1 - 2\Delta}$ $a < r < b$
Core diameter	$2a$ $<10\ \mu m$	$2a$	$2a$	$2a$			$2a$ $>80\ \mu m$	$2a$ $>50\ \mu m$
Clad diameter	$26 > 10a$						$b/a \sim 2$	$b/a \sim 2$ $b - a > 10\lambda$
Numerical aperture	0.1–0.3						0.3–0.6	0.2–0.3
Application	Long-haul fiber sensor	Dispersion control	Dispersion shift and control	Long-haul	Polarization control	Polarization control	Short-haul	Short-haul dispersion reduced

classified as a SMF. Multimode fibers have values of $V > 2.405$ and can support many modes simultaneously. The magic number 2.405 is actually the first zero of Bessel function [$J_o(x) = 0$ at $x = 2.405$], which appears in the solution of wave equation for the fundamental mode in a cylindrical lightguide. The wavelength corresponding to this value of V is known as the *cutoff wavelength* λ_c of the lightguide above which higher modes cannot propagate; and $\lambda_c = \lambda V/2.405$. Single-mode properties can be realized by decreasing the core diameter and/or decreasing the index difference Δ such that $V < 2.405$.

To design a SMF, it is critical that one know the cutoff λ_c (of the higher-order modes) and operate the fiber at a wavelength above λ_c. For example, if $n_1 = 1.5$ (silica glass, $\Delta = 0.002$, from Eq. (2-2) $\lambda_c = 0.298(2a/2.405) = 0.124(2a)$. For a 6.8-$\mu$m fiber ($2a = 6.8$ μm), $\lambda_c = 0.84$ μm. The operating wavelength should be larger than 0.84 μm or, say, at 0.85 μm.

The core diameter of a SMF is directly proportional to the wavelength. If the operating wavelength is increased to 1.3 μm, the core diameter is $2a = 12.8$ μm, making it much easier to handle.

To determine the cladding diameter $2b$ of a SMF, one has to look into the field distribution of the mode and the power dissipation in the cladding. Typically, for $V = 2.2$, about 20% of the light power will be dissipated in the cladding [3]. Cladding must be thick enough to absorb this power. Usually cladding that is 10 times larger than its core diameter in a SMF is sufficient [4].

The small core diameter of a SMF makes it very difficult to handle and to splice. To alleviate this problem, a w-fiber was initially designed to satisfy this requirement [5]. Its core is surrounded by two layers of cladding, an inner and an outer cladding with a barrier layer in between. The index profile is shown in type b in Table 2.1.

This scheme makes it possible to design a SMF with a relatively larger core dimension, aiding in fiber handling and splicing.

Further investigations have shown that the additional barrier provides a very large and efficient control of the dispersion properties of a fiber without simultaneously affecting its loss characteristics. This is because two additional design parameters, $\Delta n' = n_3 - n_2$ (in addition to the original $\Delta n = n_1 - n_2$) and the new diameter ratio b/a have been added to enhance the flexibility in design. It therefore gives added advantages beyond manufacturing reasons [6]. Other configurations, including raised inner cladding and depressed inner cladding, have also been investigated with interesting results [7].

A more complicated design, known as a *quadrupole clad design*, is listed as type c, where the profile is more complex. The additional parameter resulting from this profile gives even more flexibility of the fiber design. A much simpler design known as the *triangle* or *T-fiber* is listed as type d. It has been verified [8] that T-fiber provides a much higher second-order mode cutoff wavelength (for LP_{11} mode) and a lower attenuation per unit length than does a simple SI fiber (a). Furthermore, it shows lower sensitivity of the dispersion compensation to variations in design parameters. Polarization-controlled SMFs are realized by either introducing physical asymmetry such as an elliptical core is in (e) or by adding regions of extra birefringence as in (h), known as the *bow-tie fiber*. Twisted fiber construction is also used for polarization control. Further discussions of fiber dispersion will be given in the next chapter.

If V is increased much above 2.405, the step-index fiber can support a large number of modes. This fiber can be classified as a MMF. The maximum number of modes N_m is found to be [3]

$$N_m \approx \left(\frac{2\pi a}{\lambda}\right)^2 n_1^2 \Delta = \frac{V^2}{2} \qquad (2\text{-}4)$$

thus for $V = 10$, $N_m = 50$.

Multimode fibers can be defined as those fibers with $V > 2.405$. Either a larger core diameter and/or a larger Δ value will result in a MMF.

When many modes propagate along a fiber, carrying the same signal but at slightly different group velocities, the result is that they interfere with each other at the receiving end and produce intermodal dispersion. The quality of transmission deteriorates as the fiber length increases. For this reason, a step-index MMF has a limited use for small-capacity, short-distance communication systems only.

To reduce the effect of intermodal dispersion, a graded-index fiber is introduced. Typical graded-index core profiles are shown in Fig. 2.1a and 2.1b. The profiles are the normalized refractive index Δ between the core and the cladding, as a function of the core radius. These two alternate designs are functionally equivalent but are made by different processes. Figure 2.1a shows a depressed-index barrier layer graded-index MMF made by the MCVD process, and Fig. 2.1b shows a graded-index MMF made by OVD process. Both the modified chemical-vapor-deposition (MCVD) process and the outside vapor-deposition (OVD) process will be discussed in the "Fiber Fabrication" section. The index profile of the core of a MMF is designed according to the following expression:

$$n(r) = n_1 \left[1 - 2\Delta \left(\frac{r}{a}\right)^g \right]^{1/2} \qquad (2\text{-}5)$$

where g is the power index. It is usually chosen to reduce the effect of intermodal dispersion, yielding a wider bandwidth. For optimum effect [9], [10]

$$g = g_{opt} = 2 - \left(\frac{12}{5}\right)\Delta \qquad (2\text{-}6)$$

The refractive index of the cladding is maintained at a constant value n_2.

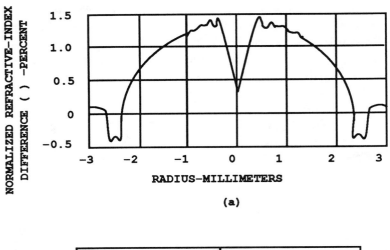

(a)

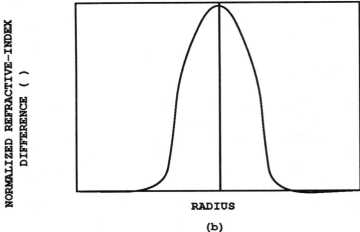

(b)

Figure 2.1 Two types of core profile for GI fibers: (a) A depressed barrier layer by MCVD method (courtesy of York Technology, Ltd.); (b) By OVD method (after A. J. Morrow et al., Chapter 2, [39], 65–94.

For a variable core index, an effective V [11] can be defined as

$$V_{\text{eff}}^2 = 2k^2 \int_0^a \left[n^2(r) - n_2^2 \right] r \, dr \tag{2-7}$$

where $k = (2\pi/\lambda)$ and the maximum mode number becomes

$$N_m = \frac{V^2/2}{1 + (2/g)} = \frac{V_{\text{eff}}^2}{2} \tag{2-8}$$

For a core material of $GeO_2–P_2O_5–SiO_2$, if $2a = 50$ μm, $2b = 125$ μm, $\Delta = 0.018$, $NA = 0.23$, $N_m = 192$ for $\lambda = 1.3$ μm. The ratio a/b can be as small as 50% without seriously affecting the operation of a fiber.

Multimode fibers have advantages as well as disadvantages compared to SMFs. The larger core diameter of a MMF facilitates launching light and splicing of similar fibers. Additionally, MMFs can be excited with LEDs, while SMFs must be excited by lasers. Light-emitting diodes are easier to fabricate and have longer life than do single-mode lasers, making their use much more desirable in some applications. Coupling efficiency η_c of the LED to a MMF can be defined as

$$\eta_c = \frac{\text{LED power into fiber}}{\text{LED power}} = \frac{(NA)^2}{1 + (2/g)} \qquad (2-9)$$

A MMF with a larger NA therefore couples more efficiently to a LED light source [12].

The disadvantages of MMFs are higher fiber losses primarily due to higher dopings and therefore higher Rayleigh scattering loss, less resistance to nuclear radiation, more complex joint and coupling problems, and higher dispersion and lower bandwidth. Graded-index-type MMFs are more expensive to fabricate. Their production yield is very poor.

Geometric perturbations of fiber dimensions, such as the core geometry variation, nonuniformity, and imperfection, may affect the final characteristics of the fiber. These effects will be treated in related subjects in later chapters.

The surface of both SMFs and MMFs need protective coating to preserve the inherent strength of the glass fiber. A nylon or polyester jacket is often applied to increase the mechanical strength of the fiber.

Materials and Doping

Fiber Materials

For obvious reasons, optical fibers are fabricated from materials that are transparent to optical frequencies. The most abundant and inexpensive material to make optical fibers is SiO_2. Silicon and oxygen form an amorphous solid, with a strong bond between Si and O. Many times, some minority metal ions in the form of metallic oxides may be included in the structure. These metallic oxides make the silicon and oxygen atoms bind less tightly and thereby change the physical properties of the glass readily. For example, the refractive index and the thermal expansion coefficient of the glass can be varied with its composition. This is the most common way to control the difference in refractive indices between the core and cladding. Other additions, such as the sodium borosilicate group, may be used to lower the softening temperature of the glass and the refractive index as well. But care must be exercised in selecting the composition for making optical fibers so as not to increase the loss factor or run into an unstable phase diagram, which might cause breakage and cracks to occur at the core–cladding interface.

An important factor to consider in choosing raw materials is to know what elements contribute to losses in optical fibers. Power losses in optical fibers occur through mechanisms such as absorption, scattering, and radiation. Aside from the intrinsic losses due to Rayleigh scattering and infrared absorption, other factors affecting losses are the inclusion of the hydroxyl radical OH, and the electronic transition of metal ions such as that of iron, copper, cobalt, and nickel. For example, 1 ppm (part per million) of OH will give rise to a 30–40-dB/km loss peak at 1.39 μm [13–15]. Other peaks due to overtones of the OH absorption occur at 0.95- and 1.25-μm wavelengths. Transition-metal ions contribute peaks at various wavelength ranges within the optical spectrum of interest [16]. High-quality pure raw materials are essential. Technology for fiber fabrication has advanced to a level where, basically, only intrinsic effects limit the losses. These losses define the transmission window, which extends from 0.7–1.8 μm wavelengths in current low-loss GeO_2-doped silica fibers.

Doping of Fiber Materials

To change the refractive index of optical fibers, the otherwise pure silica is often doped with metallic dopants in the form of oxides or chlorides. Adding germanium of phosphorous can result in an increase in the refractive index, while adding boron reduces it. The amount of dopant added to the glass must be carefully considered because a variety of trade-offs may occur. For example, adding more dopant to the core glass can increase the numerical aperture NA of the fiber, which will also increase the number of modes to be guided along the fiber. This is profitable if LEDs are used to excite the fiber. Large-NA fibers also help to reduce bending losses [16]. However, high doping increases compositional fluctuations, which increase losses via scattering. This, in turn, increases group delay to reduce bandwidth, making the fabrication more difficult due to mismatches in the physical properties of core and cladding glasses. A compromise between these contradictory factors must be worked out.

The most commonly used dopants in fiber fabrication are listed in Table 2.2. Dopants can be classified into two groups: those where the refractive indices decrease with the mole percentage and those that increase with the added whole percentage. Typical changes in n versus mole percent (mol %) are shown in Fig. 2.2a and 2.2b.

Fibers must possess high strength and flexibility to withstand the rough handling incurred during installation and the varying environmental conditions found during operation. Fabrication processes must therefore include coating and jacketing operations. Finally, one must look for materials that permit easy handling and are least costly.

Although future fiber fabrication may utilize other materials, espe-

Table 2.2 Dopants Used in Fiber Fabrication

	Core	Deposited Clad
MMF	$GeO_2-B_2O_3-SiO_2$	$B_2O_3-P_2O_5-SiO_2$
	$GeO_2-P_2O_3-SiO_2$	$B_2O_3-P_2O_5-SiO_2$
		$F-P_2O_5-SiO_2$
	SeO_2-SiO_2	SiO_2
SMF	GeO_2-SiO_2	$F-P_2O_5-SiO_2$
		SiO_2
		$P_2O_5-SiO_2$
		$F-SiO_2$
	SiO_2	$F-P_2O_5-SiO_2$
		$F-SiO_2$

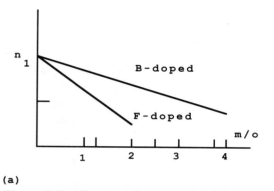

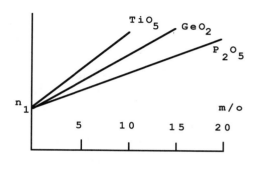

(a)   **(b)**

Figure 2.2 Sketches of two groups of dopants that affect the refractive indices of silica glass. n decreases (a) and increases (b) with mole percent of dopant. [After M. Nakahara *et al.*, *Appl. Phys.*, **50**, 1006–1020, (1981).]

cially for longer-wavelength operations, present-day technology has been developed chiefly around silica glass technology.

Fabrication processes will differ depending on how the fiber will be used. For instance, in medical applications, where only shorter fiber lengths are involved, the fiber can be made of plastics or a glass core with plastic cladding. This combination offers greater strength and ease in handling.

Materials for Long-Wavelength Glass Fibers

The useful range of wavelength of a conventional glass fiber is restricted from 0.6 to 1.8 μm. Rayleigh scattering loss limits the fiber's usefulness below 0.6 μm. Rayleigh scattering decreases inversely as the fourth power of the wavelength so that at the upper end of the wavelength range, around 1.8 μm, it becomes very small. But the infrared (IR) absorption at this wavelength is already large and increases very rapidly with increasing wavelength. Beyond 1.8 μm, no silica-based fiber can work efficiently. Silica-based glass fiber achieves a minimum attenuation at about 1.55 μm and a zero-dispersion point at about 1.27 μm.

It was estimated that a fiber loss of the order of 0.01–0.001 dB/km would be desirable for long-haul transmission so that repeaters can be spaced thousands of kilometers apart. Researchers are searching for these materials. They are aimed to find materials that have (1) low IR absorption loss at mid-IR wavelengths where Rayleigh scattering loss would be low, (2) adequate refractive-index range (1.35–1.6) for building lightguides, (3) high chemical stabilities with environmental and time changes, and (4) safe and easy processing. The search began in three groups: heavy-metal oxides, heavy-metal halides, and chalcogenides. A brief summary follows.

The heavy-metal oxides consist of oxide mixtures of metals such as aluminum (Al), barium (Ba), calcium (Ca), germanium (Ge), potassium (K), lanthanum (La), lead (Pb), tin (Sn), tantalum (Te), tungsten (W), and zinc (Zn). They can be formed, doped, and drawn into fibers in the same manner as silica fibers. Although the theoretical intrinsic attenuation of these fibers was estimated as 0.1 dB/km at 2–3-μm wavelength, the best practical loss measured experimentally was about 4 dB/km. The minimum chromatic dispersion was measured at 1.7 μm [17]. It seems that drastic improvement in fiber characteristics by this approach is not expected.

The heavy-metal halides are a combination of the heavy metals with nonmetallic elements of group VII of the periodic table: fluorine (F), chlorine (Cl), iodine (I), or alabamine (A). Glasses formed from these groups are nonoxide glasses. Potentially, they are expected to have low absorption losses in the mid-IR region. Fluoride glasses are the most experimented glasses for the 2–5-μm region. Their theoretical minimum attenuation ranges from 0.03 to 0.003 dB/km [18].

Two groups of fluoride glasses are developing: the binary system and the ternary system. In the binary system, compositions of MF_3–AlF_3 and MF_3–ZrF_4 are fabricated into fiber by rapid quenching techniques (here M represents Ca, Sr, Ba, or Pb). These fibers have a minimum loss in the 3–4-μm range. In the ternary system, the composition is RF_3–BaF_2–ZrF_4, where R represents La, Ce, Nd, Gd, or Lu for the same wavelength range. The ternary system produces comparatively better materials. They have better optical qualities, more stable chemical properties against devitrification, and lower deliquescence and toxicity than do binary-system materials [19, 20].

Vapor-phase fabrication techniques for making fluoride glasses are being developed. Although the fiber loss is not yet low enough (6.8 dB/km) for practical application, desirable chromatic dispersion properties have been revealed. It is found that the slope of the dispersion is three to four times smaller than that of silica glass. That means the difference of the source wavelength from the zero dispersion wavelength becomes less critical, thus making a very large bandwidth over a very broad spectral range possible.

Halide glasses have also been evaluated for energy transmission capabilities. Thallium bromoiodide (KRS-5) fibers have been fabricated that are capable of transmitting 20-W CW CO_2 laser beams (10.6 μm) over a distance of several meters successfully. This property makes them very attractive as a flexible waveguide in many laser-processing machines in the industry. The fiber diameter can be as large as 1 mm [21].

Chalcogenide glass fibers contain arsenic (As), germanium (Ge) phosphorus (P), sulfur (S), selenium (Se), and tellurium (Te). These fibers are prepared by zone refining or crucible techniques. The theoretical attenuation of these glasses was estimated as 0.01 dB/km at 3−5 μm. The slope of the dispersion is even smaller than the halide glasses. As a result of fabrication difficulties, however, these materials seem much less promising than the halide glasses [22].

Plastic Fibers

In certain fiber applications, as in some sensors and medical applications, the fiber length used is so short (less than a few meters) that fiber loss and fiber dispersion are of no concern. Instead, good optical transparency, adequate mechanical strength, and flexibility are the required properties. Plastic or polymer fibers may be the choice.

A good candidate for plastic fiber is made with polymethyl methacrylate polymer and its copolymer. The processing of plastic fibers is much simpler and less demanding than that for all-glass fibers. The mechanical properties of the polymer allow the fiber to be made without an additional protective layer.

All-plastic fibers are exclusively MMFs with a step-index profile and large core diameter. A core diameter as large as 200−600 μm and a cladding diameter of 450−1000 μm are not unusual. This makes the numerical aperture and the acceptance angle large ($NA = 0.5-0.6$, $\theta = 70°$). The coupling to light sources such as LEDs becomes very easy and efficient. Inexpensive and useful systems can be built, albeit with high fiber loss (350−1000 dB/km). The transmission spectrum of plastic fibers resembles that of glass fibers. Their attenuation, however, is as much as several hundred decibels higher. The refractive index can be adjusted to within a range of 1.35−1.60. A plastic fiber can withstand a rougher mechanical treatment displayed in relatively unfriendly environments [23].

Other advantages of all-plastic fiber, include less weight and lower cost compared to glass fibers. Plastic fiber weighs 40% and costs a fraction of that of glass fiber. The limitation is that its operating temperature range is from only −30 to +80°C, compared to up to 1000°C for a glass fiber.

Fiber Fabrication

Fibers can be drawn simply and economically directly from melts of silica powders in crucibles. However, it is very difficult to obtain homogeneous and high-purity glass fibers directly from powders. To produce fibers with extremely low loss, a two-step process is usually used. First, a preform is prepared with its desired index profile and the correct core/cladding-diameter ratio. Fibers are drawn from the preform [24,25]. Coating and jacket processes follow immediately.

Preform Preparation

Many techniques have been developed to prepare preforms. The main objective of these techniques is to develop a clean process that is (1) low in OH concentration, (2) low in metallic-ion contamination, (3) high in homogeneity, (4) reproducible, and (5) low in cost.

The following are a few well-developed techniques currently in use.

Outside Vapor-Deposition Process [26]

This process is also called the "soot process" first worked out by Corning Glass Works in America. Figure 2.3 shows a schematic drawing of the setup. The starting materials such as $SiCl_4$, $GeCl_4$, BCl_3, and O_2 are reacted in a hot flame to produce small particles of very high purity glass of the desired composition.

A stream of hot glass soot is emitted from the flame and directed toward a small diameter rotating and traversing ceramic rod (the target rod) held in a lathe approximately 15 cm away. The glass soot sticks to the rod in a partially sintered state, and layer by layer, a cylindrical glass preform is built up. By properly controlling and sequencing the metal halide vapor stream composition during the soot deposition process, it is possible to radially build into this porous preform the desired glass composition, refractive index, and dimensions for both the core and cladding regions. Both step-index and graded-index fiber designs can be readily achieved using this approach. Excellent-quality core–cladding interface, roundness, and concentricity can be obtained.

When the soot deposition is completed, the porous preform is slipped off the reusable target rod and then zone-sintered to a solid bubble-free glass blank by passing it though a furnace hot zone at approximately 1500°C in a controlled atmosphere, such as helium.

During the sintering process, essentially all of the hydroxyl impurity is removed from the porous glass using a chlorine gas treatment.

The center hole, if not closed, may result in breakage of sintered blanks due to tensile stresses that have developed at the center hole surface from thermal expansion mismatch of core and cladding glass

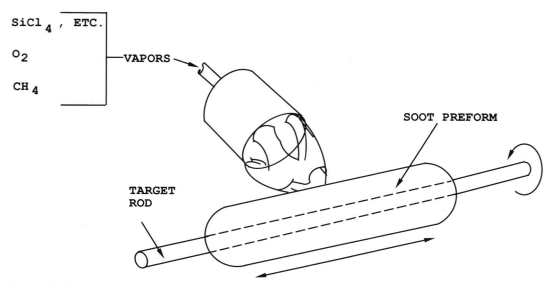

Figure 2.3 The outside vapor-deposition (OVD) process or soot process. [After P. C. Schultz, [27] © IEEE, (1980).]

composition. This problem was later eliminated by process modification [27]. OVD is actually a chemical vapor-deposition system where the reaction takes place outside a silica mandrel.

The process described above was used in Corning Glass Works and became the breakthrough for producing a glass fiber bearing a transmission loss of 20 dB/km. Since then, many improvements have been incorporated into this process to make ultra-low-loss optical fibers.

Modified Chemical Vapor Deposition

The MCVD process, employed by Bell Laboratories, is a modified chemical vapor deposition system in which vapor constituents are made to react inside a silica tube instead of outside, as in the OVD system. A sketch of this system is shown in Fig. 2.4.

A fused-quartz tube is mounted on a glassworking lathe and slowly rotated while reactants, usually $SiCl_4$, and dopant reactants, such as $GeCl_4$ and BCl_3, flow through it in an oxygen stream. Here they react to form particles that deposit down stream. An oxy-hydrogen burner is slowly traversing along the outside of the tube to provide simultaneous deposition and fusion and sintering of a layer of the reacting materials. Approximately 50 layers are deposited by multiple passes of the burner. The flow of the constituents in the vapor and the temperature of the burner and its traverse speed are closely controlled to satisfy the required growth rate and profile of the preform.

At the conclusion of deposition, the temperature of the burner is raised to collapse the tube into a solid preform [28].

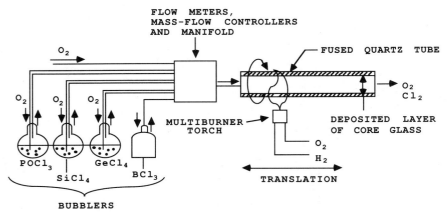

Figure 2.4 The modified chemical vapor-deposition (MCVD) process. (After
J. B. MacChesney, [28], Proc. IEEE, Vol. 68, pp. 1181–1184, © IEEE, 1980.)

The MCVD technique offers many important advantages. It is an
inherently clean process because deposition results from reaction between gaseous constituents that take place inside the silica support
tube. External contaminants, either particles or gases, are excluded.
The reactants ($SiCl_4$, $GeCl_4$, etc.) are themselves highly pure and combine directly with oxygen in the absence of hydrogen-containing
components. This produces low OH concentrations in the core and/
or cladding glass thus formed. Also, because the initial tube is col-
lapsed in the final step of preform preparation, mismatch of expansion coefficients of different glass compositions does not result in
cracking of the preform, even when high dope compositions are used
to form high numerical aperture fibers. Both step-index and graded-
index preforms can be deposited by this process.

Vapor-Phase Axial Deposition [29]

The vapor-phase axial deposition (VAD) process is distinguished
from the MCVD process as a continuous fabricating process of fiber
preform in the axial direction, which completely avoids the collaps-
ing procedure found in the conventional processes. Continuous preform fabrication is easily realized because the deposition and con-
solidation steps are arranged in sequence. Since long large-diameter
preforms can be prepared by this process, a significant cost reduction
in fiber production can be realized.

The equipment for the VAD process is arranged schematically in
Fig. 2.5. Raw materials such as $SiCl_4$, $GeCl_4$, $PoCl_3$, and BBr_3 are fed
from the bottom into oxy-hydrogen torches and glass particles produced by the flame hydrolysis reaction are deposited onto the end
surface of a starting silica glass rod, which acts as a seed. A porous
preform is then grown in the axial direction and consolidated into a

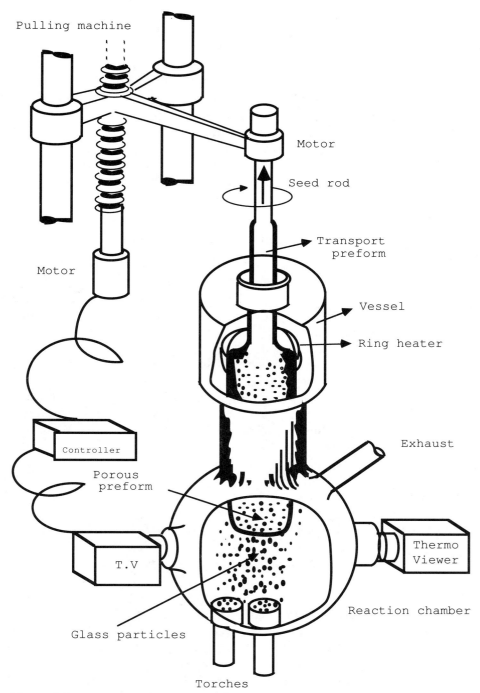

Figure 2.5 Equipment for the vapor-phase axial deposition (VAD) process. (After T. Izawa and N. Inagaki, Proc. IEEE, Vol. 68, pp. 1184–1187, © IEEE, 1980.)

transparent preform by zone melting using a carbon ring heater. A glass chamber is used to enclose the reaction and deposition regions to achieve a clean environment. The starting rod is pulled upward and rotated in the same fashion as that used to grow single crystals. A symmetrical cylinder of preform with a constant diameter can be produced. The flow rates of raw materials are precisely regulated. Flame temperature and surface temperature of the growing preform and its rotating speed are computer-controlled. A videocamera is used to sense and control the positioning of the preform such that the preform is kept at a fixed position with an accuracy of within 50 μm. This avoids variation of preform diameter. This system is preferred for mass production.

Plasma Chemical Vapor Deposition

The plasma chemical deposition (PCVD) process uses a nonisothermal plasma operated at microwave frequencies to promote the chemical reactions that result in deposition of thin layers of glass on the walls of a silica tube of a MCVD system. An inert gas, usually argon, is excited and sustained by a high-power microwave generator. The plasma discharge promotes the chemical reaction instead of a gas torch or burner as is used in the other processes. Up to 2000 thin layers of glass can be deposited on a silica tube. This plasma-augmented reaction can initiate the growth at relatively low temperatures, thus eliminating the possibility of tube deformation. However, the deposition rate is lower than that which can be achieved by other processes.

Fiber Drawing and Coating

The performance of optical fibers—specifically, the transmission losses, bandwidth, and strength—is strongly influenced by the fiber drawing and coating processes. These processes must be treated as an integrated operation designed to impart desired properties to the resulting fiber. The essential components of the whole process are shown schematically in Fig. 2.6. These consist of (1) a feed glass, (2) a heat source, (3) a fiber-diameter monitor, (4) a coating applicator, (5) a curing apparatus, and (6) a fiber puller and winding mechanism. Except for the fiber puller and winding mechanism, other components must be enclosed to provide a clean environment. The components are arranged in a vertical position to take advantage of gravitational force.

The feed glass is usually a glass preform, which typically measures 10–25 mm in diameter and 60–100 cm in length. Preform is fed at speeds of 0.002–0.03 cm/s, depending on the heat source, preform diameter, and draw speed. Continuous lengths of as much as 40–300 km have been drawn from such a preform in one drawing [30].

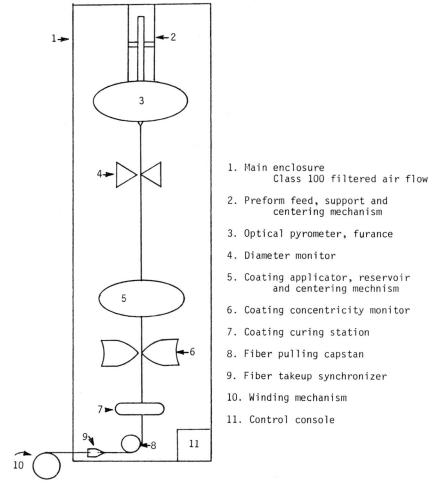

1. Main enclosure
   Class 100 filtered air flow

2. Preform feed, support and
   centering mechanism

3. Optical pyrometer, furance

4. Diameter monitor

5. Coating applicator, reservoir
   and centering mechnism

6. Coating concentricity monitor

7. Coating curing station

8. Fiber pulling capstan

9. Fiber takeup synchronizer

10. Winding mechanism

11. Control console

Figure 2.6   Schematic of glass fiber lightguide draw and coating apparatus.

Most of the silicate glasses used as lightguides have relatively high softening temperatures and, therefore, require heat sources capable of providing temperatures sufficient for drawing preforms in the range of 1950–2250°C. Inductively heated zirconia-type furnaces are widely used [31].

Fiber drawing begins by heating the tip of the preform to a molten state and allowing it to extend under the force of gravity downward while shrinking in diameter into a fine-diameter filament without the use of any die mechanism. A pulling and winding mechanism sustains a drawing force. The fiber diameter is monitored continuously as the fiber passes through the furnace. Any variance of the fiber's diameter is detected by a control device that automatically readjusts the temperature and/or the speed of pulling to compensate for the initial changes.

The strength of the fiber can be affected significantly by the quality of the starting glass tubing used for preparing the preforms. Any foreign inclusions or other defects can cause strength diminution. The preform is subjected to a process known as *fire polish* to purify it before drawing. Cleanliness and temperature capability of the draw furnace also affect the fiber strength. The drawing process is particularly vulnerable to atmospheric particle contamination during drawing because upward flow of air through the furnace by convection can cause particles to fuse to the surface of the molten glass and thus greatly reduce the tensile strength of the fiber. It has also been established that greater strengths can be attained when relatively high drawing temperatures are used [32]. Carbon dioxide lasers have been used successfully as heating sources for drawing. They are particularly advantageous to use because they are capable of providing an especially clean thermal environment that is conducive to achieving high-strength lightguides [33].

When the fiber exits the furnace, it enters the coating applicator immediately. A clean environment is maintained at this junction to reduce the risk of contaminating the fiber surface. Coating must be applied concentrically in sufficient thickness and solidified very rapidly before the fiber reaches the take-up spool [34]. Polymer coating with low moduli and low glass transition temperatures is needed to protect against bending, impact, and crushing. Liquid prepolymers, such as silicone rubber, make an effective coating material. The coating is then cured by a thermal and/or ultraviolet radiation-activated curing apparatus. Because silicone rubber is a rather poor abrasion-resistant material, it is necessary to use an abrasion-resistant jacket over the rubber coating. Nylon is the preferred material used to protect the coating [35].

Improper coating technique may result in excessive fiber loss. Coating defects such as voids, lumps, and uncoated sections should be avoided. Poor coating geometry can produce nonuniformly distributed stresses on the fiber as a result of density changes on solidification or thermal-expansion mismatches, thus inducing microbending losses in the fiber.

Soft coatings help to reduce the stresses on fibers, thus reducing the microbending. But hard coatings are needed to protect against impact and crushing forces in either the manufacturing processes and/or the installation. A dual coating consisting of a soft inner plus a hard outer coating can be used advantageously.

Organic coatings will allow the diffusion of ambient moisture into the fiber, which contributes to slow crack growth on the fiber surface by a stress–corrosion mechanism. Several inorganic hermetic coating techniques have been developed. One method involves the application of a thin metallic coating of pure indium before polymer coating is applied [36]. The metallic coating provides a shield against moisture, and the polymer coating provides flexibility and strength.

Optical Fiber Cables

The transmission properties of the optical fiber exit from the drawing and coating machine described in the preceding section may be well designed for communication applications. Yet, the fiber still needs to undergo a cabling process to improve its mechanical properties. Because of brittleness of the glass fibers, an elaborate protective structure must be improvised to support the fiber properly. We wish first to identify the stresses and strains that a fiber cable may experience during the manufacturing processes and also to loads induced on it during the installation processes and while in service. Then we offer some remedies to these effects.

Sources of Mechanical Stress on a Fiber

The major sources of fiber stresses may appear as tensile stress, torsional stress, and bending stress or flexuous stress.

Tensile Stress

Tensile stress occurs when the fiber is axially strained. This happens whenever the fiber is pulled during either the fabrication process or during installation. Although tensile strength of glass fibers is only slightly lower than that of steel wire [14 GPa (gigapascals) for glass fiber and 20 GPa for steel wire], glass fiber under an applied stress does not deform plastically as metallic wires do. Instead, it will extend elastically up to only 1% of its length. Beyond that limit, a glass fiber breaks. Furthermore, glass fibers are usually plagued with surface flaws or microcracks that distributed along the surface randomly. The statistical median strength is thus limited by its fracture strength, which is only 700–3500 MPa for glass fiber.

Torsional Stress

Torsional stress occurs when the fiber is twisted. This could happen during the installation process or during service as a result of thermal expansion or contraction. The tolerance of glass fiber to torsional stress is very small.

Bending Stress

Bending stress occurs when the fibers are subjected to a bending radius. The cable may have to fit into a duct that turns in direction. Or, during cable installation, the worker may have to bend the cable in order to fit it into the duct. The band radius must be carefully controlled.

Static Fatigue

When a glass fiber is stressed beyond its limit, it simply breaks. This is called *instantaneous failure*. The break can be detected immediately. There is another kind of failure that is slow in developing, but nevertheless is very important, because it can actually grow with time under the most unfavorable environment. This is called *static fatigue*. Static fatigue is related to the slow growth of preexisting flaws in the glass fiber under humid conditions and under tensile stress. High reliability of a fiber can be achieved only when the static fatigue failure rate is under control.

Mechanical Design Considerations

Mechanical design considerations of fiber cables have these objectives [37]: to provide physical protection against tensile loading during laying and suspending, to reduce microbending that results from loading, to accommodate thermally induced dimensional changes so as not to apply undue stress to the fiber and cause microbending, and to provide hermetic sealing against moisture diffusion. Thus, a good mechanical design will minimize the above mentioned effects on fiber cables.

The following is a list of the design considerations:

1. Protective coating of the individual fiber string. The surface of glass fibers must be adequately protected against scratches and moisture diffusion. Mechanical scratches decrease the tensile strength of the fiber under stress. Moisture diffusion along the fiber surface under tension induces static fatigue. Both cause fiber failures. Coating the fiber with polyethylene immediately after the drawing process is an effective way to provide these protections.

2. Structure member must be added to take up the tensile strength. Glass fiber is too brittle to withstand the tensile strength during installation and also during the service life of the cable. Fiber elements of a cable are usually built around a core foundation structure member made of a steel wire. Individual fiber units are stranded around the structure member with a low pitch. Cable elongation must be limited to the 0.3−0.8% range of its length.

3. The stranding of fibers around the structure member must have a sufficiently short lay in comparison to the expected bend radii such that the average position of all component fibers approximates the neutral axis to minimize the bending force on the cable. Special stranding techniques are developed to ensure that the cable-making process does not contribute a torsional stress.

4. Buffering is a technique used in conjunction with cabling of op-

tical fibers to prevent ingress and axial migration of moisture along the fiber. Either tight or loose buffering can be used. In tight buffering, usually a harder plastic or nylon coating is extruded on the fiber during drawing stress forms an annular layer to a diameter of $0.5-1.0\ \mu m$. In loose buffering, an oversized cavity is provided in the cable structure to house the fiber. The cavity is filled with soft filling compound specially designed to resist moisture intrusion. Material such as petroleum-based silicon is widely used. Buffering decouples the fiber from external forces such as bending and handling, which cause stress and strain. It also eases the stress caused by contraction or expansion of the cable due to temperature fluctuation. Various designs that are used for any particular application will be pointed out in a later section.

5. Finally, the cable is provided with an outer sheath to ensure that it is ready for operation under any working conditions. For example, in duct installation for local distribution inside a plant, a single jacket of low-friction, abrasion-resistant material such as polyethylene is a good choice. For aerial cable installation where size and weight must be minimized, a nonmetallic sheath has proved more than adequate. However, for buried land and submerged marine cables, more elaborate protection such as steel-type armoring over a cushioning inner jacket and a moisture sealed outer jacket may be necessary.

The overall consideration of cable design is always a balance between the reliability one needs and the cost of protection the money can buy. Overprotection is not cost effective, while under protection may become shortsighted.

Some Practical Cable Design

A Typical Six-Fiber Cable

Figure 2.7 is a cross-sectional view of a typical six-fiber telecommunication cable. Each fiber consists of a core fiber surrounded by cladding. The unit is then coated with an insulating jacket. In the center of the assembly is an insulated steel cable for providing tensile strengths. Six more insulated 22-gauge copper wires are distributed around the assembly to fill the gaps. These wires may also serve as a conducting path for direct-current (dc) power supply for the repeaters if needed. The fibers are then stranded and wrapped with polyester tape to bind the assembly. This is then pressure-fitted over a corrugated, bonded, and longitudinally applied aluminum shield. A polyethylene jacket is then applied over the top. The assembly can withstand a pulling force of 500 kg. It is moisture proof and can withstand temperature variations from -50 to $+60°C$. Many similar designs and constructions are available from different manufacturers.

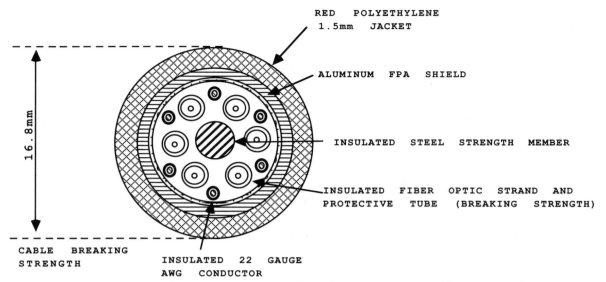

Figure 2.7 A cross-sectional view of a typical six-fiber telecommunication cable (*not to scale*).

Undersea Cable [38]

Undersea cables are generally envisaged to use only SMF. This is because for long-haul and high-bit-rate transmission, SMF operating on longer wavelengths offers lower loss and minimum dispersion and, therefore, longer repeater spacing. But the requirements of sustaining high tensile strength and accommodating high hydrostatic pressures are more stringent. It is also true that the fiber should be kept isolated from outside pressure and water intrusion, even in the case of cable rupture. Such a cable should consist of (1) a central optical core assembly that is hermetically sealed, (2) surrounding steel wires that provide most of the tensile strength and pressure resistance, (3) a metallic conductor for power feeding the repeaters, and (4) an outside insulating plastic jacket. The central optical core assembly consists of a high-tensile-strength steel wire surrounded by the fibers, which are either tightly buffered with plastic materials or embedded inside a grease coating that protects the fibers from water intrusion. A typical example of undersea optical cable structure is shown in Fig. 2.8 [38].

On the left side of Fig. 2.8, a cross-sectional view of an undersea cable assembly is displayed. It consists of a central optical fiber assembly which contains six SMF fibers arranged in containers as shown in five different diagrams. In cable assembly 1 (CI), each optical fiber is loosely fitted into a plastic tube filled with grease and arranged around a steel strength member coated with plastics. A plastic tube is used to enclose the assembly. Similar structure is used in cable assembly 5 (CV), except the fibers are now wrapped with nylon coatings and a metallic tube is used to enclose them. In cable assembly 4 (CIV), the fibers are immersed in elastomer and are en-

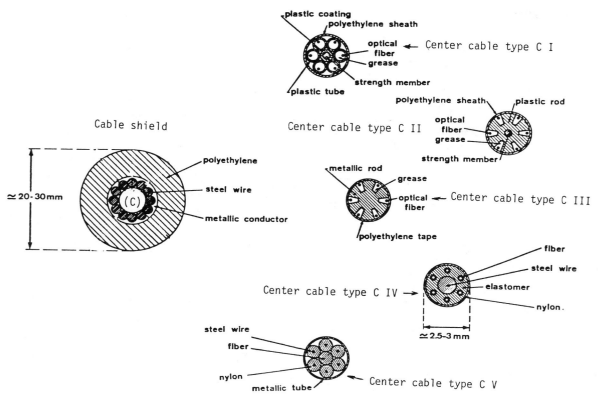

Figure 2.8 Typical examples of undersea cable structures.

closed by a nylon tubing. In cable assemblies 2 and 3, the fibers are fitted loosely in cavities around either a plastic rod (CII) or a metallic rod (CIII) and the cavities are filled with grease. Both outside tubings are polyethylene tapes. The outside diameter of the plastic sheath or the metallic tube is about 2.5–3 mm. This tube is surrounded by a layer of stranded steel wires, which are again enclosed in a sealed metallic sheath. Polyethylene sheath is then coated around the metallic sheath until the outside diameter of the assembly builds up to 20–30 mm.

Conclusion

In this chapter, classification of fiber types is introduced. Brief descriptions of the fiber fabrication techniques follow. For detailed discussions, the readers should consult the original papers cited in the reference. The author wishes to call attention to a recent book edited by Li [39] that gives a comprehensive discussion and information pertaining to materials and processes for the fabrication of optical fibers.

References

1. A. W. Snyder, Asymptotic expressions for eigenfunctions and eigenvalues of a dielectric or optical waveguide. *IEEE Trans. Microwave Theory Tech.* **MIT-17,** 1130–1138 (1969).

2. R. D. Maurer, Introduction to optical fiber waveguides. In *Introduction to Integrated Optics* (M. Baruoski, ed.), Chapter 8. Plenum, New York, 1974.

3. D. Gloge, Weakly guiding fibers. *Appl. Opt.* **10,** 2252–2258 (1971).

4. D. Gloge, Optical fibers package and its influence on fiber straighteners and losses. *Bell Syst. Tech. J.* **54,** 245–262 (1985).

5. S. Kawakami and S. Niskida, Perturbation theory of a double clad optical fiber. *IEEE J. Quantum Electron.* **QE-11,** 130–138 (1975).

6. S. Kawakami and S. Niskida, Anomalous dispersion of new doubling clad optical fiber. *Electron. Lett.* **10,** 38–40 (1974).

7. M. Monerie, Propagation in doubly-clad single-mode fibers. *IEEE J. Quantum Electron.* **QE-18,** 452–458 (1982).

8. K. I. White, Design parameters for dispersion shifted triangle profile single-mode fiber. *Electron. Lett.* **18,** 725–726 (1982).

9. D. Gloge and E. A. J. Marcatili, Multimode theory of graded-core fibers. *Bell Syst. Tech. J.* **52,** 1563–1578 (1973).

10. R. Olshousky and D. B. Keck, Pulse broadening in graded-index optical fibers. *Appl. Opt.* **15,** 483–491 (1976).

11. T. Li, Structure, parameters, and transmission properties of optical fibers. *Proc. IEEE* **68,** 1175–1180 (1980).

12. V. Daido, The estimated coupling efficiency of light emitting diode with large core high N. A. fibers. *Fujitsu Sci. Tech. J.* **14,** 25–36 (1978).

13. R. Olshousky, Propagation in glass optical waveguides. *Rev. Mod. Phys.* **51,** 341–367 (1979).

14. S. E. Miller and A. G. Chynoweth, eds., *Optical Fiber Telecommunications.* Academic Press, New York, 1979.

15. D. B. Keck, R. D. Maurer, and P. C. Schultz, On the ultimate lower-limit of attenuation in glass optical waveguides. *Appl. Phys.* **22,** 307–309 (1979).

16. P. C. Schultz, Progress in optical waveguide process and materials. *Appl. Opt.* **18,** 3654–3693 (1979).

17. J. A. Harrington, Infrared fiber optical waveguides. *Tech. Dig. Conf. Laser Electro-Opt. Syst.,* p. 96 (1980).

18. J. Lucas, M. Changhansinh, M. Poulain, and M. J. Weber, Preparation and optical properties of neodymium flouziconate glasses. *J. Non-Cryst. Solids* **27,** 273–283 (1978).

19. M. G. Drexhage and K. P. Quinlan, Fluoride glasses for visible to mid-IR guided-wave optics. In *Physics of Fiber Optics,* 57–73 (B. Bendos and S. S. Mitra, eds.). American Ceramic Society, Columbus, Ohio, 1981.

20. S. Takahashi, S. Shuibata, T. Kanamori, S. Mitachi, and T. Mainabe, New fluoride glasses of IR transmission. In *Physics of Fiber Optics* 74–83 (B. Bendos and S. S. Mitra, eds.). American Ceramic Society, Columbus, Ohio, 1981.

21. S. Sakuragi, K. Imagawa, M. Saito, and H. Kotani, Ir transmission capabilities of thallium halide and silver halide optical fibers. In *Physics of Fiber Optics,* 84–93 (B. Bendos and S. S. Mitra, eds.). American Ceramic Society, Columbus, Ohio, 1981.

22. J. A. Savage, Optical properties of chalcognide glasses. *J. Non-Cryst. Solids* **47** (1), 101–116 (1982).

23. R. M. Glen, A review of plastic optical fibers. *Chemtronics* **1,** 98 (1986).

24. D. Gloge, Bending loss in multi-mode fibers with graded and ungraded core index. *Appl. Opt.* **11,** 2506–2513 (1972).

25. D. Marcuse, Gaussian approximation of the fundamental modes of graded-index fibers. *J. Opt. Soc. Am.* **68,** 103–109 (1978).

26. K. J. Beals, C. R. Day, W. J. Duncan, J. E. Midwinter, and G. R. Newns, Preparation of sodium borosilicate glass fiber for optical communications. *Proc. Inst. Electr. Eng.* **123,** 591–596 (1976).

27. P. C. Schultz, Fabrication of optical waveguide by the outside vapor deposition process. *Proc. IEEE* **68,** 1187–1190 (1980).

28. J. B. MacChesney, Materials and processes for preform fabrication. *Proc. IEEE* **68,** 1181–1184 (1980).

29. T. Izawa and N. Inagaki, Materials and processes for fiber preform fabrication-vapor phase axial deposition. *Proc. IEEE* **68,** 1184–1187 (1980).

30. U. C. Paek, High-speed high-strength fiber drawing. *Jour. IEEE Lightwave Tech.* LT-4**(8),** 1048–1060 (1986).

31. A. D. Pearson, Fabrication of single-mode fiber at high rate in very long lengths for submarine cable. Tech Dig. 3rd Integer. Opt. and Opt. Fib. Conf., WA-3, 86–87, San Francisco (1981).

32. R. E. Jaeger, A. D. Pearson, J. C. Williams, and H. M. Presby, Fiber drawing and control. In *Optical Fiber Telecommunications* (S. E. Miller and A. G. Chynoweth, eds.), Chapter 9. Academic Press, New York, 1979.

33. F. V. DiMarello, D. L. Brownlow, and D. S. Shenk, Strength characterization of multikilometer silica fiber. In *Conf. Proc. Int. Conf. Opt. Fiber Commun. Integr. Opt.*, MG-6, 26–27 (1981).

34. U. C. Paek and C. M. Schroedes, High-speed coating of optical fibers with UV curable materials at a rate of greater than 5 m/s. *Appl. Opt.* **20,** 4028–4034 (1981).

35. L. I. Blyler, Jr. and F. V. DiMarcello, Fiber drawing, coating, and jacketing. *Proc. IEEE* **68,** 1194–1198 (1980).

36. U. C. Paek and C. M. Schroeder, High speed coating of optical fibers with UV curable materials. In *Appl. Opt.* **20,** 4028–4034 (1981).

37. M. Sato, *Fukudo, Dig. Integrated Optics and Optical Fiber Communication '81*, San Francisco, Paper MG.4.

38. M. I. Schwartz, P. F. Gagen, and M. R. Santana, Fiber cable design and characterization. *Proc. IEEE* **68,** 1214–1219 (1980).

39. T. Li, ed., *Optical Fiber Communications*, Vol. 1. Academic Press, Orlando, Florida, 1985.

3

Wave Propagation in Lightguides

Introduction

A lightguide is a dielectric waveguide in which an optical wave, usually in the visible to infrared spectrum, can be transmitted along the dielectric medium. The structure of a lightguide may consist of a thin dielectric slab sandwiched between two similar slabs as in a planar lightguide or a thin-core fiber surrounded by a dielectric cladding as in a cylindrical optical fiber. If n_1 and n_2, the refractive indices of the core and cladding, respectively, and the condition that $n_1 > n_2$ is maintained, lightwave transmission along the guide is possible.

In this chapter, we investigate the following:

1. The propagation properties of two types of lightguides, the planar and cylindrical optical fiber
2. Methods of treatment for multimode fibers
3. The transmission loss mechanisms in dielectric lightguides
4. The dispersion properties of a lightguide.

Planar Lightguide

Although optical fibers used in communication systems are usually cylindrical in shape, planar lightguide plays an increasingly important role in modern optical fiber systems as integrated optics is introduced into the development. Moreover, the geometry of a planar lightguide simplifies the analysis. The results may even help us to visualize the more complicated propagation analysis of a cylindrical optical fiber.

Planar Lightguide Geometry

A typical dielectric lightguide may have a geometric configuration as shown in Fig. 3.1. The center core is a dielectric slab of thickness $2a$ and with a refractive index n_1. It is sandwiched between two slabs (the cladding slabs) with refractive index n_2 (assume symmetrical slabs). The thickness of the cladding slabs is assumed to be infinite, and $n_1 > n_2$. The guide is oriented such that the direction of wave propagation is along the z-axis, while the core and cladding thickness direction is along the x-axis. Both y- and z-direction dimensions are extending to infinity. Indices n_1 and n_2 are both uniform and lossless.

Assume also that the z-direction variation of the field components in time and z are sinusoidal and can be expressed as $\exp[i(\omega t - \beta z)]$, where ω is the angular frequency and β is the propagation phase constant of the wave motion.

Introducing this simplified geometry to Maxwell's equations reveals the existence of two independent modes, the TE and TM modes in the solution for planar lightguides [1]. For the TE mode, there exist the E_y, H_x, and H_z component fields and for the TM mode, there are H_y, E_x, and E_z field components. The E_y component of the TE mode can be obtained by solving the wave equation

$$\frac{d^2E_y}{dx^2} + (k_0^2 n_j^2 - \beta^2)E_y = 0 \qquad (3\text{-}1)$$

where $k_0 = 2\pi/\lambda_0$ is the free-space propagation constant, λ_0 is the free-space wavelength, and n_1 is the refractive index; $j = 1$ for the core,

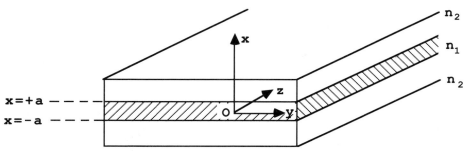

Figure 3.1 Example of a dielectric slab waveguide consisting of a sheet of material with refractive index n_1 and thickness $2a$ between two layers of material with n_2, and n_2 lower than n_1.

$j = 2$ for the cladding. Other field components can be obtained from Maxwell's equations as

$$H_x = -\frac{\beta}{\omega\mu_0}E_y \tag{3-2}$$

and

$$H_z = -\frac{1}{i\omega\mu_0}\frac{\partial E_y}{\partial x} \tag{3-3}$$

where μ_0 is the material permeability.

For the TM mode, we solve the $1 - d$ wave equation

$$\frac{d^2H_y}{dx^2} + (k_0^n n_j^2 - \beta^2)E_y = 0 \tag{3-4}$$

and obtain the other components as

$$E_x = \frac{\beta}{\omega\varepsilon}H_y$$

$$\tag{3-5}$$

$$E_z = \frac{1}{i\omega\varepsilon}\frac{\partial H_y}{\partial x}$$

where ε is the dielectric constant and is related to n as $n^2 = \varepsilon/\varepsilon_0$. The solution of the wave equations is as follows. Within the core slab,

$$\left.\begin{array}{l} E_y = A_e \cos k_x x + A_o \sin k_x x \\ H_y = B_e \cos k_x x + b_o \sin {}_x x \end{array}\right\} \quad x \geq \pm a \qquad \begin{array}{l}(3\text{-}7) \\ \\ (3\text{-}8)\end{array}$$

where A_e, A_o, B_e, and B_o are constants. The subscript e denotes even, subscript o denotes odd, and

$$k_x = \pm\sqrt{k_0^2 n_1^2 - \beta^2} \tag{3-9}$$

Outside of the core, we have

$$\left.\begin{array}{l} E_y = C_1 e^{-k_c x} \\ H_y = C_2 e^{-k_c x} \end{array}\right\} \quad x \geq \pm a \qquad \begin{array}{l}(3\text{-}10) \\ \\ (3\text{-}11)\end{array}$$

where C_1 and C_2 are constants and

$$k_c = \pm\sqrt{\beta^2 - k_0^2 n_2^2} \tag{3-12}$$

Introducing a new set of parameters used frequently in fiber optics, let

$$u^2 = a^2(k_0^2 n_1^2 - \beta^2) \tag{3-13}$$

$$w^2 = a^2(\beta^2 - k_0^2 n_2^2) \tag{3-14}$$

and

$$V^2 = u^2 + w^2 = a^2 k_0^2(n_1^2 - n_2^2) = \left(\frac{2\pi}{\lambda_0}\right)^2 a^2(n_1^2 - n_2^2) \tag{3-15}$$

Table 3.1 Field Component Equations

	TE Mode		TM Mode	
	Even	Odd	Even	Odd
Inside the guide region $n = n_1$	$E_y = A_e \cos \dfrac{u_x}{a}$	$E_y = A_o \sin \dfrac{u_x}{a}$	$H_y = B_e \cos \dfrac{u_x}{a}$	$H_y = B_o \sin \dfrac{u_x}{a}$
	$H_x = -\dfrac{\beta}{\omega\mu_o} A_o \cos \dfrac{u_x}{a}$	$H_x = -\dfrac{\beta}{\omega\mu_o} A_o \sin \dfrac{u_x}{a}$	$E_x = \dfrac{\beta}{\varepsilon_o n_1^2 \omega} B_e \cos \dfrac{u_x}{a}$	$E_x = \dfrac{\beta}{\varepsilon_o n_1^2 \omega} B_o \sin \dfrac{u_x}{a}$
	$H_z = \dfrac{1}{i\omega\mu_o} \dfrac{u}{a} A_e \sin \dfrac{u_x}{a}$	$H_z = -\dfrac{A_o}{i\omega\mu_o} \dfrac{u}{a} \cos \dfrac{u_x}{a}$	$E_z = \dfrac{iuB_e}{a\varepsilon_o n_1^2 \omega} \sin \dfrac{u_x}{a}$	$E_z = \dfrac{B_o}{i\varepsilon_o n_1^2 \omega} \dfrac{u}{a} \cos \dfrac{u_x}{a}$
In the cladding region $n = n_2$		$E_y = C_1 e^{-w_x/a}$		$H_y = C_2 e^{-w_x/a}$
		$H_x = \dfrac{\beta}{\omega\mu_o} C_1 e^{-w_x/a}$		$E_x = \dfrac{\beta C_2}{\varepsilon_o n_2^2 \omega} e^{-w_x/a}$
		$H_y = \dfrac{1}{i\omega\mu_o} C_1 e^{-w_x/a}$		$E_z = \dfrac{i\omega}{a} \dfrac{C_2}{\varepsilon_o n_2^2 \omega} e^{-w_x/a}$

The field components are solved and listed in Table 3.1. The component field equations will be used to obtain the eigenvalue equations, field distributions, and power relations in the subsequent sections.

The Eigenvalue Equations and Mode Designation

To find the eigenvalue equation of a particular mode, we match the field components at the boundary. For example, for the even solution of the TE mode, we match the tangential E field and H field at $x = \pm a$. The tangential E and H fields must be continuous. Thus

$$A_e \cos u = C_1 e^{-w} \tag{3-16}$$

and

$$A_e u \sin u = C_1 w e^{-w} \tag{3-17}$$

Dividing, we obtain

$$u \tan u = w \tag{3-18}$$

Following the similar process, Table 3.2 is generated. Notice that as $n_1 \sim n_2$, actual plots for TE and TM modes are very close. We choose

Table 3.2 Eigenvalue Equations of a Planar Lightguide

TE Mode		TM Mode	
Even	Odd	Even	Odd
$u \tan u = w$	$u \cot u = -w$	$u \tan u = \left(\dfrac{n_1}{n_2}\right)^2 w$	$u \cot u = -\left(\dfrac{n_1}{n_2}\right)^2 w$

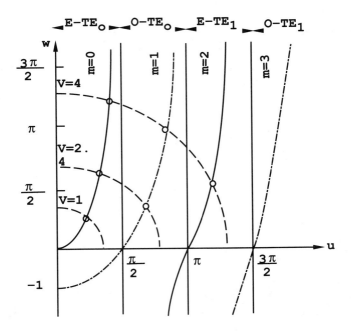

$$\underline{\hspace{1cm}} : \quad w = u \ tg \ u$$

$$\underline{\cdot\underline{}\cdot\underline{}} : \quad w = -u/tg \ u$$

$$\underline{}\underline{} : \quad u^2 + w^2 = V^2$$

Figure 3.2 Graphical solution to Eqs. (4-9) and (4-10) for determining u and w of TE modes for constant values of V in a planar dielectric slab lightguide.

to plot the TE mode as in Fig. 3.2. Here w is plotted against u. For $w = u \tan u$, or $w = -u \cot u$.

The equation $V^2 = u^2 + w^2$ is plotted as arcs of circles for constant V values. The intersections of the circles with the solid or dot–dash lines represent the possible solutions of this set of equations.

The following information becomes obvious:

1. For a circle drawn with a radius corresponding to $V < \pi/2$, there is only one intersection with the solid curve marked $m = 0$. This becomes the only solution for the guiding TE mode, although there exists a similar intersection for the TM mode. To reiterate for the TE mode, this solution signifies the only guiding mode with two polarizations for $V < \pi/2$.

2. No matter how small V becomes, even at $V = 0$, there is always an intersection or a possible solution. This means that this mode has no cutoff frequency.

3. For a circle drawn with a radius corresponding to $V = \pi/2$, the circle intersects two points: one on the $m = 0$ line at $u \cong \pi/3$ and $w = \pi/10$ and the other on the $m = 1$ line at $u = \pi/2$ and $w = 0$.

The first intersection signifies one possible guiding mode while the second intersection indicates the cutoff of a second mode.

4. For $\pi/2 \leq V \leq \pi$, typically for $V = 2.4$, there are two intersections: one with the $m = 0$ line (even TE_o) at $u = \frac{3}{8}\pi$, $w = \frac{2}{3}\pi$ and one with the $m = 1$ line (odd TE_o) at $u = \frac{2}{3}\pi$, $w = \frac{1}{3}\pi$. For $V = \pi$, a third mode is at cutoff and is ready to appear if V increases further.

5. As V increases, the number of modes increases.

6. Cutoff occurs at $u = V$ and $w = 0$. Solving the equation for $w = 0$, one obtains $\beta = k_0 n_2$ and $u = a\sqrt{k_0^2 n_1^2 - k_0^2 n_2^2}$, and this occurs at $u = m(\pi/a)$. Equating $m(\pi/2) = ak_0\sqrt{n_1^2 - n_2^2}$, one can solve for $2a$, the thickness of the guide layer necessary to retain the guiding property as

$$2a = \frac{m\lambda}{2(n_1^2 - n_2^2)^{1/2}} \tag{3-19}$$

where $m = 0, 1, 2$.

For the TM modes, a similar graph may be drawn and similar conclusions may be obtained.

The Propagation Constant β

Once u and w are determined from the graph (Fig. 3.2), the propagation constant β can be calculated from the defining equations. A qualitative discussion may serve as a guide for obtaining the β values. Figure 3.3 shows the $\beta-\omega$ plot of a planar lightguide. Lines $\beta = k_0 n_1$ and $\beta = k_0 n_2$ are two straight lines originating from the origin. These lines form the boundary between which lie all the possible values of β. For a particular mode, $\beta(\omega)$ approximates $k_0 n_2$ at zero frequency (or cutoff frequency), while at infinite frequency (or for frequencies far from cutoff), $\beta(\omega)$ approximates $k_0 n_1$. Thus, a $\beta(\omega)$ curve can be sketched within these boundary lines as shown in Fig. 3.3. The non-

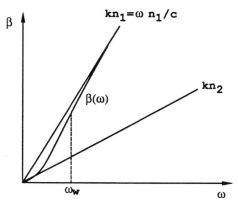

Figure 3.3 The $\beta-\omega$ diagram showing the nonlinear nature of the curve qualitatively.

linearity of the $\beta(\omega)$ curve is obvious. Nonlinearity in $\beta(\omega)$ gives rise to dispersions. We shall discuss these effects later in this chapter.

The Field Shape

The electric and magnetic field distributions of various modes in a planar lightguide are listed in Table 3.1. Field distribution can be evaluated using this table. For example, Fig. 3.4 shows the normalized field distributions E_y/A_e for the first three modes of the even TE modes of a planar lightguide with the following parameters: $n_1 = 3.590$, $n_2 = 3.385$, $2a = 1\ \mu m$, and $\lambda = 0.9\ \mu m$. Notice that for the even modes $m = 0,2$, the maximum field is at the center, while for $m = 1$, the center field is zero. Also notice that at the boundaries between the layers, $x = \pm a$, the field is far from zero. This means that the light field is actually not confined within the boundaries. The existence of field outside the guiding layer excites the evanescent or even the radiating modes, and power is lost through these modes. It is also possible that field outside of the guiding layer may provide couplings between the existing guiding modes and affect the transmission properties of the lightguide.

The light intensity I is proportional to the magnitude of the Poynting vector $\mathbf{p}$ defined by

$$\mathbf{p} = \mathbf{E} \times \mathbf{H} \tag{3-20}$$

The time average of the Poynting vector along the z-axis is given by

$$p_z = \frac{1}{2} \int (E_x H_y - H_x E_y)\, dx \tag{3-21}$$

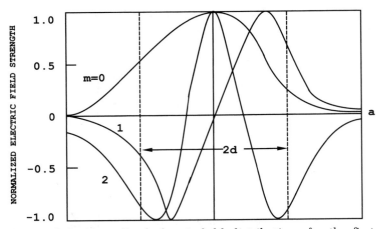

Figure 3.4 Normalized electric-field distributions for the first three low-order modes of the even TE mode of a planar lightguide: $n_1 = 3.590$, $n_2 = 3.385$, $2a = 1\ \mu m$, $\lambda = 0.9\ \mu m$.

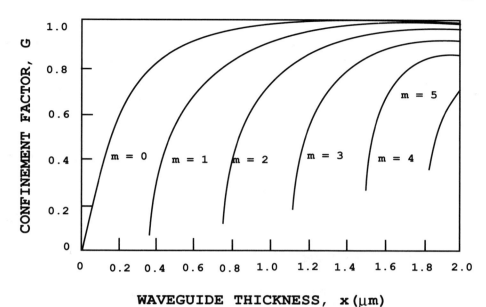

Figure 3.5 Variation of the confinement factor G as a function of the waveguide thickness $2a$ for various modes in a planar lightguide.

For TE modes, both E_x and H_y vanish. This reduces P_z to

$$P_z = \frac{1}{2} \frac{\beta}{\mu\omega} \int E_y^2 \, dx \qquad (3\text{-}22)$$

Thus, the intensity distribution within the guide is proportional to the square of the electric field strength.

A factor indicating the fraction of power that can be guided is defined as the ratio of the intensity inside the lightguide, I_{core}, to the total intensity, which is the sum of the intensities within the guiding core layer and that outside the guiding layer in the cladding. Since

$$I_{core} = \int_{-a}^{a} \frac{\beta A e^2}{\mu\omega} \cos^2 \frac{u_x}{a} \, dx \qquad (3\text{-}23)$$

and

$$I_{cladding} = 2 \int_{a}^{\infty} \frac{\beta A e^2}{\mu\omega} \cos^2 u \exp\left\{ -2 \frac{w}{a} (x - a) \right\} dx \qquad (3\text{-}24)$$

The confinement factor G is

$$G = \frac{I_{core}}{I_{core} + I_{cladding}} = \left[1 + \frac{\cos^2 u}{w[a + (\sin u \cos u)/u]} \right]^{-1} \qquad (3\text{-}25)$$

Figure 3.5 shows the G factor as a function of waveguide thickness for various modes in an even TE mode of a planar lightguide. The following observations can be made:

1. For each mode, G increases as the thickness increases, first rather rapidly and then much more slowly as it reaches saturation.
2. For a very thin lightguide, as $2a$ approaches zero, G approaches zero; in other words, if there is no guidance, all power is lost in the cladding.
3. The G factor varies significantly with m. For each m, there is a minimum thickness to achieve some guidance.
4. For low-order modes, saturation occurs at a smaller thickness.
5. For high-order modes, most of the power is lost in the cladding.

Similar graphs can be constructed for the TM modes, but these results reveal no additional information.

Cylindrical Optical Fibers

Cylindrical Optical Fiber Geometry and Mode Nomenclature

A cylindrical optical fiber is a dielectric fiber consisting of a core fiber of refractive index n_1 surrounded by a concentric dielectric cladding of index n_2 as described in Chapter 2. Cylindrical coordinates are used to describe its geometric configurations. The direction of propagation is along the z-axis. The core–cladding cross section lies in a plane $(r - \theta$ plane$)$ perpendicular to z. Angular symmetry is usually assumed.

For an optical fiber to transmit light efficiently, it is essential that $n_1 > n_2$. If n_1 is uniform across the core, it is called a *step-index* (SI) *fiber*. By the ray-tracing method, one is able to track the ray paths along the fiber. Incident light parallel to the z-axis is known as the *axial ray*, which takes the shortest path and therefore the least time to traverse the fiber length to the receiver. Light incident at different angles smaller than a critical angle encounters multiple reflections along the core–cladding interface, thus taking a longer path and more time to reach the destination. Light incident at angles greater than the critical angle is refracted out of the core area and lost through the cladding. Each ray path may be considered as a mode of propagation. The fiber is a multimode fiber (MMF). Since different rays of the MMF carry the same signal but arrive at the receiver at different times, signal dispersion result. Dispersion limits the bandwidth and ultimately the length of fiber usable for communication. To reduce the differential time delay, graded-index (GI) fibers were designed. The index profile of the core of a GI fiber is designed such that off-center rays that travel longer distances will be accelerated by traveling into mediums of gradually lowering refractive indices to compensate for the lost time.

The ray-tracing method gives a sketching description of the propagation properties of a fiber. It fails to prescribe the parameters desperately needed to design a better fiber. A more exact analytical method is sought. The effort may be rewarding but is not without limitation. Thus far, theoretical analysis of optical fibers yield useful results only when the simplest type (SI fiber), ideally structured, and with single-frequency operation is attempted. Even this was not possible without simplifying assumptions (such as $n_1 \sim n_2$). Some of the results will be highlighted in the next section.

Mode Designation in Cylindrical Optical Fibers

For an ideal optical fiber of the SI type operating at a single frequency, Maxwell's equations can be applied to derive a wave equation in cylindrical coordinate. The solution of the wave equation is complicated by the fact that the boundary conditions of an optical fiber contribute to the existence of all six components of electric and magnetic fields. The resulting mode structure becomes extremely complex. It includes the desired propagating modes along the guide length and the extraneous radiating and evanescent modes elsewhere. Worse yet, all these modes can be coupled to each other through any irregularities and nonsymmetrics introduced along the fiber.

Table 3.3 [2] shows the complexity of the various propagating modes in a cylindrical optical fiber. The first column indicates the range of V, the normalized frequency within which the mode will appear. The second column gives the name of the mode(s), and the third column indicates the total number of propagating modes. Modes having approximately the same cutoff frequency and propagation constants are grouped together although they may have entirely different field distributions. The LP modes are the linearly polarized designation intended to simplify the nomenclature.

The total number of modes shown on the right side of this table arises because each hybrid mode has a twofold degeneracy with clockwise and counterclockwise polarizations, thus doubling the combinations and increasing the total propagating modes. As V increases, the total number of modes increases rapidly.

Table 3.3 Order of Appearance of Various Modes in a Cylindrical Optical Fiber

Range of V	Additional Modes		Total Number of Propagating Modes
0–2.4048	HE_{11}	(LP_{01})	2
2.4048–3.8317	TE_{01}, TM_{01}, HE_{21}	(LP_{11})	6
3.8317–5.1356	HE_{12}, (LP_{02}) EH_{11}, HE_{31}	(LP_{21})	12
5.1356–5.5201	EH_{21}, HE_{41}	(LP_{31})	16
5.5201–6.3802	TE_{02}, TM_{02}, HE_{22}	(LP_{12})	20
6.3802–7.0156	EH_{31}, HE_{51}		24
7.0156–7.5883	HE_{13}, EH_{12}, HE_{32}		30
7.5883–8.4172	EH_{41}, HE_{61}		34

The propagating waves that are confined to the core of the fiber sustain a variety of modes. Those with a strong E_z field compared to the magnetic H_z field along the direction of propagation are designated as the EH modes. Likewise, those with a stronger H_z field than E_z field are called the HE modes. These are the hybrid modes, consisting of all six field components and possessing no circular symmetry. Occasionally, some circularly symmetrical TE and TM modes can also exist. The propagating modes are found to be discrete in nature. They require two indices to identify a given mode, such as HE_{lm} and EH_{lm} modes. Here l, an integer, is the constant introduced in the analysis to separate the variables in the scalar wave equation, and m, another integer, indicates the mth roots of the Bessel functions, J_l and K_l, which are the Bessel functions of the first kind and the modified Bessel functions, respectively. (Solutions of wave equations in cylindrical coordinates involve Bessel functions.) Thus, for each l, there are m possible roots, $m = 1, 2, \ldots, m_{max}$, with one exception. For $l = 0$, there exist two linearly polarized sets of modes, the TE_{0m} and TM_{0m} modes, where either the E or H field in the direction of propagation becomes zero. These are modes with circular symmetry.

The lowest order of mode is not TE_{01} or TM_{01} mode, which have a cutoff frequency corresponding to the first zero of the $J_0(u)$ function at $u = 2.405$, where u is a parameter containing the fiber dimension a and the free-space wavelength λ defined in Chapter 2. The lowest mode is the HE_{11} mode, which is the fundamental mode of a cylindrical lightguide fiber. Its cutoff frequency is zero, corresponding to the first zero of the $J_1(u)$ function at $u = 0$. When $0 < u < 2.405$, there exists only the HE_{11} mode. This is the only region in which the lightguide can be considered as a single mode guide. However, the electric field of the HE_{11} mode has two polarizations orthogonal to each other. One rotates clockwise and the other counterclockwise as they propagate down the guide length. Only by carefully decoupling these polarizing modes can one achieve a truly single-mode, single-polarization transmission free from intermodal interferences [3, 4].

The next-higher-order modes consist of a group of modes designated as the TE_{01}, TM_{01}, and HE_{21} modes. All these modes have approximately the same cutoff frequency ($u = 2.405$) and an almost identical propagation constant β_{01}. With a twofold degeneracy of the HE_{21} mode, there exist six possible modes for $2.405 < u < 3.832$. At $u = 3.832$, the next-higher-order modes set in.

The next-higher-order modes have a cutoff frequency of $u = 3.832$. These consist of HE_{12}, EH_{11}, and HE_{31}. Each of these modes can have a twofold degeneracy, raising the total possible modes to 12 for $3.832 < u < 5.136$. Higher-order mode groups have correspondingly larger combination modes. It is therefore absolutely necessary to limit u, or the size of the fiber, to within $7-10$ μm in order to have a single-mode structure.

The designation of the modes from the characteristic equation was only possible after we made an approximation to the general characteristic equation. Furthermore, by assuming that Δn $(= n_1 - n_2)$ is very small, Gloge [5,6] has shown, at least for the lower-order modes, that the combination modes have the electric field configuration resembling a linearly polarized pattern. Thus, Gloge named these the linearly polarized LP_{lm} modes. The fundamental HE_{11} mode is named the LP_{01} mode, the TE_{01}, TM_{01}, and HE_{21} combination modes as the LP_{11} mode, and so on. Because of the relative simplicity of this notation, it has been universally adopted for mode designation for the fiber guides.

The Characteristic Equation

The actual mathematical analysis follows in the routine manner by solving the wave equation in a cylindrical coordinate system. The direction of propagation is assumed to be along the z-axis.

The vector fields $\bar{H}(r, \phi, z)$ and $\bar{E}(r, \phi, z)$ are expressed as

$$\bar{E}, \bar{H} = \bar{E}(r, \phi), \bar{H}(r, \phi) \exp[i(\omega t - \beta z)] \tag{3-26}$$

where ω is the signal frequency of propagation and β is the propagation constant.

Following the classical routine for solving the wave equation, we first obtain the solution of the z-components of the field as

$$E_z = A \frac{J_l(ur/a)}{J_l(u)} \exp[il\phi] \quad \text{for} \quad 0 \le r \le a \tag{3-27}$$

$$= A \frac{K_l(wr/a)}{K_l(w)} \exp[il\phi] \quad \text{for} \quad r > a \tag{3-28}$$

and

$$H_z = B \frac{J_l(ur/a)}{J_l(u)} \exp[il\phi] \quad \text{for} \quad 0 \le r \le a \tag{3-29}$$

$$= B \frac{K_l(wr/a)}{K_l(w)} \exp[il\phi] \quad \text{for} \quad r > a \tag{3-30}$$

where A and B are arbitrary constants to be determined; u and w are parameters defined in Eqs. (3-13)–(3-15), respectively; and l is the constant used to separate the variables in solving the cylindrical wave equations and is an integer having values ranging from 0 to $l_{max}(0, 1, 2, 3, \ldots, l_{max})$.

The transverse field components E_r, E_ϕ, H_r, and H_ϕ can be expressed in terms of E_z and H_z by using Maxwell's equations. The characteristic equation can be obtained by writing out the tangential field component equations and applying boundary conditions and the continuity relationships.

Snyder and Gloge [5, 7] recognized the fact that the practical fabrication methods such as MCVD or VAD give a small change in index. Also, to keep the pulse dispersion small, the index difference must be very small. In practice, if n_1 is in the vicinity of 1.5, n_2 must be chosen such that the difference $n_1 - n_2$ is on the order of $0.001-0.02$. In other words, let $\Delta = n_1 - n_2$ be small.

If the approximation $\Delta \sim 0$ or $n_1 \sim n_2$ is used, a simplified characteristic equation is obtained as

$$\pm \left\{ u\, \frac{J_1(u)}{J_{1\mp1}(u)} \right\} = w\, \frac{K_1(w)}{K_{1\mp1}(w)} \quad \text{for} \quad n_1 \sim n_2 \tag{3-31}$$

or, alternatively, as

$$u\, \frac{J_{1\mp2}(u)}{J_{1\mp1}(u)} = +w\, \frac{K_{1\mp2}(w)}{K_{1\mp1}(w)} \quad \text{for} \quad n_1 \sim n_2 \tag{3-32}$$

The solutions to their characteristic equations [Eq. (3-31) or (3-32)] yield useful information about the transmission properties of the fiber, as we will discuss in the following sections.

This approximation is justified as practical SMF fibers usually have Δ values of $0.001-0.02$.

We notice that the characteristic equation carries $\pm$ signs. These signs are used to catalog the hybrid modes. The upper sign associates with the HE_{lm} modes and the lower sign, with the EH_{lm}. We use double subscripts l and m for mode designation because for each l value, there are m possible solutions because of the periodic nature of Bessel functions. To catalog the HE_{lm} modes, we choose $J_{l-1,m}$ and $K_{l-1,m}$ as functions; for EH_{lm} modes, we choose $J_{l+1,m}$ and $K_{l+1,m}$.

For example, if $l = 1$, the characteristic equation Eq. (3-31) for the HE_{lm} modes is

$$u\, \frac{J_1(u)}{J_0(u)} = w\, \frac{K_1(w)}{K_0(w)} \tag{3-33}$$

We can find the cutoff frequency of these modes by letting $w = 0$ and $u = V$. The first zero of $J_1(V)$ occurs at zero for $m = 1$. This is the HE_{11} mode. It has no cutoff frequency. Similarly, for $m = 2$, the HE_{12} mode is cut off at $V = 3.832$, and so on. For $l = 2$, the characteristic equation for the HE_{2m} mode becomes, using Eq. (3-30),

$$u\, \frac{J_0(u)}{J_1(u)} = w\, \frac{K_0(w)}{K_1(w)} \tag{3-34}$$

The first zero of $J_0(V)$ is at $V = 2.405$ for the HE_{21} mode. For $m = 2$, that is, for the HE_{22} mode, the cutoff is at $V = 5.52$. For $l = 0$, the circular symmetrical modes TE_{0m} and TM_{0m} exist. The cutoff frequencies of these modes correspond to the zeroes of $J_0(V)$ at $V = 2.405$, 5.52, and so forth. Thus, for $2.405 < V < 5.52$, three modes exist: the TE_{01}, TM_{01}, and HE_{21} modes. All have the same cutoff frequency at $V = 2.405$.

Using the small-index differences approximation, Snyder and Gloge have developed another simplified characteristic equation for designating the linearly polarized LP_{lm} modes:

$$u \, \frac{J_{l-1}(u)}{J_l(u)} = -w \, \frac{K_{l-1}(w)}{K_l(w)} \qquad (3\text{-}35)$$

The LP_{lm} modes appear as a group of modes and can be counted as a single mode. The HE_{11} modes appear as LP_{01} modes followed by the LP_{11} modes, which contain the group of modes designated as TE_{01}, TM_{01}, and HE_{21}. The LP_{02} mode is actually the sum of the TE_{02}, TM_{02}, and HE_{22} modes, and the LP_{21} mode is the sum of the HE_{31} and EH_{11} modes, and so on. The modes are so grouped because within this designated mode, the field distribution can be shown to be linearly polarized. This is possible since the fields in the longitudinal direction have been neglected as a result of the small-index-difference approximation.

Evaluation of the Propagation Constants

The information of β, the propagation constant of the propagating modes, is contained in the parameters u and w. By solving the characteristic equation of a certain mode together with the equation $u^2 + w^2 = V^2$, we can obtain the information of $\beta(V)$. For example, for fundamental LP_{01} (HE_{11}) mode, we solve equations $u^2 + w^2 = V^2$ and $u[J_1(u)/J_0(u)] = w[K_1(w)/K_0(w)]$ to obtain, say, u as a function V. The value of β can be calculated by substituting values of u and V in the defining equations. A plot of u versus V for various groups of modes is shown in Fig. 3.6.

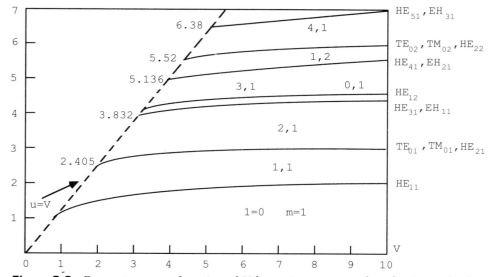

Figure 3.6 Parameter u as a function of V for various groups of modes in a cylindrical dielectric lightguide.

Field Distribution and the Confinement Factor

Once u and w are obtained, we can calculate the field distributions in the fiber. For the approximation that $\Delta n = n_1 - n_2$ is small, Gloge recognized that the field expressions are considerably simpler in appearance if they are expressed in Cartesian coordinates rather than in cylindrical coordinates [5]. Assuming that the transverse field components are essentially linearly polarized for the LP modes, these field components can be expressed as

$$E_y = \frac{E_1}{J_1(u)} J_1\left(\frac{ur}{a}\right) \cos 1\theta \quad \text{for} \quad r < a \qquad (3\text{-}36)$$

and

$$= \frac{E_1}{K_1(w)} K_1\left(\frac{wr}{a}\right) \cos 1\theta \quad \text{for} \quad r > a \qquad (3\text{-}37)$$

where E_1 is the electric field strength at the core–cladding boundary.

The choice of using $\cos 1\theta$ instead of $\sin 1\theta$ is entirely arbitrary. Again, using Maxwell's equations, we obtain the following equations by letting $\beta = k_0 n_1$ or $k_0 n_2$:

$$H_x = \begin{cases} -\dfrac{n_1}{z_0} E_y & \text{for} \quad r < a & (3\text{-}38) \\[2ex] -\dfrac{n_2}{z_0} E_y & \text{for} \quad r > a & (3\text{-}39) \end{cases}$$

where

$$z_0 = \frac{\omega\mu}{k_0}$$

where z_0 is the plane-wave impedance in a vacuum. In this approximation, the E_x and H_y components are very small compared to E_y and H_x.

The lowest mode is designated as the HE_{11} or LP_{01} mode, which is the fundamental mode. It is the dominant mode within the range $0 \le V \le 2.405$. The fields of the HE_{11} mode, from Eqs. (3-36) and (3-37), are given by

$$E_{y,x} = \frac{z_0}{n_2} H_{x,y} = E_0 \begin{cases} \dfrac{J_0\left(\dfrac{ur}{a}\right)}{J_0(u)} & \text{for} \quad 0 \le r \le a & (3\text{-}40) \\[3ex] \dfrac{K_0\left(\dfrac{wr}{a}\right)}{K_0(w)} & \text{for} \quad r \ge a & (3\text{-}41) \end{cases}$$

where E_0 is the normalized amplitude.

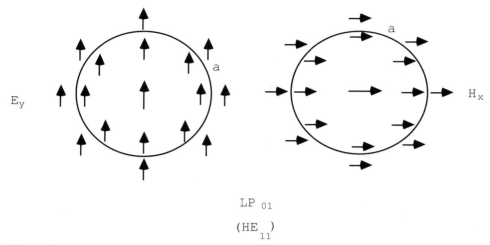

LP $_{01}$

(HE$_{11}$)

Figure 3.7   Electric- and maganetic-field distributions for the LP_{01} or HE_{11} mode.

The corresponding E_y and H_x fields are distributed in the plane of a circular cross section as shown in Fig. 3.7. Similarly, high-order field distributions can be calculated.

The total power carried by a particular fiber mode along the z-direction is given by the following integral [8]:

$$P_t = \frac{1}{2} \int_0^\infty \int_0^{2\pi} \text{Re}[\vec{E} \times \vec{H}^*] \cdot \mathbf{1}_z r \, dr \, d\phi \qquad (3\text{-}42)$$

where Re indicates the real part, $\times$ indicates the vector product, $\cdot$ denotes the scalar product, $*$ indicates the complex conjugate, and $\mathbf{1}_z$ is a unity vector in the z-direction. Using the values of E and H for the particular mode, one can integrate to obtain the total power, including the power transmitted into the core and the power lost in the cladding.

The fractional power carried by the fiber core and that carried by the cladding can be obtained by changing the limit of the first integral of Eq. (3-42) from $\int_0^\infty$ to $\int_0^a$ and $\int_a^\infty$, respectively, and then dividing it by P_t. For the LP_{01} mode, we obtain

$$\frac{P_{core}}{P_t} = 1 - \left(\frac{u}{V}\right)^2 \left[1 - \left(\frac{K_0(w)}{K_1(w)}\right)^2\right] \qquad (3\text{-}43)$$

and

$$\frac{P_{clad}}{P_t} = \left(\frac{u}{V}\right)^2 \left[1 - \left(\frac{K_0(w)}{K_1(w)}\right)^2\right] \qquad (3\text{-}44)$$

Adding Eq. (3-43) to Eq. (3-44) results in $P_t = P_{core} + P_{clad}$, as is expected.

The expression $P_{clad} \neq 0$ indicates that a certain amount of power is

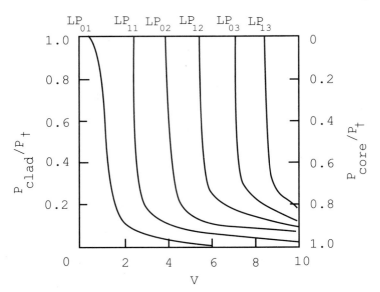

Figure 3.8 A plot of the normalized optical power in the core and cladding as a function of V for several low-order LP modes.

lost. Power is lost either through radiation or as heat loss through evanescent modes. Figure 3.8 is a plot of P_{clad}/P_t and P_{core}/P_t as a function of V for various lower-order modes. Note that for larger values of V increasing amounts of power will be carried onto the core. As $V = (2\pi a/\lambda)(NA)$, a larger V means that the fiber is gathering more light. This happens only in multimode fibers as a/λ is large. For simple-mode fibers, V is usually best kept small. Notice that below cutoff values of V, all power is consumed in the cladding. For example, for the LP_{01} mode, if $V = 1$, 80% of the power is lost in the cladding. To keep 90% of the power in the core, V must be chosen to be about 5. This could set a limit on how small the core radius can be designed for a SMF. For the LP_{01} mode, about 30–50% of the fundamental power is carried by the cladding.

Further power loss in fibers can be introduced by fiber irregularities. In fact, any dielectric waveguide will radiate if it is not absolutely straight. In a bend, for example, its radius of curvature can affect how much radiation loss occurs [9–22]. For the radius of curvature $R > a$, fundamental power is lost through coupling to higher-order modes and/or radiation modes.

Multimode Cylindrical Lightguides

A multimode fiber has a relatively large core and can support many modes of propagation. This makes the study of a multimode fiber by electromagnetic analysis more difficult because it is impractical to track down each mode individually. Instead, ray optics, WKB (Wentzel–Kramers–Brillouin) analysis, or coupled-mode theory are used. Ray optics, or geometric optics, on the other hand, can provide much valuable information about light propagation in MMF by treating the combination of modes as a light ray. This treatment is accurate provided that the core and the length scale governing refractive-index fluctuations are both large compared to the light wavelength. An approximate method, known as the *WKB method*, will also be discussed. Although the WKB method does not give us more accurate results than ray theory, it provides an approximate eigenvalue equation for the mode propagation constant β, which remains undetermined by ray theory. It enables us to count the total number of guided modes and provides some insight into the interesting leaky modes.

Optical fibers are usually treated theoretically as if they were ideal and perfect. Practical fibers are never perfect. Any existing geometric unsymmetry or irregularity will promote couplings between modes. Coupled-mode theory can provide an effective and convenient way to treat these problems. This method will also be treated briefly.

Ray Optics

Ray Path

Consider a planar lightguide consisting of a core slab of refractive index n_1 and thickness d, bounded by top and bottom slabs of refractive indices n_2 and n_3. The thickness of these slabs may be infinitely thick. Assume $n_1 > n_2$ or n_3. For simplicity, let $n_2 = n_3$. The ray paths of the lower orders, say, the LP_{0m} group with $m = 0, 1, 2, 3$, are as shown in Fig. 3.9 [5]. For $m = 0$, the ray path is close to the horizontal. Let its path length be L. As m increases, rays take off at larger angles and the paths are totally reflected at the boundaries and back down to the guide only to be reflected from the lower boundary again. Thus

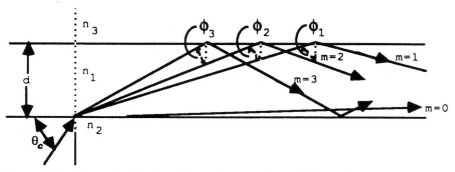

Figure 3.9   Ray paths for the LP_{0m} modes of a planar lightguide: $m = 0,1,2,3$ are the mode order; $\phi_1, \phi_2, \ldots, \phi_m$ are the angles of total reflection of the m ray.

the ray paths are zig-zag as shown. This accounts for the larger group delay, which contributes to signal dispersion. The larger m is, the larger the number of reflections and the longer the path length. Of course, if the entrance angle of the ray is larger than θ_c, rays will be refracted into the cladding and lost. The path length of the propagating rays is increased by a factor $1/\cos \phi_m$, where ϕ_m is related to n_1 and θ by Snell's law ($\sin \theta = n_1 \cos \phi_m$ for $\theta < \theta_c$). Also, $\cos \theta_m = \sqrt{n_1^2 - n_2^2}/n_1 = NA/n_1$, where NA is the numerical aperture in optics.

In cylindrical fibers, the ray paths are more involved [13]. The shortest path follows the axial ray path, which remains on axis. Other rays that enter the fiber obliquely but within an acceptance angle θ_c will travel in helical paths, also being subjected to multiple reflections at the core–cladding boundary. They traverse longer paths and give rise to dispersion. To find the ray path in cylindrical fibers, the general ray equation should be used.

Ray Equation

The ray equation is expressed as

$$\frac{d}{ds}\left(n \frac{d\mathbf{r}}{ds} \right) = \nabla n \qquad (3\text{-}45)$$

where $\mathbf{r}$ is a position vector on the ray path, s is the ray path, and ∇n is the gradient of the refractive index n. To simplify the expression, let us consider only the axial ray. This allows us to replace ds by dz, and let $\tan \theta \approx \sin \theta \approx \theta$ and $\cos \theta \sim 1$. Equation (3-45) can then be written in component forms, enabling us to solve these equations to yield $r(z)$ and $\phi(z)$, which become the parametric equation of the ray path. One can also relate the differential path length to the group velocity and calculate from it the pulse delay and the dispersion effect.

The WKB Method [14]

Although the ray method gives expressions to calculate the dispersion effect, it does not yield the propagation constant β and other parameters essential to guide fiber design. The WKB method is an approximate wave analysis method. It allows us to evaluate the modal propagation constant β and to estimate the total number of guided modes, however, and even provides insight into the tunneling effect known as the *leaky modes*.

For a graded-index MMF with a refractive index profile given by

$$n(r) = \begin{cases} n_1 \left[1 - 2\Delta \left(\dfrac{r}{a} \right)^g \right]^{1/2} & \text{for} \quad r < a \\[2ex] n_1 (1 - 2\Delta)^{1/2} & \text{for} \quad r \geq a \end{cases} \qquad (3\text{-}45a)$$

we can assume that (1) the core radius is large so that many modes will be propagating, (2) the index variation is small over distances of a wavelength so that plane wave approximation can be applied, and (3) the index difference is small so that only transverse *EM* waves exist.

The following wave equation can be written [14]:

$$\frac{d^2\psi}{dr^2} + \frac{1}{r}\frac{d\psi}{dr} - \left[n^2(r)k_0^2 - \beta^2 - \frac{l^2}{r^2}\right]\psi = 0 \qquad (3\text{-}46)$$

Assuming a solution of the form

$$\psi(r) = A(r)e^{i\phi(r)} \qquad (3\text{-}47)$$

we can proceed to solve Eq. (3-46) in the usual manner.

To obtain the WKB solution, we must assume that $\phi(r)$ is a much more rapidly varying function than the amplitude function $A(r)$. This assumption allows us to neglect the second derivative of $A(r)$ relative to the square of the first derivative of $\phi(r)$ and to the second derivative of $\phi(r)$. Making the necessary mathematical manipulations, we obtain the amplitude function $A(r)$

$$A(r) = \frac{C}{\left[\left(n^2k_0^2 - \beta^2\right)r^2 - l^2\right]^{1/4}} \qquad (3\text{-}48)$$

and the phase function

$$\phi(r) = \int_0^r \left[\left(n^2k_0^2 - \beta^2\right) - \frac{l^2}{r^2}\right]^{1/2} dr \qquad (3\text{-}49)$$

where C is an integration constant.

Notice that in both Eqs. (3-48) and (3-49), the argument

$$\left(n^2k_0^2 - \beta^2\right) - \frac{l^2}{r^2} = 0 \qquad (3\text{-}50)$$

may have special significance. For the r values satisfying Eq. (3-50), the amplitude function goes to infinity and the phase function yields imaginary results. Both suggest that within two values of r, r_{max} and r_{min}, the solution is real and the function is oscillatory. Outside these bounds, no oscillation is possible and the resulting exponential functions indicate decay. The limiting boundaries can be found graphically by plotting the functions $n^2(r)k_0^2 - \beta^2$ *and* l^2/r^2 as a function of r as shown in Fig. 3.10. The solid curve is for $n^2(r)k_0^2 - \beta^2$, the dotted curve is for l^2/r^2. The intersections of these curves at r_{min} and r_{max} determine the limits. Thus, when the solid curve is above the dotted curve, the field is oscillatory. When it is below the dotted curve, we have evanescent behavior of the field. Outside the boundaries r_{max} and r_{min}, the field decays exponentially into the r direction.

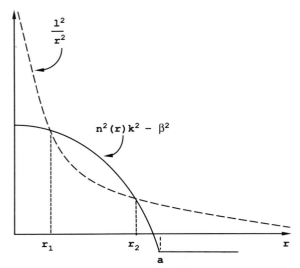

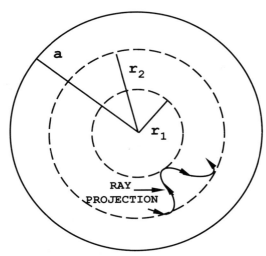

Figure 3.10 Graphical representations of the solutions of the WKB method. The field is oscillatory between the turning points r_1 and r_2; $r_1 = r_{min}$, $r_2 = r_{max}$ in the text. [After D. Gloge and E. A. J. Marcatili, [14] *Bell Syst. Tech.* **52**, 1563–1578 (1973).]

Figure 3.11 Projections on the fiber cross-section of a trapped ray in a graded-index fiber. [After D. Gloge and E. A. J. Marcatili, *Bell Syst.* **52**, 1563–1578 (1973).]

A projection of the ray trajectory on the cross-sectional plane of a fiber is sketched in Fig. 3.11. In this figure, the wave is trapped and oscillating between two circles whose radii are $r = r_{min}$ and $r = r_{max}$, respectively. The limiting boundaries may also be regarded as the turning points of the ray where the total internal reflections take place.

The Eigenvalue Equation

As we noted earlier, one of the conditions for establishing guiding modes in an optical fiber is that on two consecutive reflections (at r_{min} and r_{max}), the total phase angle must be an integer multiple of 2π. Let this integer be m (m is associated with the mth radial mode number). Then Eq. (3-49) yields

$$m\pi = \int_{r_{min}}^{r_{max}} \left[k_0^2 n^2(r) - \beta^2 - \frac{l^2}{r^2} \right]^{1/2} dr \qquad (3\text{-}51)$$

This equation enables us to determine the possible values of the propagation constant β and is the principal result of our discussion. If $n(r)$ is defined, β_{lm} can be calculated. Equation (3-51) is the eigenvalue equation. Figure 3.9 can be extended to explain the physics of a leaky mode.

The turning points r_{min} and r_{max} shown in Fig. 3.10 exist inside the fiber core, that is, $r_{min} < r_{max} < a$. However, the crossing points on the graph actually depend on the values of β and l. For given values of $n(r)$ and k_0, by increasing the l value, the dotted curve moves upward, causing the turning points to move closer together. For even larger values of l, the dotted curve may not touch the solid curve at all. In this case, no guided modes exist. On the other hand, if l remains unchanged while β is increased, this forces the solid curve to move downward, and again it will disengage the dotted curve and create a condition for nonguiding modes. Modes that fail to be guided by the fiber are said to be *cutoff*.

A cutoff condition of quite a different nature occurs when $\beta < n_1 k_0$. When this condition is satisfied, $k_0^2 n^2(r) - \beta^2$ becomes positive for $r > a$. The solid curve in Fig. 3.10 has thus been replaced by a new solid curve shown in Fig. 3.12. For $r > a$, the horizontal portion of the $k_0^2 n^2(r) - \beta^2$ curve is now above the axis. It is then possible that the solid curve can intersect the dotted curve at three places: at $r_{\min}$, $r_{\max}$, and r_3. For $r < r_3$, the solid curve lies above the dotted curve, again indicating oscillating solutions for the field. This causes energy to

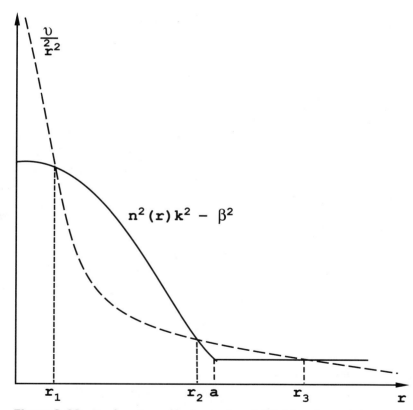

Figure 3.12 Explanation of leaky modes. A third turning point appears at r_3. Power can tunnel out between points r_2 and r_3. [After D. Gloge and E. A. J. Marcatili, *Bell Syst. Tech. J.* **52**, 1563–1578 (1973).]

radiate away in the radial direction into the cladding region, forming the leaky modes [15]. This effect, known as the *tunnel effect* in quantum mechanics, indicates that light energy tunnels through the region between the turning points r_{max} and r_3 and emerges in the cladding at $r > r_3$ as radiant energy. The amount of energy lost per fiber length due to tunneling depends on the length of the tunneling regions $r_3 - r_{max}$. A long tunneling region causes very little leakage of light power; a short tunneling region may lead to high losses. However, inside the core the fields differ only slightly from truly the guided modes.

Total Number of Modes and the Propagation Constant β

The total number of modes and the propagation can be evaluated from Eq. (3-51). First, by integrating over the *l* number and substituting the index profile and integrating over *r*, we obtain

$$M(\beta) = \frac{g}{g+2} (n_1 k_0 a)^2 \Delta \left[\frac{n_1^2 k_0^2 - \beta^2}{2 n_1^2 k_0^2 \Delta} \right]^{(g+2)/g} \tag{3-52}$$

Solving this equation for β yields

$$\beta = n_1 k \left[1 - 2\Delta \left(\frac{M(\beta)}{N} \right)^{g/(g+2)} \right]^{1/2} \tag{3-53}$$

where

$$N = \frac{g}{g+2} (n_1 k_0 a)^2 \Delta \tag{3-54}$$

where N is identified as the total number of guided modes while M is the total number of modes; M values may include those with propagation constants that are larger than β.

Using the important parameter V introduced in Chapter 2, we can write the total number of guiding modes as

$$N = \frac{g}{2(g+2)} V^2 \tag{3-55}$$

Thus, for a parabolic index profile fiber with $g = 2$,

$$N = \frac{1}{4} V^2 \tag{3-56}$$

For a step-index fiber, $g = \infty$, and

$$N = \frac{V^2}{2} \tag{3-57}$$

This indicates that the number of modes in a graded-index fiber is only about one-half of that in a step-index fiber of identical core diameter.

Coupled-Mode Method

The coupled-mode method is an approximate method intended for solving problems involving imperfections and irregularities in regular fiber lightguides. To take advantage of simplified mathematical manipulations, weakly coupled elements are assumed [16]. For those not weakly coupled, more complex procedures are required [17].

Coupled-Mode Equations: Normal-Mode Equations

The initial step in using this method is to convert the system differential equations into coupled-mode form. This involves first neglecting the coupling (or considering each element independently), and writing the decoupled equations. The wave amplitudes are normalized so that they represent energy or power carried by the modes. The coupling is then taken into account by adding a term containing the effect of the presence of the coupled element through the use of coupling coefficients (as yet unknown). The coefficients of coupling are then determined either by direct comparison with the exact equations or by computing the power transfer between modes and applying the principle of energy conservation. These equations can then be simplified further if some of the modes can be neglected. The final step is to determine the propagation constants for the modes. When the boundary conditions are finally applied, a complete solution is obtained.

For example, the coupling effect between two transmission lines can be written as

$$\frac{da_1}{dz} = -i\beta_1 a_1 + C_{12}a_2 \tag{3-58}$$

$$\frac{da_2}{dz} = -i\beta_2 a_2 + C_{21}a_1 \tag{3-59}$$

where a_1 and a_2 are the normal mode amplitudes of the first and second transmission lines, respectively, and C_{12} and C_{21} are the coupling coefficients per unit length. Other components of a values can be written similarly.

In terms of matrix representation,

$$\mathbf{A'} = \mathbf{KA} \tag{3-60}$$

where

$$\mathbf{A} = \begin{bmatrix} a_1 \\ a_2 \end{bmatrix}, \qquad \mathbf{A'} = \begin{bmatrix} a_1' \\ a_2' \end{bmatrix}, \qquad \mathbf{K} = \begin{bmatrix} -i\beta_1 & C_{12} \\ C_{21} & -i\beta_2 \end{bmatrix} \tag{3-61}$$

The prime refers to the z-derivation of $\mathbf{A}$. Coefficients C_{12} and C_{21} are assumed to be independent of length. For weak coupling, it is assumed that C_{12} and C_{21} are small compared to β_1 and β_2.

The modes or wave amplitudes are normalized so that $2|a_{1,2}|^2$ represents approximately the average power carried by each mode.

Solution of Eq. (3-60) is of the form $e^{\gamma z}$, where γ is given by

$$\gamma_{1,2} = \pm \sqrt{\mp|C_{12}|^2 - \left(\frac{\beta_1 - \beta_2}{2}\right)^2} - i\left(\frac{\beta_1 + \beta_2}{2}\right) \qquad (3\text{-}62)$$

The upper sign of $|C_{12}|^2$ is to be used if the group velocities of the uncoupled modes are in the same direction and the lower sign of $|C_{12}|^2$ is taken if the group velocities are in opposite directions. The γ_{12} are the propagation constants of the coupled system. The complete solution of a_1 and a_2 consists of a linear combination of the two modes, plus the constants of integration to be determined by the boundary conditions.

Determination of the Coupling Coefficients

An approximate way to determine the coupling coefficients is to take the ratio of powers carried by the modes. In this manner, C_{ij}, the coupling coefficients that couples mode i to mode j, can be defined as

$$C_{ij} = \left(\frac{P_{ji}}{P_i P_j}\right)^{1/2} \qquad (3\text{-}63)$$

Here, P_i and P_j represent the power carried by the mode i and mode j, respectively; P_{ji} is the power converted from mode i to mode j. Thus the coupling coefficients can be evaluated only when a particular coupled system is considered.

We add here a consideration that allows us to simplify the computation. It is the principle of energy conservation:

$$\frac{d}{dt}\left[|a_1|^2 + |a_2|^2\right] = [(C_{12} - C_{21}^*)a_1^* a_2 + (C_{12}^* + C_{21})a_1 a_2^*] = 0 \qquad (3\text{-}64)$$

For this to be true,

$$C_{12} = -C_{21}^* \qquad (3\text{-}65)$$

The individual average power in modes 1 and 2 are, respectively,

$$P_1(z) = 2|a_1(z)|^2, \qquad P_2(z) = 2|a_2(z)|^2 \qquad (3\text{-}66)$$

Couplings in Nonconventional Fibers

Nonconventional fibers contain irregularities and imperfections. Equation (3-61) can be modified to include these effects. The square matrix of $\mathbf{K}$ is put into the following form:

$$\mathbf{K} = i \begin{bmatrix} -\beta_1 + \Delta\beta_1 & K_1 + iK_1 \\ K_1 - iK_2 & -\beta_2 + \Delta\beta_2 \end{bmatrix} \qquad (3\text{-}67)$$

where the attenuation has been ignored. Also, we consider the case $\beta_1 = \beta_2 = \beta$; K_1 and K_2 are coupling coefficients, and $\Delta\beta_1$ and $\Delta\beta_2$ are the changes in phase constants due to coupling.

According to Qian [18], couplings of ideal modes in deformed fibers are generally of two kinds. For the first kind, the couplings are characterized by imaginary coupling coefficients (i.e., jK_1, K_1 real) as well as by a change in phase constants $\Delta\beta_1$ and $\Delta\beta_2$. These deformations, such as ellipticity and bending, appear symmetrical on the two sides of a fiber cross section. For the second kind, the deformations appear unsymmetrical on the two sides of a fiber cross section. These are characterized by real coupling coefficients (i.e., K_2) and appear to have no change in the phases.

Solutions of problems involving nonconventional fibers usually require several kinds of modes and their suitable couplings to describe the complete field structure. Marcuse [20] developed a coupled-mode theory for these fibers in which ideal modes and local modes are linked. His derivations start from Maxwell's equations. However, because he uses vector notations and complex mathematical manipulations, the physical interpretations are often obscured. Huang [21] suggested a different approach. He considers the coupling of ideal modes, local modes, and super local modes as a representation of the electromagnetic field in the actual fiber. Although the coupled mode formulation is by no means unique, as the choice of the reference waveguide is arbitrary, the above-mentioned three sets of modes are found useful in solving problems involving nonconventional fibers such as those with twists and elliptical core. Huang uses matrix theory to relate different sets of modes. For the weakly coupled case of multimode fibers, the matrix equations are admirably simple in actual calculations. Readers should refer to his original work to appreciate the elegance of his approach [23].

Optical Fiber Loss Mechanisms

Optical fibers suffer transmission losses just as any other transmission medium. But the mechanism responsible for the loss may be quite different. Figure 3.13 shows the general shape of the fiber loss in decibels per kilometer and its constituents as a function of the wavelength. First, we notice that the fiber loss is wavelength-dependent. The estimated (theoretical) loss decreases with increasing wavelength in the visible to infrared range (1.0–1.7 μm) but increases sharply for $\lambda > 1.7$ μm. The experimental curve shows more humps in between.

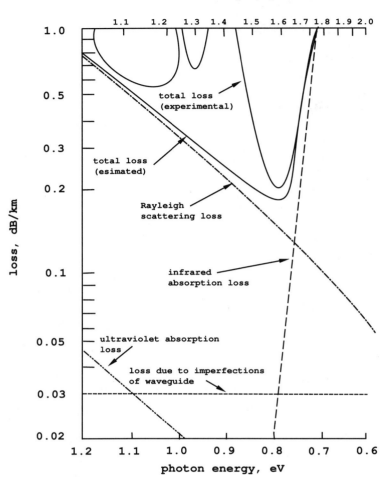

Figure 3.13 A typical loss spectrum of a GeO$_2$-doped silica glass.
[After T. Miya *et al.*, *Electron. Lett.* **15**, 116–118 (1979).]

Analysis of the mechanism shows that the theoretical curve is dominated by two mechanisms, Rayleigh scattering and IR absorption losses. At low wavelengths, Rayleigh scattering dominates, and is inversely proportional to the fourth power of $\lambda(1/\lambda^4)$. Above 1.7 μm, the infrared absorption loss takes over, increasing rather sharply with λ. In the experimental curve, it is found that besides Rayleigh scattering and IR absorption there are many loss peaks due primarily to the resonance absorption of impurities and contamination in the fiber material.

In the following sections, we try to understand the loss mechanism and establish a criterion for building a low-loss optical fiber.

Classification of Fiber Losses

First let us divide the loss mechanism into two categories, the inherent and the induced fiber losses. Inherent losses include Rayleigh scattering, ultraviolet (UV), and IR absorption losses. These losses will persist even though other losses could be reduced to zero. They are the limiting losses in fibers. The induced and/or introduced losses include the inclusion of contaminating atoms or ions and those affected by geometric irregularities, bending, microbending, and radiation. Design and processing of fibers are aimed at reducing these losses as much as possible. A short discussion of the characteristics of these loss mechanisms follows.

Intrinsic Losses

Scattering

Because of the granular appearance of the atoms or molecules of the glass fiber, light power transmitted through the fiber suffers scattering loss. This is known as *Rayleigh scattering loss*. The power loss through scattering can be expressed in terms of a loss coefficient defined as the ratio of the total power that is lost per unit length to the power carried by the incident wave. The power loss coefficient of Rayleigh scattering of SiO_2 glass can be expressed [22] as

$$2\alpha = \frac{A}{\lambda^4} \tag{3-68}$$

where $1/\lambda^4$ is the characteristic wavelength-dependence factor of Rayleigh scattering and A is the parameter geared to the material properties and the temperature of glass. For a single-component glass such as SiO_2, $A = (8\pi^3/3)n_0^8 p^2 \beta kT$, where n_0 is the refractive index, p is the photoelastic coefficient, β the thermal compressibility, k is Boltzmann's coefficient, and T is the absolute temperature of the sample. For compound glass, the compositional fluctuations can also enter into the picture and increase the power loss.

The addition of dopants into the silica glass increases Rayleigh scattering loss because the microscopic inhomogeneities become more important [24, 25]. It has also been observed that the loss increases linearly with the index difference for GeO_2-doped fibers [24]. For other dopants, the influence of the index difference on fiber loss seems to be slightly less.

Absorption (Intrinsic)

The absorption process is closely related to the resonant mechanism. Whenever an applied optical energy causes a quantum transition between different energy levels of the electrons of the fiber material or of different vibrational states of the molecules, the system is said to be in resonance with the source. Energy is absorbed by the system and is eventually lost as heat or in other forms. Light absorption occurs at a wavelength λ for which the relation $\lambda = hc/(E_2 - E_1)$ holds. Here, h is Planck's constant, c is the velocity of light in free space, and E_1 and E_2 are the respective initial and final energy states of either the electronic or vibrational states of the fiber material. Strong electronic absorption occurs at UV wavelengths, while at IR wavelengths, the vibrational absorption becomes dominating.

Ultraviolet absorption leads to an absorption tail in the wavelength range of interest below 1 μm. It decreases exponentially with increasing wavelength. Typically, values of 1 dB/km at 0.62 μm and 0.02 dB/km at 1.24 μm have been reported [26]. In comparison with the Rayleigh scattering loss within this wavelength range, it is less significant and is often negligible in comparison [see the ———--——--——— (labeled ultraviolet absorption loss) line in Fig. 3.13].

Infrared absorption behaves quite differently. Izawa et al. [27] showed that the IR absorption tail of silica occurs around 9–13 μm as a result of the strong resonance of the silica tetrahedron. The peak of the absorption in this wavelength range amounts to 10^{10} dB/km. Overtones and combinations of these fundamental vibrations lead to various absorption peaks at shorter wavelengths. The important peaks are around 3 μm with an intensity of 5×10^4 dB/km and at 3.8 μm with a 6×10^5-dB/km intensity. The tails of these various absorption peaks result in typical values of 0.02 dB/km at 1.55 μm, 0.1 dB/km at 1.63 μm, and 1 dB/km at 1.77 μm. This trend of IR loss is responsible for the long-wavelength cutoff of the transmission in high-silicon optical fibers around 1.8 μm. The dashed line in Fig. 3.13 refers to the pure-silica IR absorption losses. The dotted–dash curve in the same figure labeled Rayleigh scattering loss corresponds to the pure-silica Rayleigh scattering loss. The relative importance of these losses in the corresponding wavelength range is obvious.

The sum of these intrinsic losses results in a solid curve marked the "estimated total loss curve." It represents the ultimate loss that can be expected from a silica glass as a function of wavelength. The wavelength at which ultralow loss can be reached is clearly indicated as around 1.55 μm. Experimental data plotted above the ideal curve shows higher loss and is extremely irregular. The cause of these irregular peaks and valleys is believed to be contamination.

Besides the GeO_2-doped silica fiber, other dopants such as B_2O_3 and P_2O_5 are used for both the cladding and the core. Dopants can be used to influence the fiber loss spectrum, particularly on the longer wavelength end of the spectrum [28, 29]. Thus boron and/or phosphorus will shift the IR absorption edge to lower wavelengths, while germanium has its influence mainly on Rayleigh scattering.

Losses Due to Contaminations

Inclusion of contaminants such as metal ions and hydroxyl radical ions (OH) is the chief cause of impurity absorption.

Transition-Metal Ions

It is very difficult to produce fused silica of perfect purity. The transition metals, such as copper, iron, nickel, vanadium, chromium, and manganese, occur in the host material and ions usually have electronic absorption in or near the visible part of the spectrum. The transition-metal ions are formed during the oxidation process in glass and leave incompletely filled inner electron shells. This gives rise to the characteristic absorptions by inducing transitions between shells. Only through proper balance between oxidation and reduction by controlling the partial pressure of the oxygen in the melt can this absorption loss be reduced to a minimum. Table 3.4 lists the ions, the absorption peaks, and the concentrations in parts per billion (ppb) that are required to keep their contribution to light absorption below 1 dB/km at the peak absorption wavelength [30]. It can easily be seen that extreme purification of the glass materials intended for optical fibers is required.

OH Radicals

Water is an important, undesirable contaminant of optical fibers. Its presence contributes to the vibrational absorption losses in fibers. The OH radical of the H_2O molecule vibrates at a fundamental frequency corresponding to the IR light wavelength of $\lambda = 2.8$ μm. Since the OH radical is slightly aharmonic (not a true harmonic oscillator), "overtones" can occur. These cause OH absorption lines to occur at $\lambda = 1.39, 0.95,$ and 0.725 μm, the second, third, and fourth harmon-

Table 3.4   Transition Metallic Ions

Ion[a]	Absorption Peak (μm)	Concentration (ppb)
Cu^{2+}	0.8	0.45
Fe^{2+}	1.1	0.40
Ni^{2+}	0.65	0.20
V^{3+}	0.475	0.90
Cr^{3+}	0.675	0.40
Mn^{3+}	0.50	0.90

[a]The ionization state of the metal for which these data apply is indicated by superscripts [30].

ics of the fundamental vibrational frequencies, respectively [31]. Broad peaks can appear.

For a concentration of 1 ppm (parts per million), the corresponding attenuations introduced by the absorption are about 1 dB/km at 0.95 μm, 3 dB/km at 1.24 μm, and 40 dB/km at 1.39 μm. It appears necessary to keep the OH concentration at levels below 0.1 ppm if ultralow losses are desired in the 1.20–1.60-μm range.

Induced Fiber Losses

Fiber losses may be introduced during manufacturing processes. These include the irregularities introduced in the fiber drawing process, microbends resulting from coating and cabling processes, curvature or bends in the installation process, joint and jacketing effects, the leaky mode effect, and similar processes. With suitable design and careful fabrication, the added losses from these causes can be made very small.

Geometrical Effects

Irregularities in fiber size include fluctuations of core diameter and perturbations in the size of the fiber as a function of z. Usually the variation of core diameter is under strict control and its effect can be kept to a minimum. But the fluctuation along the length direction is difficult to control. Marcuse [32] and others [33] found that transfer of power from guided modes to radiating modes takes place in a fiber length of approximately 5 cm for a 0.1% change in core width. A root-mean-square (rms) deviation of one of the waveguide walls of 9 Å will cause radiation loss of 10 dB/km.

Present preform fabrication and fiber drawing techniques are under sufficient control to render the losses from these perturbations negligible in both SMF and MMF. Avoidance of these losses is usually achieved by maintaining a uniform-fiber cross section, aided by surface tension in the flowing glass as it is drawn, and having residual core size changes occur with very long periods.

Bending Losses

Any dielectric waveguide will radiate if it is not absolutely straight. The simplest qualitative description of the bending losses in a fiber can be obtained by assuming that in the bent fiber, the field is not significantly changed compared to that of a straight fiber. The plane wavefronts associated with the guided mode are pivoted at the center of curvature of the bent fiber, and their longitudinal velocity along the local fiber axis increases with the distance from the center of curvature. As the phase velocity in the core is slightly smaller than that of a plane wave in the cladding, which is assumed to be extending to infinity, there must be a critical distance from the center of curvature, above which the phase velocity would exceed that of a plane wave in the cladding. The electromagnetic field resists this phenomenon by radiating power away from the guide, causing radiation losses. The radiation attenuation coefficient has the form [34]

$$2\alpha_r = C_1 \exp(-C_2 R) \qquad (3\text{-}69)$$

in which C_1 and C_2 are independent of R. The attenuation coefficient is thus exponentially related to the bend radius R. Equation (3-69) allows us to conclude that (1) the bending losses will increase dramatically when the radius of curvature decreases; (2) as the field decays in the cladding exponentially, decreasing the radius of curvature quickly increases the radiation power; (3) a mode close to cutoff will be affected more than that far from cutoff; and (4) a high index difference will be of primary importance for decreasing the bending losses.

For a permissible bending loss $B(v)$, the bending radius R is related to Δ and λ by the expression [16]

$$R = \frac{B(v)\lambda}{\Delta^{3/2}} \qquad (3\text{-}70)$$

where $B(v)$ is a function of the normalized frequency V. Letting $B(v) = 0.1$ dB/km, the permissible radius of curvature of the fiber bend can be calculated and plotted as a function of Δ for various values of λ. For a step-index profile, this is plotted in Fig. 3.14. The relation is useful in fiber structure design.

The refractive index difference Δ must be made large so that the radius of curvature, which is produced in cabling and fiber handling, does not become smaller than the permissible bending radius [35].

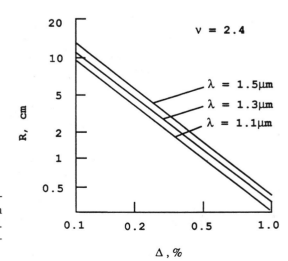

Figure 3.14 The permissible radius of curvature R of an optical fiber bend as a function of the refractive-index difference Δ. [After H. Heiblam and J. H. Harris, [34] *Quantum Electron.* **QE-11**, 75–83 (1975) © IEEE 1975.]

Microbending Losses

Microbending losses are induced primarily during the process of fiber jacketing and cabling. In these processes, the fiber is subjected to microscopic deviations of the fiber axis from the straight condition. This causes a random oscillation of the radius of curvature $R(z)$ around its normal straight position. The problem becomes serious when R of the bend is itself small but large compared with the radius of the fiber. The bending curvature may change slowly and continuously with the fiber length. As a result of bending, the fundamental mode power is eventually lost through coupling to high-order modes and/or radiation modes [36].

In general, microbending losses decrease rapidly with increasing Δ. Therefore, it is very important to use a single-mode, step-index fiber with a numerical aperture as high as possible for long distance telecommunication systems.

In MMF, microbending loss is found to vary as $a^{2\rho}/\Delta^{\rho+1}$, $\rho \geq 1$ [37]. It can be controlled by decreasing the fiber core size and/or increasing its numerical aperture NA. In practice, proper cable design can keep microbending loss below 0.2 dB/km [38–40].

Cladding and Jacketing Effects

Fibers often have cladding materials with higher losses than the core materials. It is inevitable that some signal power will be scattered into the cladding. Reconversion of the power from cladding modes to core-guided modes would have undesirable effects on the group delay characteristic for the guided mode. It is therefore important to provide a loss in the cladding to dissipate the scattered power. Jackets of very large lossy material are used to cover the cladding to reduce cross-talk between fibers in a cable and to suppress unwanted cladding modes. Since the losses from these effects are proportional to the fractions of the modal field intensities that reside in these regions, high-order modes tend to be affected much more than low-order modes.

Marcuse [32] found that a proper core/cladding radius ratio a/b is also important to fiber loss. For SMF, the ratio $(b - a)/a = 5$ will yield an attenuation loss factor of approximately 0.2 dB/m or 200 dB/km added to the HE_{11} mode loss due to the lossy jacket. Thus, the cladding thickness of five times the core radius is not sufficiently large. However, a cladding thickness of 10 times the core radius yields a million times smaller loss for the HE_{11} mode by this computation. For multimode fibers, typical cladding thickness is about 1.5 times the core radius.

Intermodal and Intramodal Dispersions

Signal dispersion in fiber communication systems can be classified as intermodal or intramodal dispersion. Intermodal dispersion, which dominates the dispersion in multimode fibers, is due to the differential time delay between modes at a single frequency. The significance of the intermodal dispersion in MMF can be made to diminish by grading the refractive index n of the core or by simply replacing them with SMF. Intramodal dispersion, known also as *chromatic dispersion*, is due to the variation of group velocity of a particular mode with wavelength. It can be subdivided into material and waveguide dispersion depending on whether the fiber material or the waveguide structure is responsible for the signal dispersion. Intramodal dispersion plays a significant role in SMF as only one mode is propagating. In practical optical fibers, both types of dispersion exist. In MMF with a graded index profile, the intermodal dispersion could be reduced. Then, however, the intramodal dispersion may become important. In SMF, when operating under zero dispersion condition, the two orthogonally polarized modes of the HE_{11} mode will travel down a birefringent fiber with a slightly different group velocity, thus making intermodal dispersion important again.

Intramodal Dispersion of a Single-Mode Fiber

Intramodal dispersion consists of the material dispersion and waveguide dispersion.

Material Dispersion

If the refractive index n of the fiber material varies with wavelength, thus causing the group velocity to vary, it is classified as material dispersion. Our first question is how n, the refractive index of fiber material, becomes wavelength-dependent.

Readers may recall that in the last section, we introduced the resonant interaction of electromagnetic wave fields with bound electrons or molecules of the fiber material that contributes to absorption loss in fibers. Since elastically bound particles tend to oscillate at characteristic frequencies, the interaction between the lightwave and bound electrons becomes frequency-dependent. This, in turn, causes the refractive index n of the glass material, and thereby the material dispersion, to be frequency-dependent.

The effect of resonance on the refractive index of glass can be expressed by Sellmeier's equation [41]. By neglecting the effect of damping and considering multiple resonances, Sellmeier's equation may be written as

$$n^2 - 1 = \sum_{j=1}^{p} \frac{\lambda^2 B_j}{\lambda^2 - \lambda_j^2} \tag{3-71}$$

where λ_j represents the jth resonant wavelength, B_j the corresponding constants, and $j = 1, 2, \ldots, p$ are integers.

Marcuse [42] used two resonances (i.e., $j = 1, 2$): one at $\lambda_1 =$

0.1 μm, the UV resonance, and one at $\lambda_2 = 9.0$ μm, the IR resonance of fused-silica SiO_2 to demonstrate the validity of Eq. (3-71). By choosing B_1 and B_2 (from experimental measurements of n), he fitted the calculated results with the experimental curve closely. In actual fibers, there may be many resonant elements to affect the refractive index. But Marcuse's demonstration clearly indicated that (1) the resonance mechanism is responsible for the wavelength-dependence of the refractive index of fibers and (2) the IR and UV resonances could be the major resonances that affect n.

Our second question is how $n(\lambda)$ affects the light propagation in fibers.

To discuss the material dispersion of plane-wave propagation in a homogeneous dispersive medium, the concept of group velocity should first be introduced. Although individual plane waves propagate with a phase velocity $v_p = \omega/\beta$, a signal travels with a group velocity $v_g = d\omega/d\beta$.

The transit time required for a pulse to travel a distance L is

$$\tau = L/v_g = L\, d\beta/d\omega = L\frac{d\beta}{d\lambda} \cdot \frac{d\lambda}{d\omega} \tag{3-72}$$

where $\beta = nk = 2\pi n/\lambda$, and $\lambda = 2\pi c/\omega$. Since n is a function of λ, therefore, the group velocity also varies with wavelength. Thus τ, the travel time per unit length, becomes

$$\tau = \frac{L}{c}\left(n - \lambda\frac{dn}{d\lambda}\right) \tag{3-73}$$

The second term on the right-hand side of Eq. (3-73) indicates the wavelength-dependence of the transit time. The difference in travel time at two extreme wavelengths $\Delta(\tau/L) = (\tau/L)_2 - (\tau/L)_1$ is a measure of dispersion, provided that the pulses of λ_1 and λ_2 have well-defined edges. In practice, the source spectra width will affect the group delay time. Our third question is how the source spectral spread affects the dispersion. For a dispersive medium, a pulse produced by a light source of spectral width $\Delta\lambda$ that arrives at the receiver, after traveling a distance L, will spread out over a time interval

$$\Delta\tau = \frac{d\tau}{d\lambda}\Delta\lambda \tag{3-74}$$

The derivative $d\tau/d\lambda$ describes the pulse spreading and is therefore more interesting than the delay time τ itself.

From Eq. (3-73), we obtain

$$\frac{d\tau}{d\lambda} = -\frac{L}{c}\lambda\frac{d^2n}{d\lambda^2} \tag{3-75}$$

The dispersion, as measured by the pulse spread, can be expressed as

$$\Delta\left(\frac{\tau}{L}\right) = -\frac{L}{c}\lambda\frac{d^2n}{d\lambda^2}\Delta\lambda \tag{3-76}$$

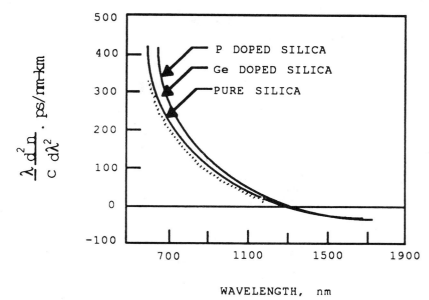

Figure 3.15 Material-dispersion measurements for pure and doped silica glasses. [After D. N. Payne and A. H. Hartog, [43].

A normalized dispersion parameter σ_m can be defined as

$$\sigma_m = \frac{\Delta\tau}{L\,\Delta\lambda} = -\frac{L}{c}\frac{d^2n}{d\lambda^2} \qquad (3\text{-}77)$$

where σ_m represents the dispersion parameter due to material dispersion. The units of σ_m are nanoseconds of pulse width increase per kilometer of path length per nanometer of spectral width of the source (ns/nm·km).

Figure 3.15 shows the effect of dispersion of most silicate fibers as a function of wavelength [43]. The dotted curve represents pure silica glass. Also shown are two solid curves, one for a P-doped and the other for a Ge-doped silica. The curves assume similar shapes with a slight shift in the zero crossing points. Below the crossing point at shorter wavelengths, σ_m is negative. This means that longer wavelength travels faster than shorter wavelength. The reverse is true for wavelengths above the crossing point. For silical fiber without doping, the crossing point is at 1.27 μm.

Waveguide Dispersion of a Single-Mode Fiber

Waveguide dispersion is the result of wavelength-dependence of the propagation characteristic β of the optical fiber. Usually, β must be obtained by solving the eigenvalue equation of the particular mode of interest. Exact solution is very difficult to obtain. However, the use of some new parameters such as u, w, and V enable us to express β in an approximate form from which the dispersion parameter can be derived. For an ideal SI fiber and by defining a new normalized propagation constant b as

$$b = 1 - \frac{u^2}{V^2} \tag{3-78}$$

we can relate b to β by the approximation $b = (\beta^2/k_0^2 - n_2^2)/(n_1^2 - n_2^2)$, or

$$\beta = n_2 k_0 (b\Delta + 1) \tag{3-79}$$

we can then express the dispersion parameter due to waveguide dispersion as

$$\sigma_w = -\frac{n^2\Delta}{c\lambda} V \frac{d^2(Vb)}{dV^2} \tag{3-80}$$

For the LP_{01} (or HE_{11}) mode, Gloge [44] has calculated b, $d(Vb)/dv$, and $Vd^2(Vb)/dV^2$ and plotted these quantities as a function of V as shown in Fig. 3.16. The justification of using b as the normalized propagation parameter can be seen from the following considerations.

At cutoff, since $w \cong 0$ and $u \approx V$, it follows that $b \cong 0$. At zero wavelength, as $V \sim \infty$, $b \sim 1$. Thus, $0 < b < 1$ completely defines the propagation characteristic of the fiber. Recall the discussion in a planar lightguide that at cutoff, $\beta = n_2 k_0$. This point corresponds to $b = 0$. At zero wavelength, $\beta = n_1 k_0$, which corresponds to $b \sim 1$; thus $k_0 n_2 < \beta < k_0 n_1$ corresponds to $0 < b < 1$.

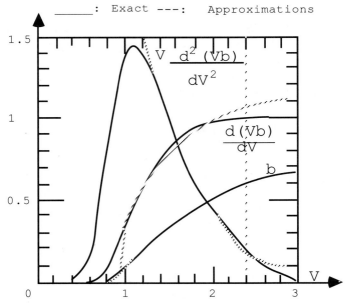

Figure 3.16 Normalized propagation constant b, normalized group delay $d(Vb)/dV$, and normalized waveguide-dispersion parameter $V[d_2(Vb)/dV_2]$ as a function of the normalized frequency V. [After D. Gloge, [44] (1971).]

The curve identified as $(d/dV)(Vb)$ is proportional to the normalized group delay time. A normalized group delay equal to zero means that the signal propagates at the group velocity of a plane wave in the cladding, whereas a normalized group delay equal to one leads to the plane-wave group velocity in the core.

Chromatic dispersion, representing the combined effect of material and waveguide dispersion, can be grouped as $C(\lambda)$, where

$$C(\lambda) = M(\lambda) - \frac{\Delta n}{c\lambda} V \frac{d^2(Vb)}{dV^2} \qquad (3\text{-}81)$$

The Zero Dispersion Point

Zero material dispersion occurs whenever $d^2n/d\lambda^2 = 0$ [4]. For pure-silica glass, this occurs at about 1.27 μm. Since V and λ are related by an expression $V = 2.405\lambda_c/\lambda$, for a single-mode pure silica fiber with a $\lambda_c = 1.18$ μm, the zero material dispersion (ZMD) point corresponds to $V \approx 2.2$. In general, from experimental observations [43], it appears that for the LP_{01} mode, above $\lambda > 1.2$ μm, material and waveguide dispersions possess opposite signs. Therefore, it is possible for these effects to cancel and achieve zero chromatic dispersion in fibers within a range of wavelengths. It would be even better if this range of wavelengths could coincide with that for minimum fiber loss. For this purpose, many researchers have been trying to shift the zero material dispersion point or the waveguide crossover point to accomplish true cancellation at the desired wavelength.

The following is a short summary of the efforts to shift the zero chromatic dispersion point.

1. *Effect of core radius on the dispersion curve.* Marcuse [45] found that there is an optimum core radius at which the effect of waveguide dispersion is very small. For fibers with a core radius larger than this optimum radius, material dispersion dominates the chromatic dispersion. For a smaller core radius the effect of waveguide dispersion on total chromatic dispersion increases very rapidly. It can even be used to shift the zero crossing wavelength slightly.

2. *Effect of doping on the material-dispersion curve.* Adams [46] found that by varying the GeO_2 doping to the SiO_2 fiber from 0 to 15% may shift the zero material dispersion point from 1.07 to 1.4 μm.

3. *Effect of the index difference Δ.* If the effect of the dopant is neglected, it is found [46] that the chromatic dispersion decreases as Δn is increased [as $V(Vb)$ increases]. The wavelength at which zero dispersion occurs shifts toward a longer one.

4. *The w-fiber.* The w-fiber [47] gives added flexibility for moving the zero-dispersion wavelength by manipulating the design parameters. These include the ratio of the depressed cladding/core ra-

dius ratio b/a and the amount of index change $\Delta n'/\Delta n$, and so on. One can either achieve a shift of the zero chromatic dispersion (ZCD) point between 1.3 and 1.7 μm or achieve a flat response within this wavelength range.

Intermodal Dispersion in a Multimode Fiber: Ideal (Uncoupled) Multimode Fibers

To qualify as an ideal multimode fiber, let us assume that (1) the power distribution among all modes is uniform, (2) the loss phenomena attenuate all modes equally, and (3) no coupling exists between modes.

Intermodal dispersion is caused by the variation in group delay time among the propagating modes at a single frequency. It is the dominating dispersion factor that affects the signal distortion in multimode fibers. It has been shown that the delay time of each mode depends slightly on wavelength, but more strongly on the mode number. This fact makes the WKB method a convenient means to compute the intermodal dispersion effect in multimode fibers.

The simplest way to compute the delay time in a step-index multimode fiber is by ray tracing. Here we compute the time difference between a ray that travels the shortest path and one that travels the longest path. Recall that in Chapter 1, we recognized that an axial ray that enters the fiber of an angle $\theta = 0$ travels the axial path and this is known to be the shortest path. In contrast, a ray that enters the fiber at an angle θ_m, where $\theta_m = \sin^{-1}(\sqrt{n_1^2 - n_2^2}/n_1)$ is taking the longest path. Assuming that only meridian rays are present, the core profile is uniform, and the phase shift accompanying each total reflection at the boundary can be neglected, we can easily write the time difference τ per unit length as

$$\tau = t_{\text{max}} - t_0 = (\sec \theta_m - 1)t_0 \approx \frac{1}{2} \sin^2 \theta_m t_0 \approx t_0 \Delta \qquad (3\text{-}81)$$

where t_{max} and t_0 are the time for traveling the maximum and axial path length, respectively. This equation suggests that the delay time is proportional to the index difference, a fact that is well recognized as true in optical fiber practice. For example, if $n_1 = 1.5$, $\Delta = 0.01$, and $n_1/c = 5$ μs/km, then $\tau = 50$ ns/km; τ is roughly on the right order of magnitude for a weakly guiding multimode fiber.

To use the WKB method to estimate the intermodal dispersion of a graded-index multimode fiber, we first recall Eq. (3-53) for β, the propagation constant of the fiber:

$$\beta = n_1 k \left[1 - 2\Delta \left(\frac{M(\beta)}{N} \right)^{g/(g+2)} \right]^{1/2} \qquad (3\text{-}53)$$

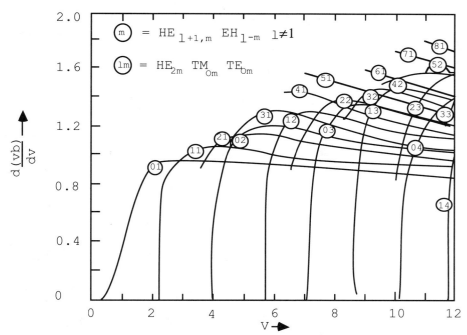

Figure 3.17 Normalized group delay as a function of V. [After D. Gloge, *Appl. Phys.* **10**, 2252–2258 (1971).]

where N is the total number of guided modes defined in Eq. (3-54) and M is the total number of modes whose propagation constants are greater than β. Modes with large β values are more tightly guided. The value of β decreases with increasing mode order and reaches the value $\beta_c = n_2 k_0$ at cutoff.

To calculate the delay time, we use Eq. (3-53) in Eq. (3-72) to obtain τ, from which an impulse response expression can be derived, and the pulse broadening can then be evaluated. Pulse broadening determines the bandwidth of the MMF. The fiber bandwidth is inversely proportional to the rms pulse width from a given pulse shape.

The mode delay expressed as $d(Vb)/dv$ for a SI multimode fiber can be calculated and plotted against V as shown in Fig. 3.17. Two important features can be observed: (1) for increasing V values, the mode delay approaches a lower asymptote. This is to say, in a large core fiber, more rays travel directly down the center of the guide; and (2) as the mode approaches cutoff, the group delay rapidly decreases and a progressively larger amount of energy is shifted from the core with the cladding.

Olshansky and Keck [48] obtained an expression for pulse broadening that contains the parameters Δ and g. It is a complex equation, and we will not repeat the derivation but instead will summarize the observations as follows:

1. The expression of the pulse width contains a leading term that is proportional to Δ. This confirms the simple finding obtained from the ray theory.

2. The fiber parameters that affect intermodal dispersion are those characterizing the refractive-index profile. In the case of a graded-index profile with power-law variation, Δ and g are the parameters of importance.

3. The optimum g value for minimum pulse width (or maximum bandwidth) can be obtained by differentiating the expression for the pulse width with respect to g and equating it to zero. If the effect of source spectral width is neglected,

$$g_{opt} = 2 - \frac{15\Delta}{5} \tag{3-82}$$

4. The effect of source spectral width on the intermodal dispersion can be demonstrated by calculating the rms pulse spreading in a graded-index fiber as a function of g for various light sources of different spectral width. The results are shown in Fig. 3.18. In the calculation three light sources [an LED, an injection laser, and a distributed feedback (DFB) laser] having spectral width values of 150, 10, and 2 Å, respectively, are used. The fiber is a titanium-doped fiber operating on a wavelength of 0.9 μm. The parameters involved are $n_1 = 1.460$ and $n_2 = 1.452$. For a wide-spectral-width

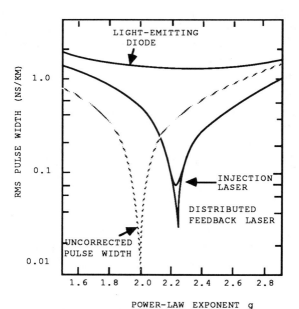

Figure 3.18 Calculated rms pulse spreading in a Ti-doped fiber as a function of the parameter g for various source spectral widths. [After R. Olshansky and D. B. Keck, [48] (1976).]

source (an LED), the rms pulse width is almost flat as g varies. A pulse broadening of less than 1.5 ns/km can be achieved if g is within 25% of the optimum value. For an injection laser, the curve shows a sharp dip at the optimum g value. Now, a g value within 5% of the optimum value will yield a pulse width of less than 0.2 ns/km. For a narrow-spectral-width source such as the DFB laser, the curve shows a very sharp dip at the optimum g value. A pulse width of less than 0.05 ns/km can be predicted at the optimum g value. The sensitivity of the dispersion to the spectral width is clearly demonstrated. The dotted, peaked curve represents the uncorrected curve, which assumes no material dispersion and negligible source spectral width.

5. The optimum g value for minimum pulse width shown in Eq. (3-82) is valid provided the host glass and dopant material have identical material dispersion. If the dispersive properties of the dopant differ from those of the host glass, the group velocity of light is different at different radial positions in the fiber core. This means the optimum g value will change with the dopant. This is called *profile dispersion*. To reduce this effect, the refractive index profile can be reshaped [49].

6. Minor deformations of the profile can reduce the peak bandwidth drastically. Very tight fabrication control is required to make a MMF of high bandwidth. The use of a composite power-law profile or a multiplicity of dopant materials has been tried to broaden the spectral range of high bandwidth of a multimode fiber.

Effects of Couplings between Modes

Practical fibers are usually not ideal. In reality, there exists coupling between modes. In fact, the neighboring modes are usually quite tightly coupled. Also different modes will most likely be attenuated differently and the power distribution among all modes will never be uniform.

Coupling among guided modes has a beneficial effect on intermodal dispersion. However, couplings to nonguiding modes such as the evanescent and radiating modes may result in additional loss.

Assuming that power is coupled between the guided modes, a portion of the pulse, which may have been traveling at more than the average speed, can couple and transfer energy to slower modes. As a result, the average spreading over all modes of the pulse width may be reduced. It has been shown that instead of spreading at a rate proportional to L, the length of the transmission, with mode coupling, the spreading rate increases only to the square root of L [50]. Mode coupling tends to average out the propagation delays associated with the modes, thereby reducing intermodal dispersion. The total delay time can vary between $\sigma_0 L$ and $\sigma_0 L^{1/2}$, where τ_0 is the intermodal dispersion per unit length.

Pulse Width and Bandwidth

The dispersion coefficient can be expressed in either of two ways: in terms of pulse spread or bandwidth. In the former case, time-domain analysis is used. A short pulse is injected at the input, and the time response at the output is measured. Because of the random nature of the interaction, the rms values are used. The total pulse spread σ_t is expressed as

$$\sigma_t^2 = \sigma_i^2 + \sigma_c^2 \qquad (3\text{-}83)$$

where σ_i is the pulse spread due to intermodal dispersion and $\sigma_c^2 = \sigma_m^2 + \sigma_w^2$ is the chromatic spread, the sum of the pulse spread due to material and waveguide dispersion. To include the effect of source spread $\Delta\lambda$, the dispersion coefficient D_λ can be expressed in picoseconds per nanometer-kilometer as

$$D_\lambda = \frac{\sigma_t}{\Delta\lambda \ L} \qquad (3\text{-}84)$$

Bandwidth is defined as the frequency at which the Fourier amplitude of the transformed pulse response falls to half of its peak value. For pulses with Gaussian temporal response, the frequency bandwidth $\Delta f_{-3d\beta}$ (electrical) in picoseconds can be related to the rms group delay time $\sigma = (\sigma_t)$ as

$$\Delta f_{-3d\beta} \text{ (electrical)} \cong \frac{1.33}{\sigma} \qquad (3\text{-}85)$$

A similar definition for an optical bandwidth is

$$\Delta f_{-3d\beta} \text{ (optical)} \cong \frac{1.87}{\sigma} \qquad (3\text{-}86)$$

Here, the frequency is between points when the optical power is reduced to half of the original value.

Notice that the bandwidth is inversely proportional to the rms delay time.

At zero-dispersion wavelength where σ becomes zero, the bandwidth can approach infinity. In reality this never happens. At this wavelength, the second-order dispersion effect which has been neglected thus far becomes important.

Sometimes a term called bit rate, B, is used in place of the frequency bandwidth and

$$B = f_{-3d\beta} \text{ (optical)} \qquad (3\text{-}87)$$

The product $B \times$ distance is often used in system comparison.

Conclusion

In this chapter we have discussed the wave propagation properties in lightguides. The treatment is different for different lightguides. In planar lightguide, we solved the planar wave equations to obtain the field expressions of the *TE*- and *TM*-mode waves. The modes are identified and propagation constants are defined. Although planar lightguides are seldom used for communication, they are increasingly useful for building couplers and switches in integrated optics. Cylindrical optical fibers are introduced next. We have avoided the massive mathematical manipulations for the analysis. Only the results of the findings are given. Multimode optical fibers reserve special treatment. We introduced those approaches to analyze the problem. Finally, the fiber loss mechanisms and dispersion effects are briefly discussed. Ample references are given for serious readers to pursue further.

References

1. J. E. Midwinter, *Optical Fibers for Transmission*. Wiley, New York, 1979.
2. A. N. Snyder, Asymptotic expression for eigenfunctions and eigenvalues of a dielectric or optical waveguide. *IEEE Trans. Microwave Theory Tech.* **MTT-17,** 1130–1138 (1969).
3. L. Eyges, P. Gianiuo, and P. Wintersteiner, Modes of dielectric waveguides of arbitrary cross sectional shape. *J. Opt. Soc. Am.* **69,** 1226–1235 (1979).
4. R. B. Dyott, J. R. Cozens, and D. B. Morris, Preservation of polarization in optical-fiber waveguides with elliptical cores. *Electron. Lett.* **15,** 380–382 (1979).
5. D. Gloge, Weakly guided fibers. *Appl. Opt.* **19,** 2252 (1971).
6. C. Yeh, Guided-wave modes in cylindrical optical fibers. *IEEE Trans. Educ.* **E-30,** No. 1 (1987).
7. A. N. Snyder, Asymptotic expression for eigenfunctions and eigenvalues of a dielectric or optical waveguide. *IEEE Trans. Microwave Theory Tech.* **MTT-17,** 1130–1138 (1969).
8. R. Olshansky, Propagation in glass optical waveguides. *Rev. Mod. Phys.* **51,** 341–367 (1979).
9. D. Marcuse, Microbending losses of single-mode, step-index and multimode. Parabolic-index fibers. *Bell Syst. Tech. J.* **55,** 937–955 (1976).
10. K. Petermann, Theory of microbending loss in monomode fibers with arbitrary refractive index profile. *Acta Electron. Ubertraqung* **30,** 337–342 (1976).
11. J. Sakai and T. Kiumura, Practical microbending loss formula for single-mode optical fibers. *IEEE J. Quantum Electron.* **QE-15,** 497–500 (1979).
12. L. B. Jeunhomme, *Single-Mode Fiber Optics*. Dekker, New York, 1983.
13. D. Marcuse, *Theory of Dielectric Optical Waveguides*. Academic Press, New York, 1974.
14. D. Gloge and E. A. J. Marcatili, Multitude theory of graded-core fibers. *Bell Syst. Tech. J.* **52,** 1563–1578 (1973).
15. S. E. Miller and A. G. Chynoweth, eds., *Optical Fiber Telecommunications*, Chapter 3. Academic Press, New York, 1979.
16. W. H. Louisell, *Coupled Mode and Parametric Electronics*. Wiley, New York, 1960.
17. H. A. Haus, Electron beam waves in microwave tubes. *Tech. Rep.—Mass. Inst. Technol. Res. Lab. Electron.* **316** (1958).

18. J. R. Qian, On coefficients of coupled-wave equations. *Acta Electron. Sin.* No. 2, p. 46 (1982).

19. R. Ulrich and A. Simon, Polarization optics of twisted single mode fiber. *Appl. Opt.* **18**(13), 224–231 (1979).

20. D. Marcuse, *Theory of Dielectric Optical Waveguides.* Academic Press, New York, 1974.

21. H. C. Huang, On local normal modes in optical fiber and film waveguides. *Sci. Sin. (Engl. Ed.)* **22**(10), 1147–1150 (1979).

22. I. L. Fabelinskii, *Molecular Scattering of Light.* Plenum, New York, 1968.

23. H. C. Huang and J. R. Qian, Theory of imperfect nonconventional single mode optical fibers. In *Optical Waveguide Science*, by H. C. Huang and A. W. Snyder. Martinus Nijhoff Publishers, The Hague, Netherlands, 1983.

24. P. Bachmann, P. Geittner, and H. Wilson, The deposition efficiency for the GeO_2-doped single-mode and step-index fibers prepared by the low pressure PCVD process. *Eur. Conf. Opt. Comm. Proc.* **8**, 614–617, Cannes (1982).

25. D. A. Pinnow, T. C. Rich, F. W. Osterwager, and M. Dominico, Fundamental optical attenuation limits in the liquid and glassy state with application to fiber optical waveguide material. *Appl. Phys. Lett.* **22**, 527 (1973).

26. D. B. Keck, R. D. Maurer, and P. C. Schultz, On the ultimate lower limit of attenuation in glass optical waveguides. *Appl. Phys. Lett.* **22**, 307–309 (1973).

27. T. Izawa, N. Shibata, and A. Takeda, Optical attenuation in pure and doped fused silica in the IR wavelength regions. *Appl. Phys. Lett.* **31**, 33–35 (1977).

28. H. Osanai, T. Shoida, T. Moriqama, S. Araki, M. Horiguchi, T. Izawa, and H. Takata, Effects of dopants on transmission loss of low-OH-content optical fibers. *Electron. Lett.* **12**, 549–550 (1976).

29. T. Miya, Y. Terunuma, T. Hosaka, and T. Miyashita, Ultra low loss single-mode fibers at 1.55 μm. *Rev. Electr. Commun. Lab.* **27**, 497–507 (1979).

30. R. D. Maurer, Glass fibers for optical communication. *Proc. IEEE* **61**, 452 (1973).

31. P. Kaiser, Spectral losses of unclad fibers made from high-grade vitreous silica. *Appl. Phys. Lett.* **23**, 45 (1973).

32. D. Marcuse, Power distribution and radiation losses in multimode dielectric slab waveguides. *Bell Syst. Tech. J.* **50**, 1817–1832 (1971).

33. S. E. Miller, Some theory and applications of periodically coupled waves. *Bell Syst. Tech. J.* **48**, 2189–2219 (1969).

34. H. Heiblam and J. H. Harris, Analysis of curved optical waveguides by conformed transportation. *IEEE J. Quantum Electron.* **QE-11**, 75–83 (1975).

35. R. W. Davies, D. Davison, and M. P. Singh, Single-mode optical fiber with arbitrary refractive-index profile: Propagation solution by the numerical method. *IEEE J. Lightwave Technol.* **LT-3**(3), 619–629 (1985).

36. D. Gloge, Bending loss in multimode fibers with graded and ungraded core index. *Appl. Opt.* **11**, 2506–2513 (1972).

37. R. Olshansky, Mode coupling effects in graded-index optical fibers. *Appl. Opt.* **14**, 935–945 (1975).

38. M. I. Schwartz, P. F. Gagen, and M. R. Santaur, Fiber cable design and characterization. *Proc. IEEE* **68**, 1214–1219 (1980).

39. D. Marcuse, Microbending losses of single-mode, step-index and multimode, parabolic-index fibers. *Bell Syst. Tech. J.* **55**, 937–955 (1976).

40. J. Sakai and T. Kimura, Practical microbending loss formula for single-mode optical fibers. *IEEE J. Quantum Electron.* **QE-15**, 497–500 (1979).

41. W. Sellmeier's equation was discussed in a paper "Dispersion and its anomalies" in Annalen der Physik und Chemie, 5th series **145**, 399–421 (J. C. Poggemdorff, ed.) (1872).

42. D. Marcuse, Pulse distortion in single mode fibers. *Appl. Opt.* **19**, 1653–1660 (1980).

43. D. N. Payne and A. H. Hartog, Determination of the wavelength of zero material dispersion in optical fibers by pulse delay measurement. *Electron. Lett.* **13,** 627–629 (1977).
44. D. Gloge, Dispersion in weakly guiding fibers. *Appl. Opt.* **10,** 2242–2445 (1971).
45. D. Marcuse, Interdependence of waveguide and material dispersion. *Appl. Opt.* **18,** 2930–2932 (1979).
46. M. J. Adams, D. N. Payne, F. M. E. Sladen, and A. H. Hartog, Wavelength-dispersive properties of glasses for optical fibers: The germanium enigma. *Electron. Lett.* **14,** 703–705 (1978).
47. D. Marcuse, Equalization of dispersion in single-mode fibers. *Appl. Opt.* **20,** 696–700 (1981).
48. R. Olshansky and D. B. Keck, Pulse broadening in graded-index optical fibers. *Appl. Opt.* **15,** 483–491 (1976).
49. H. M. Presby and I. P. Kamikow, Binary silica optical fibers: Refractive index and profile dispersion measurements. *Appl. Opt.* **15,** 3029–3036 (1976).
50. D. Gloge, Optical power flow in multimode fibers. *Bell Syst. Tech. J.* **51,** 1767–1783 (1972).

4

Auxiliary Components for Optical Fiber Systems

Introduction

Practical implementation of optical fiber systems requires the use of interconnecting devices. Fiber length needs to be extended to cover a required distance, optical fibers need to be coupled to the light source and detector, and other equipment may have to be connected or disconnected to the fiber as required. Splices, connectors, and couplers form an integral part of the system design. Since losses introduced by each item can be summed to significantly increase the total system loss, the design and use of these components becomes equally important as the fiber and devices themselves. We discuss these components in this chapter.

Splices

A splice is employed to join cabled fibers permanently when the system span is longer than the available continuous fiber cable length. Because of the small cross section of the fibers, it is extremely difficult to make a perfect joint. Since misalignment of fibers results in loss at joints, a study of the loss mechanism of joints is warranted. To accurately align the cores of two fibers being joined, two sets of parameters require attention: the intrinsic and the extrinsic parameters. Intrinsic parameters include the core diameter differences, numerical aperture differences, core–cladding eccentricity, and index profile mismatching for the graded-index fibers. Extrinsic parameters include

84

the transverse offset between fiber cores and the separation and axial tilt for multimode graded-index fibers.

To maintain the joint within an acceptable tolerance, many techniques have been developed. Intrinsic misalignment can usually be minimized by choosing the fiber parameters (the core and cladding diameters and refractive index profile) carefully. Fibers of the same kind are used to connect together. Extrinsic misalignments, on the other hand, can easily be introduced at any stage of the process. These include (1) transverse offset or lateral misalignment of the axis of the fibers, (2) angular misalignment, (3) end separation, and (4) non-smooth end surfaces. These misalignments need more attention.

The Loss Mechanism Due to Disalignment

Transverse Offset

In Fig. 4.1a, two fibers of the same diameter are jointed end-to-end but are offset laterally by a distance x. Figure 4.1b shows the computed loss curve assuming uniform light illumination. Loss expression for the calculation is shown in the graph. It is derived from the calculation of the coupling efficiency defined as the ratio of the common area a through which light power is transmitted to the core end area πa^2.

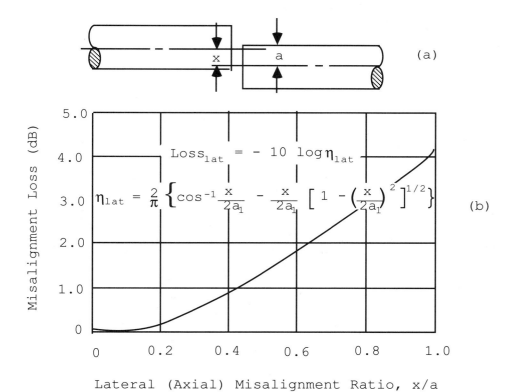

$$\text{Loss}_{lat} = -10 \log \eta_{lat}$$

$$\eta_{lat} = \frac{2}{\pi} \left\{ \cos^{-1} \frac{x}{2a_1} - \frac{x}{2a_1} \left[1 - \left(\frac{x}{2a_1} \right)^2 \right]^{1/2} \right\}$$

Figure 4.1 Lateral (axial) misalignment loss for SI fibers. [After D. Gloge, [1] (1976).]

For graded-index fibers, the calculation of the coupling loss due to lateral offset is more complex since the numerical aperture varies across the fiber end face. For a parabolic GI fiber, if the axial misalignment x is small compared to a, say, $x/a < 0.4$, then [1, 2]

$$L_{\text{offset(GI)}} \approx -10 \, \log \left(1 - \frac{8x}{3\pi a} \right) \qquad (4\text{-}1)$$

This loss expression is only approximately correct because mode coupling has been neglected. Mode coupling tends to change the power distribution and affect the common area. In a GI multimode fiber (MMF), the index profile mismatch tends to complicate the power distribution even more. Fortunately, the power loss given by Eq. (4-1) is on the conservative side. One can use these results as upper limits for guidelines.

For a single-mode fiber (SMF), the lateral loss expression becomes a function of x/w instead of x/a because the spot width w is now determining the common area of transmission.

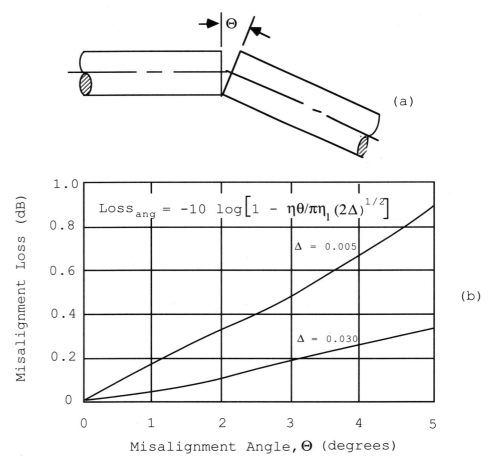

Figure 4.2 Loss due to angular misalignment. [After D. Gloge, [1] (1976).]

Angular Misalignment

Figure 4.2a shows the geometry of angular fiber misalignment. The loss curve and loss expression are shown in Fig. 4.2b. Here η_1 is the refractive index of the fiber core, η is that of the medium filling the gap, and θ is the angular misalignment expressed in radians. Loss curves for two values of Δ, the index difference between the core and cladding, are shown. It is observed that for a specific angular misalignment, the losses are greater for fibers with small values of Δ (or low numerical aperture NA).

End Separation

When fiber ends do not contact to each other but leave an intervening gap, light power is lost in transmission. This is because light beams diverge at the end. As the receiving fiber is further away, it intercepts proportionately less of the cone of light projected by the transmitting fiber, thus giving rise to end separation loss (see Fig. 4.3).

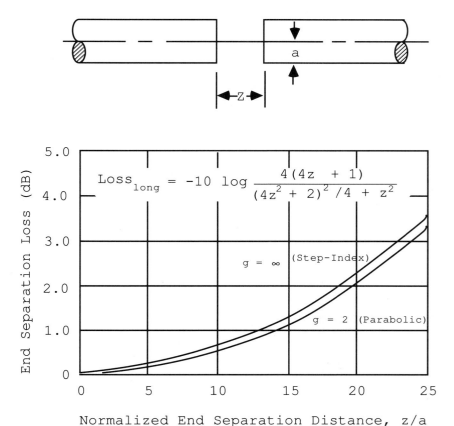

$$\text{Loss}_{\text{long}} = -10 \log \frac{4(4z + 1)}{(4z^2 + 2)^2/4 + z^2}$$

Figure 4.3 End separation losses for single-mode fibers. [After D. Gloge, [1] (1976).]

Nonsmooth End Surfaces

Roughness of the fiber end surfaces also introduces insertion loss. This loss can usually be minimized by filling the gap between the end surfaces of the fibers with index-matching liquid or epoxy.

Among the four types of misalignment loss, the most dominating loss is the lateral misalignment. Other misalignments are usually small by comparison. In practice, a 5% lateral misalignment is considered as tolerable.

Splice Techniques

Splicing is considered a permanent joint. This can be accomplished by either mechanical adhering or fusion splicing [3].

Methods of fiber splicing are classified as the V-groove, fusing and sleeve methods. These are shown in Fig. 4.4.

The simplest mechanical splice uses a V-block to align the bare fibers whose end surfaces have been prepared and apply epoxy resin to hold them in place. Protective layers are then applied to reinforce the structure. Other methods include the tight-sleeve fitting, loose tube (square tubing), and 3-rods to align and secure the fibers in joint.

Fusion splicing is the most prominent technique for joining individual multimode fibers. It is accomplished by applying localized heating at the interface between two butted, prealigned fiber ends, causing the fibers to fuse together. Electric arc or laser heating can be used as an energy source [4, 5].

The average splice insertion loss for MMF ranges from 0.1 to 0.2 dB per splice. Even with careful handling, the tensile strength of a fused fiber can be as low as 30% that of the original fiber. The completed splices are usually packaged so that little or no tensile loading is added. This includes a provision of an accurately produced, rigid metal, or other material alignment number into which the fibers are permanently bonded. Refer to Bisbee [4] for detailed design.

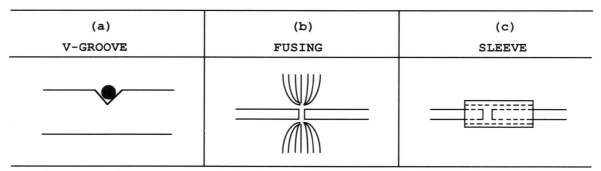

(a)	(b)	(c)
V-GROOVE	FUSING	SLEEVE

Figure 4.4   Fiber splicing methods.

Single-mode fibers typically have core diameters of 5–10 μm and require more accurate transverse offset control (typically <1 μm) to achieve a low splice loss.

Fusion splicing of a SMF has produced an average splice loss of 0.2 dB with well-matched fibers at 1.3 μm wavelength. Other splicing techniques have been reported in the literature [6, 7, 8].

Connectors

A connector is a demountable device used to connect and disconnect fibers whenever it is necessary to do so. A removable connector may have a different set of requirements as a splice joint. Besides the requirement of low insertion loss, a connector must be easy in construction and simple to mount. It must be interchangeable with other connectors, stable in insertion losses after a great number of connect and disconnect actions, low in cross-talk among multiple connectors, and, of course, low in cost. Also, the coupling loss must remain constant with temperature variations and rotational angle.

Connectors may be divided into two basic categories, the lens-coupled and butt-coupled (without lenses).

Lens-Coupled Connectors

A liquid-lens connector is shown schematically in Fig. 4.5. It is designed for a 50-mm-core, graded-index fiber [9]. The center piece is a

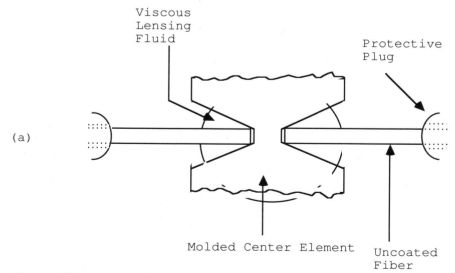

(a)

Figure 4.5 A lens-coupled connector. [After J. F. Dalgleish, (1980) © IEEE; [15].]

molded lens structure. It is composed of a molded, plastic biconical center element with an optical fluid in each of the two concave cavities. The geometric and optical lens parameters are selected to optimize the coupling for a specific fiber. The lens system requires accurate transverse alignment of the fibers, which can be achieved by precision modeling of the biconical center element. Each fiber is mechanically retained in a retractable plug. The insertion loss of this connector is about 1 dB.

Butt-Coupled Connectors

A butt-coupled connector generally consists of a ferrule for each fiber, a precision sleeve into which the ferrules fit, and a cap to maintain the connection. Several types are available [9, 10]. Figures 4.6, 4.7, 4.8, and 4.9 show four varieties: (1) tube alignment, (2) tapered sleeve, (3) jewel bushing, and (4) resilient ferrule.

Tube Alignment Connector

Prepared fiber ends (cut and polished) are inserted into a jack and plug pair of the correct length (Fig. 4.6). The jack and plug are then pushed together to make the connection. The alignment hole in the plug is provided for initial alignment. The device is held in position by screwed caps. This type of connector can never be of high quality and the connector loss is usually high. Since the tube dimension must be large enough to accept the fiber easily, lateral misalignment is the consequence. Gap separation between fiber ends is unavoidable. Also, any contamination in the alignment hole will be pushed forward and pile up between the fibers.

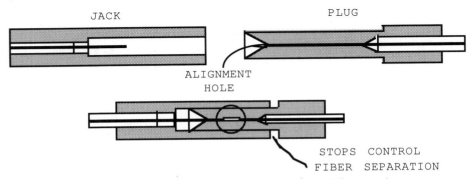

Figure 4.6 A tube alignment connector. Courtesy of AMP Corporation.

Tapered-Sleeve Connector

A tapered sleeve connector uses an accurately machined tapered sleeve to replace the tube and uses plastic buttons on the ends (Fig. 4.7). The fibers are molded into the plastic plugs during manufacture and are maintained in position very precisely.

Jewel Bushing Connector

A jewel bushing connector makes use of precision watch jewel bearings to align the fiber (see Fig. 4.8). Fibers are fitted into stainless-steel ferrules supported by the jewel bearings. Two mated pairs are then slid into a concentric sleeve from both ends for perfect alignment. They are held in position by caps.

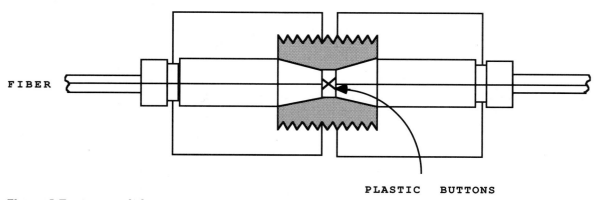

Figure 4.7 A tapered sleeve connector. Courtesy of AMP Corporation.

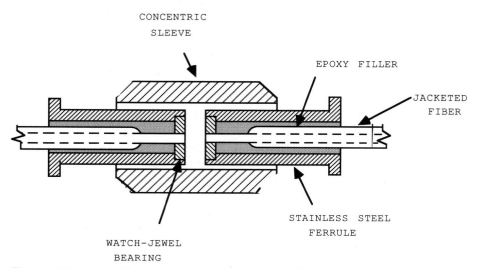

Figure 4.8 A jewel bushing connector. Courtesy of ITT Corporation.

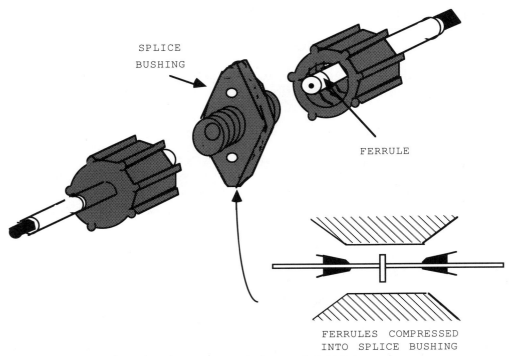

Figure 4.9 A resilient ferrule connector. Courtesy of AMP Corporation.

Resilient Ferrule Connector

A resilient ferrule connector consists of three parts: two plugs and a splice bushing (Fig. 4.9). Each plug is a tapered ferrule through which the fiber is threaded (a manufacturing process). The ferrule holds the fiber and serves as a resilient part of the alignment mechanism. The plug is provided with a crimp ring to pressure-hold the fiber. Two plugs, one from each end, are screwed into the center splice bushing, which is also tapered. The taper compresses the ferrules and seals the fiber ends for a tight fit.

This type of connector can easily be modified to fit a source or a detector bushing.

Couplers

While splices and connectors are two-port devices, couplers [11–13] require a multiport configuration. Multiport optical couplers are the basic interconnecting elements between various networks of modern optical communication systems. These networks can be an omnidirectional broadcast from a single light source to multiple receivers or bidirectional as in local-area networks (LANs). Multipost couplers are the essential optical components that perform the efficient multiplexing and demultiplexing of the light sources into and out of the trunk fiber. In the majority of networks, it is the performance of the couplers that ultimately limits the network performance.

An Ideal Coupler

An ideal coupling element should have the following characteristics:

1. It should distribute the light power among the branches of the network without scattering losses. Finite scattering loss at the couplers could limit the number of branches usable in the network.
2. It should be insensitive to the wavelength within the window for which it is designed.
3. Its performance should not be affected by either the light-power distribution among the fiber modes or the state of polarization.
4. It must be flexible in function to permit the layout of different network configurations for the optimal usage of cable or available optical power.
5. It should also provide the flexibility for connecting to a number of branches even with varying fiber types.
6. It should be able to select the branching power ratios as desired.

No coupler could possess all these attributes. Many designs necessitate compromise among various requirements. In Fig. 4.10, we show an n-port coupler and define the functions of different couplers symbolically. The I_i values indicate the multiinput ports and the O_i values, the multioutput ports ($i = 1, 2, \ldots, n$). The dots indicate possible paths of couplings. Most functions can be described according to the data in Table 4.1. As one can see, most devices have at least three ports.

Generally, a coupler may have a number of input and output ports, labeled $I_1, I_2 \ldots I_n$ and $O_1, O_2, \ldots, O_n$, respectively. They can be arranged to perform two types of function, the tap and the star. A simple tap or T-coupler consists of one input port I_1 and two outputs O_1 and

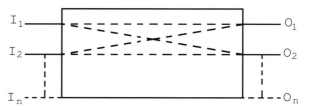

Figure 4.10   A multiport (n-port) coupler.

Table 4.1 Function of Couplers

Type	Input Port(s)	Output Port(s)	Features
Tap	I_1	O_1	Through channel
		O_2	Local output
Directional coupler	I_1	O_1	Through channel
	I_2	O_2	Local output
		O_3	Local output
Star coupler	I_1	$O_1 \cdots O_n$	Equal strength
	I_2	$O_1 \cdots O_n$	on all output
	.	.	ports
	.	.	
	.	.	
	I_n	$O_1 \cdots O_n$	
Bidirectional coupler	I_1	O_1	Half-duplex
	O_1	I_2	operation
WDM multiplexer	$I_1(\lambda_1)$	$O_1(\lambda_1 \cdots \lambda_n)$	
	$I_2(\lambda_2)$		
	.		
	.		
	$I_n(\lambda_n)$		
WDM demultiplexer	$I_1(\lambda_1 \cdots \lambda_n)$	$O_1(\lambda_1)$	
		$O_2(\lambda_2)$	
		.	
		.	
		$O_n(\lambda_n)$	

[After J. F. Dalgleish, *Proc. IEEE* **68**, 1229 (1980).]

O_2. One of the output ports may be the preferred one, usually called the *throughput*, and the other, the tap output. The T-coupler is used to branch out a portion of the light power for monitoring or detecting purposes or to input other sources to the mainstream of the busbar. Of course, if we design the coupler such that $O_1 = O_2$, we have a power splitter. In fact, any power dividing ratio between O_1 and O_2 can be achieved. A more general structure may include another input, say, I_2, to become a four-port coupler, provided isolation between I_1 and I_2 is good.

By interchange of one set of the input and output ports, a bidirectional directional coupler is obtained. This is possible because a coupler is a passive system; any port can be used as input or output.

The star coupler is a multiterminal coupler which has two sets of ports, a set of N transmitting (or input) ports, and a set of N receiving (or output) ports. The star coupler distributes power equally to each of the receiving ports from any one of the transmitting ports.

WDM multiplexers and demultiplexers are couplers permitting multiplexing and demultiplexing of wavelength-division multiplexing of wideband communication systems. Details of these couplers will be discussed in Chapter 13.

Loss Characteristics in Couplers

The optical coupler is a passive device. It introduces insertion losses, which affects the distribution characteristics of a communication system. The following terms are frequently used to characterize couplers: throughput loss, tap loss, excess loss, power-splitting ratio, and directivity. To simplify the presentation, let us limit the example to a four-port coupler, that is, let $N = 4$ in Fig. 4.10. We name port 1 as the *input port*, port 2 as the *preferred output port*, port 3 as *auxiliary port* (for monitoring or other usages), and port 4 as an *isolation port*. Let P_1, P_2, P_3, and P_4 be the light power appearing at each port, respectively. Then we define the following expressions:

Throughput loss $\qquad L_{12} = -10 \log \dfrac{P_2}{P_1}$ $\qquad\qquad\qquad$ (4-2)

Tap loss $\qquad\qquad L_{13} = -10 \log \dfrac{P_3}{P_1}$ $\qquad\qquad\qquad$ (4-3)

Excess loss $\qquad\quad L_{ex} = -10 \log \dfrac{P_1 - P_2 - P_3}{P_1}$ $\qquad$ (4-4)

Power-splitting ratio $\quad R_{23} = \dfrac{P_2}{P_3}$ $\qquad\qquad\qquad\qquad$ (4-5)

Directivity $\qquad\qquad L_{14} = -10 \log \dfrac{P_4}{P_1}$ $\qquad\qquad\qquad$ (4-6)

In an ideal coupler (where $P_1 = P_2 + P_3$ and $L_{ex} = 0$) if no power appears at port 4, or $P_4 = 0$, then the isolation is perfect. A coupler is usually named after the tap loss. Thus a 3-dB coupler implies $L_{13} = 3$ dB, or $P_2 = P_3$, which gives a power splitting ratio of $1:1$. A 10-dB coupler will have an $L_{13} = -10$ dB and a power ratio of $9:1$ and $L_{12} = -0.46$ dB. In terms of tap loss, the throughput loss expression becomes

$$L_{12} = -10 \log (1 - 10^{-L_{13}/10}) \qquad\qquad (4-7)$$

Practical couplers have access losses, but seldom in excess of 1 dB. Other loss expressions have to be adjusted accordingly.

Techniques for Constructing Couplers

Many techniques can be used to construct couplers. Some basic configurations are shown in Figs. 4.11, 4.12, and 4.13. A biconical tapered coupler [11] is shown in Fig. 4.11. Two multimode step or graded-index fibers are twisted together. They are heated and pressed

Separate Fibers
on ends

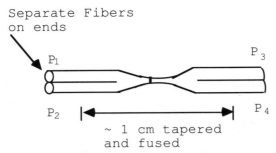

Figure 4.11 A biconical tapered coupler. [After J. F. Dalgleish, 1980 [15] © IEEE, 1980.]

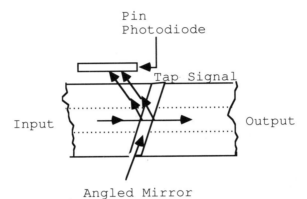

Figure 4.12 A beam splitter tap. (After J. F. Dalgleigh [15].)

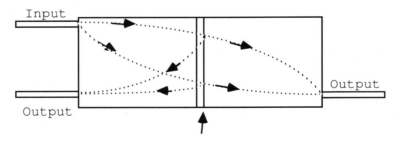

Figure 4.13 A 3-dB beam splitter. (After J. F. Dalgleish, [15].)

to form a double-tapered section. The high-order modes entering port 1 are coupled into the cladding by the input taper, thereby becoming common to the cladding of both ports 3 and 4. This cladding light is recoupled into the cores by the output tapers. Coupling ratios of 3−20 dB have been realized with both graded and step-index fibers. A typical 20-dB graded-index fiber tap has an insertion loss of 1 dB. By increasing the cladding thickness, one can reduce the insertion loss.

Figure 4.12 shows a beam-splitter tap. The beam splitter is created by splitting a fiber into two parallel, angled fiber faces separated by an airgap. Reflections at the two fiber–air interfaces direct light to the photodetector. A beam splitter of a 10-dB coupler with a 0.6-dB insertion loss has been reported [12].

Figure 4.13 illustrates an approach for a 3-dB splitter. A greater pitch self-focused lens, joined across a half-mirrored interface to a second lens, causes the input beam to be expanded onto the interface. One reflected and one transmitted beam are then refocused into the output fibers. The coupling ratio can be changed by varying the thickness of the mirror layer [14].

Other coupling structures can be seen in a review article by Dalgleish [15].

The star coupler [16] is useful as a data distribution system. The construction of a star coupler is remarkably simple. A bundle of bared fibers are wound around one another and fused while under tension, and we have a star coupler. Star couplers can generally be grouped into the transmissive and reflective categories. An 8 × 8 transmissive star coupler and an eight-port reflective star coupler are shown in Fig. 4.14. For the transmissive star (Fig. 4.14a), power put into any port on one side of the coupler emerges from all ports on the other side, divided equally. The ports on the same side of the couplers are isolated from each other. The insertion loss L ideally is

$$L = -10 \log \frac{1}{N} \qquad (4\text{-}8)$$

In practice, excess losses have to be added. Excess losses include scattering loss, absorption loss, connection losses, cable losses, and any other losses incurred in the coupling. Excess losses should be kept as small as possible. By looping one side of the device as shown in Fig. 4.14b, a reflective star is formed.

The reflective star couples light from any one port to all the ports, including the input. Thus every fiber connected to the star carries both transmitted and received data, a directional coupler is needed to separate the two signals at each terminal. Thus, a reflective star coupler suffers all the additional excess losses in the transmissive star plus an additional 3 dB due to power division.

Both T-couplers and star couplers are used extensively in data bus and large distribution network systems. T-Couplers can be connected in series to link many stations, thus cutting the number of fibers compared to using dedicated lines. However, T-couplers in series experience an increased vulnerability to breakage. A break at any point will cut communication. A data bus that uses a star coupler broadcasts the signal to all terminals. Breaking any one line only affects that particular terminal, other terminals will not be affected.

In systems using T-couplers in series, the transmitting light power suffers losses through each coupler. If a portion t of the signal enters

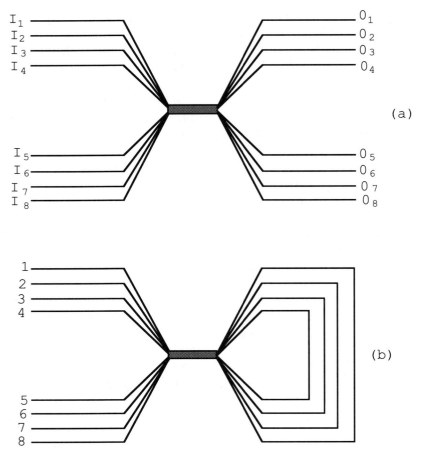

Figure 4.14   Star couplers: (a) a transmissive star; (b) a reflective star.

each T-coupler, the power through the second coupler becomes t^2. After going through N T-couplers, the remaining power becomes t^N. Thus, the throughput loss L_T is

$$L_T = -10 \log t^N = -10 N \log t \qquad (4\text{-}9)$$

Equation (4-9) indicates that L_T is linearly increasing with N. A star coupler has a throughput loss expressed by Eq. (4-9) that is related to N logarithmatically. For small N, T-couplers in series may offer small loss. For large N, however, a star coupler should outperform T-couplers. The cross point will depend on the actual loss in couplers.

Coupling to Source and Detector

Fiber systems require a source and detector to operate [17]. Light power needs to be coupled to the fiber at the sending end and coupled back to a detector at the receiving end. While maximizing the power transfer may be desirable, it is usually more important to maximize the coupling efficiency in most communication systems involving optical fibers. However, the requirements for couplers at the transmitting and receiving ends are different.

Source-to-Fiber Coupling

Coupling efficiency may be defined as

$$\eta = \frac{P_f}{P_s}; \quad \text{coupling loss} = -10 \log \eta \quad (4\text{-}10)$$

where P_f is the power in the fiber and P_s is the source power. In fiber optics, source-to-fiber coupling is usually very inefficient for several reasons:

1. The emitting surface of the source and the fiber end are butt-joined together at the coupler. Any intervening airgap becomes the origin of reflection loss. Reflection loss can be minimized if these surfaces are joined together with a matching-index epoxy.
2. The emitting area A_s may differ from the fiber core cross-sectional area A_c. If $A_s > A_c$, the coupling efficiency is reduced by a ratio A_c/A_s. If $A_s < A_c$, this loss does not exist.
3. In multimode fibers, the coupling efficiency could be proportional to the square of the numerical aperture. For a step-index fiber, the coupling loss due to a small numerical aperture could be very large.
4. The problem of lateral and angular alignment within the coupler between the emitter and the fiber can be very acute.

To improve coupling efficiency at the transmitting end, we can do the following.

Match the Emission Pattern of the Source to Fibers
For a Lambertian source, which emits light omnidirectionally, the radiation pattern can be approximated by a power loss variation of intensity of the form $I = I_0 \cos \phi$, where I_0 is the axial intensity and ϕ is the angle of emission.

Many LEDs are Lambertian sources. Many solid-state light sources have emission patterns that are non-Lambertian. This intensity pattern can be approximated by $I = I_0 \cos^m \phi$, where m is a numerical value characteristic of the source.

For the fibers, the intensity pattern of a SMF has a Gaussian distribution that has a peak at its center, as in curve c in Fig. 4.15 (for the HE_{11} mode).

Couplers with Lenses
Some designs of optical fiber couplers may include lenses to improve the coupling efficiency. It is intended to match the source emitting

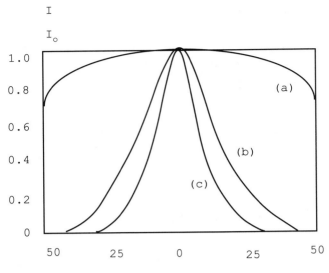

Figure 4.15 Radiation pattern of (a) Lambertian source, (b) non-Lambertian source, and (c) Gaussian source. [After T. Uchida and S. Sugimoto, (1978).]

area to the fiber core area for good coupling efficiency. If the source emitting area is larger than the core area, usually no lens is needed. Otherwise, a lens system can provide a magnification of the emitting area to match exactly the core area (the end face) of the fiber. Lens designs such as image sphere (large sphere), nonimage sphere (small sphere), and cylindrical lenses, have been developed [13].

Multiport couplers can make use of the integrated optics to great advantage. Integrated couplers offer direct coupling between light sources and system components without the necessity of energy conversion. This makes the device very compact and less costly. This will be discussed further in Chapter 13.

Pigtailed Devices
A pigtailed device is simply a source–fiber combination assembly that has been aligned and permanently locked into place. It avoids the necessity of mounting alignment thus improving the coupling efficiency. A fiber-to-fiber connector is then used to complete the connection.

Fiber-to-Detector Coupling

The coupling of an optical fiber to an optical power detector is more easily done because its geometry is much more favorable. High coupling efficiency between fiber and detector can be achieved even if the area of the detector is many times larger than the cross-sectional area of the fiber. Alignment is not critical. It is necessary only to butt the fiber against the active area of the detector. Reflector loss should be minimized. If the geometry does not allow this direct coupling, lenses may be used to focus the light onto the detector.

Multiplexing Couplers

Multiplexing couplers are passive components that perform efficient multiplexing and demultiplexing of a light source into and out of the trunk fiber.

In WDM, optical sources of different wavelengths are combined to propagate in the same optical fiber. This requires a set of sharp wave filters of different wavelengths. The passband insertion loss of each filter at λ_n must be low while the interband attenuation must be high. These properties ensure good transmission and minimize cross-talk. There are, in general, two common approaches to building these couplers: (1) the interference filter, in either highpass, lowpass, or bandpass configurations and (2) the diffraction grating of different variations. Some of these couplers will be discussed in Chapter 13.

In time-division multiplexing (TDM), messages are digitized and sent over the optical fiber in a time-sharing fashion; that is, each channel is allowed to transmit and receive the message in a short duration assigned to the channel only, allowing other channels to use the trunk in successive sequence. This coupler requires a switching mechanism that allows synchronized switching between paired transmitter and receiver use of the trunk for a specified time interval. Figure 4.16 illustrates the general idea.

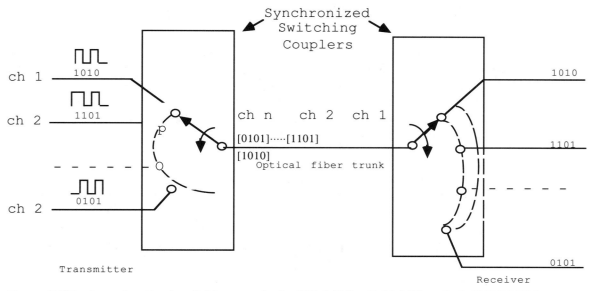

Figure 4.16 A synchronized switching coupler for TDM. [After C. M. Miller *et al.*, [18] (1980).]

Switches

Switches [18] are used to reroute the optical signals.

The performance of a switch is determined by its characteristic properties. These include the insertion loss, cross-talk, reproductivity, and speed. As in any three-port network, a two-position switch can have an input/output with two positions, "on" and "off." Assign the "on" position as port 2 and the "off" position as port 3; then the insertion loss can be written as

$$L_{12} = -10 \log \frac{P_2}{P_1} \qquad (4\text{-}11)$$

A loss less than 1 dB is desirable for any switch.

Cross-talk measures the isolation between the input and the off port,

$$L_{13} = -10 \log \frac{P_3}{P_1} \qquad (4\text{-}12)$$

A well-isolated switch should have L_{13} in excess of 40−60 dB.

Switching is subject to repeated activation. It should maintain the same insertion loss after many switchings.

How fast a switch can be activated or deactivated is crucial in many applications. The actual speed depends on the mechanism involved in switching.

Mechanical Switching. Examples of mechanically activated optical switching include a movable mirror prism placed in the path of a beam to deflect the light. This method is simple to implement, but its speed of operation is often too low for usable applications.

Electronic Switching. Although electronic circuitry can be designed to switch optical signals fast enough for modern communication applications, the complexity involved in electronic-to-optical conversion techniques may limit the use of electronic switching to optical fiber communication. Besides, optical switching offers a more direct switching in a more elegant way.

Optical Switching. Optical switching makes use of the properties of an optical waveguide, usually the planar lightguide and electrooptic and acoustooptic phenomena to actuate the switching. Its speed can be very fast. The insertion loss can be low and isolation high. As the switching device is usually built with integrated optics, we shall study these in Chapter 13.

Attenuators and Isolators

Optical Attenuator

An optical attenuator [19] is a passive device inserted into a transmission path to control the insertion loss. Both fixed or adjustable attenuators are available.

Fixed attenuators make use of an absorbing layer axially aligned between two segments of fibers. Because of the small cross-sectional area of the fibers, this type of attenuator cannot handle high light levels. Other elaborated methods use GRIN-rod lenses to collimate the light between fibers and then refocus the light into the downstream fiber. The absorbing layer is sandwiched between the two lenses.

Adjustable Attenuators

Instead of a solid piece of absorption material inserted between fibers, a wedged-shaped absorber whose position can be adjusted accurately is used to build an adjustable attenuator.

Optical Isolators

An optical isolator [20] is a device used to prevent return reflections along a transmission path.

The nonreciprocal nature of Faraday rotation property can be used to construct an isolator. A light beam through a polarizer (say, a vertical polarizer) is sent into a Faraday rotator that rotates the light polarization clockwise by 45°. When it is sent through a second polarizer set at +45°, the light passes in the forward direction with little insertion loss. The reflected light that is oriented with the second polarizer will be rotated counterclockwise 45° in the Faraday rotator and will thus cross with the first polarizer and will not pass through.

Material for the rotator is chosen to provide low absorption loss and strong Faraday rotation at the desirable wavelength of operation. Paramagnetic glass terbium aluminum garnets have been used. For longer wavelengths, iron-garnet can be used.

This will be discussed further in Chapter 13.

Recent Development in Fiber Optic Components

The continuous expansion of fiber applications promotes rapid growth and increased sophistication in the development of fiber optic components. This is evidenced by the frequent conferences on the "Components for Fiber Optic Applications" sponsored by SPIE, IEEE, and other professional organizations. Many new developments have been reported through their proceedings. We wish to cite just a few interesting ones here.

Tapered-Fiber and Fused-Fiber Couplers

The purpose of a tapered coupler is to provide field access required for coupling to a second fiber in a simple and efficient way. A simple tapered coupler is made by pulling a fiber coupler under heated conditions. In a recent paper by Black et al. [21], the theory of how such a coupler works was reviewed. The propagation mechanism for a single-fiber, biconical taper, the dependence of the model properties on the transverse index profile of the cladding diameter, and the evolution of the field along the guide as a function of the taper slope are explained. The Black et al. paper should promote better understanding of tapered couplers and help in the design of these couplers.

Motivated by the requirement to provide a broad range of power splitting applications such as filtering, wavelength division multiplexing, and switching, we can implement fused multifiber couplers. Some of these devices will be discussed in Chapters 8 and 13.

In Chapter 3, Section 3 entitled "Coupled-Mode Analysis," where coupled modes were first discussed, we recall that when a pair of wires carrying electromagnetic wave is coupled, there is a critical length at which power in one line can be transferred completely to a second line. This length is called a *beatlength*. A longer coupled length will cause power to be transferred back and forth periodically between the lines. This is also true in fused multifiber couplers. On the basis of these findings, a multichannel wavelength multiplexer using SMF has been developed [22].

A two-channel fused coupler is referred to as a *wideband* (WB) coupler when the coupler has been pulled through one or two beatlengths during fabrication. The multiplexed wavelengths of this coupler are widely separated, say, greater than 200 nm. If the pulling process were allowed to continue for several beatlengths, however, the coupling ratio would continue to oscillate back and forth with increasing frequency. Thus, the coupling ratio would vary more rapidly with wavelength change than in the wideband case. This is called a *narrowband coupler,* whose separation can be narrowed down to several nanometers. Multichannel multiplexing in SMF can be achieved by combining narrowband and wideband WDM in a tree structure. Up to 1/16 channel couplers have been reported.

A Fiber Optic Rotary Connector

Many fiber optic applications require the transfer of data and power between a rotary and a stationary point across a rotary interface. A simple single-channel on-line fiber optic rotary connector built by TRW is shown schematically in Fig. 4.17a and 4.17b. In Fig. 4.17a, two fiber ends are aligned on-line without a lens in between. The coupling is through the optical flux in the gap. Both fibers can be rotated. The loss through the gap can be minimized by introducing a pair of lenses as shown in Fig. 4.17b. This connector, also called an *optical slip ring* [23], is bidirectional.

Multichannel fiber optic slip rings can be built using off-axis optical bundles as shown in Fig. 4.18. By rotating one end relative to the other, channels can be connected to whichever pair one desires. This is also a *bidirectional coupler.* Coupling loss due to the gap can be reduced by filling the gap with index-matching liquid. But the set experiences loss due to the eclipsing effect caused by the light modulation when the connector is rotated.

Multichannel slip rings can be used in place of the expensive multichannel multiplexing couplers as in WDM.

ON-AXIS BIDIRECTIONAL FIBER OPTIC SLIP RING USING SINGLE FIBERS

a

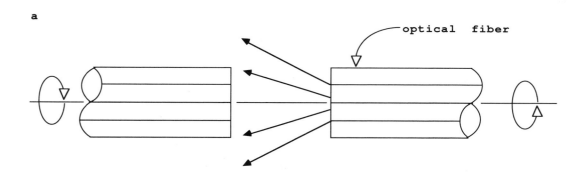

b

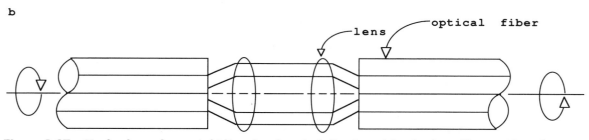

Figure 4.17 Single-channel on-axis bidirectional optical slip rings: (*a*) using bare fibers, with no focusing lenses; (*b*) using focusing lenses. [After J. A. Speer and W. W. Koch, [23] (1988).]

OPTICAL FIBER BUNDLE SLIP RING

On-axis bidirectional slip ring using concentric bundle

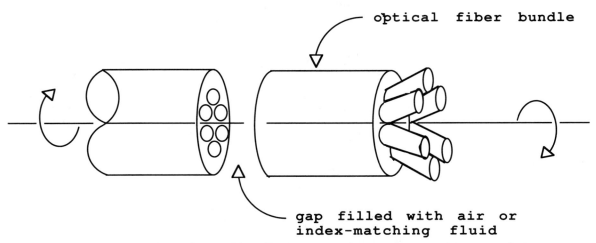

Figure 4.18   A multiple-channel optical bundle slip ring. [After J. A. Speer and W. W. Koch, [23]

Conclusion

This chapter describes the auxiliary components needed to complete the installation of an optical communication system. These include connectors for fiber-to-fiber, source-to-fiber, fiber-to-detector connection; couplers; and switches. Brief comments follow each device. Others are left for discussion in later chapters.

A list of fiber component manufacturers is included indicating the worldwide interest in this industry.

Fiber Components Manufacturer Companies

Cabloptic
   Costallorid, Switzerland
Canadian Institute and Research, LTD
   Mississauga, Ontario, Canada
Cansfar
   Scarborough, Ontario, Canada
CLTO
   Bezons, France
CTEC Research Laboratory
   Wembley, United Kingdom
Farukawa Electric Company
   Japan

General Optronics
Edison, New Jersey USA
Gould, Inc.
Glen Barrie, Maryland USA
Hitachi Cable America, Inc.
Tokyo, Japan
ITT
Roanoke, Virginia USA
JDD Optics, Inc.
Ottawa, Ontario, Canada
Kaptron
Palo Alto, California USA
Phalo Optical System
Manchester, New Hampshire USA
NEC Electronics
Mountain View, California USA
STC Components
London, England
York Tech, LTD
Hampshire, England

References

1. D. Gloge, Offset and tilt loss in optical fiber splices. *Bell Syst. Tech. J.* **55**, 905–916 (1976).
2. T. C. Chu and A. R. McCormick, Measurement of loss due to offset end separation, and angular misalignment in graded-index fibers excited by incoherent source. *Bell Syst. Tech. J.* **57**, 595–602 (1978).
3. D. Gloge, A. H. Cherin, C. M. Miller, and P. W. Smith, Fiber splicing. In *Optical Fiber Telecommunications* (S. E. Miller and A. G. Chynoweth, eds.), pp. 456–461. Academic Press, New York, 1979.
4. D. L. Bisbee, Splicing silica fibers with an electric arc. *Appl. Opt.* **13**, 767 (1974).
5. K. Eqashira and M. Kobayashi, Optical fiber splicing with a low-power CO_2 laser. *Appl. Opt.* **16**(6), 1636 (1977).
6. C. M. Miller and G. F. DeVean, Signal high-performance mechanical splice for single mode fibers. *Opt. Fiber Commun.*, San Diego, *1985*, pp. 26–27 (1985).
7. C. H. Gartside, III and J. L. Baden, Single mode ribbon cable and array splicing. *Opt. Fiber Commun.*, San Diego, *1985*, pp. 106–107 (1985).
8. M. H. Hodge, Fiber optic-interconnection: A new approach. Proc. 12th Annual Connector Symp., p. 190, Oct. 1979.
9. M. A. Holzman, A detachable connector for multimode graded index optical waveguides. *Annu. Connector Symp. Proc.* **11**, 194 (1978).
10. J. Minowa, M. Saruwatari, and N. Suzuki, Optical componentry utilized in field trial of IEEE transmission. *J. Quantum Electron.* **QE-18**, 705 (1982).
11. B. S. Kawasaki and K. O. Hill, Low-loss access coupler for multimode optical fiber distributed network. *Appl. Opt.* **16**(7), 1974 (1977).
12. M. A. Karr, T. C. Rich, and M. Dimenico, Jr., Lightwave fiber tap. *Appl. Opt.* **17**(14), 2215 (1978).
13. T. Uchida and S. Sugimoto, Micro-optic devices for optical communications. *Proc. Euro. Conf. Opt. Commun. 4th, 1978*, p. 374 (1978).

14. J. G. Ackenhausen, Microlenses to improve LED-to-optical fiber coupling and alignment tolerance. *Appl. Opt.* **18,** 3694–3696 (1979).

15. J. F. Dalgleish, Splices, connectors and power couplers for field and office use. *Proc. IEEE* **68**(10), 1226–1232 (1980).

16. J. C. Williams, S. E. Goodman, and R. L. Coon, Fiber-optic subsystem considerations of multimode star coupler performance. *Prof. Conf. Opt. Fiber Commun. Conf.,* New Orleans, *1984,* Paper WCG (1984).

17. M. K. Barnoski, Coupling components for optical fiber waveguides. In *Fundamentals of Optical Fiber Communications* p. 147–185 (M. K. Barnoski, ed.), 2nd ed. Academic Press, New York, 1981.

18. C. M. Miller, R. B. Kammer, S. C. Metterr, and D. N. Ridgway, Single-mode optical fiber switch. *Electron. Lett.* **16,** 783 (1980).

19. E. G. Hanson, Polarization-independent liquid crystal optical attenuator for fiber optics applications. *Appl. Opt.* **21,** 1342 (1982).

20. K. Shiraishi, S. Sugaya, and S. Kawakami, Fiber Faraday rotator. *Appl. Opt.* **23,** 1103 (1984).

21. R. J. Black, E. Gonthier, S. Lacroix, J. Lapierre and J. Bures, Tapered fibers: An overview. *Proc. SPIE—Int. Soc. Opt. Eng.* Vol. **839,** 2–19 (1988).

22. P. M. Kopera, K. L. Sweeney, and K. M. Schmidt, Multichannel wavelength multiplexing in single mode optical fibers. *Proc. SPIE—Int. Soc. Opt. Eng.* **839,** 25–30 (1988).

23. J. A. Speer and W. W. Koch, The diversity of fiber optic rotary connectors (slip rings). *Proc. SPIE—Int. Soc. Opt. Eng.* **839,** 122–129 (1988).

5

Optical Fiber Measurements

Introduction

The parameters to be measured in optical fibers include the core and cladding dimensions, the refractive-index profile, the loss characteristic, and the dispersion properties. Accurate measurements are needed to assess the result of changing parameters to assure a better fiber. Measurements are sometimes also used to help fabrication stay within the bounds of tolerance and to guide users to specify parameters for new system designs.

A good measurement technique must be simple in both sample preparation and technique and be easily accessible for obtaining data. The principle of measurement must be clear and easy to interpret, and the method must be nondestructive and/or noncontact. None of the measurement methods described here can satisfy all these criteria. A proper choice must be made in accordance with the actual existing situation.

In this chapter, we describe optical methods for measuring the refractive index, using either the scattering, reflection, refraction, or interference method; the power losses by radiation and absorption methods; and the dispersion properties by the pulse and bandwidth measurements. Fiber dimension measurements are not included. They are usually specified by the manufacturers. Under each category, the principle of measurement, method of gathering data, computation of the results, and finally, interpretation of the data are described. Sometimes alternative methods are introduced to cover a broad scope of the measurement technique.

The key parameters that need to be determined may differ for single-mode and multimode fibers (SMF and MMF) and, therefore, the measurement methods will differ. For example, in MMF, the modal properties, mode coupling, and differential modal attenuation are the important parameters, while in SMF, the cutoff wavelength of the second mode LP_{11} is more important. Most examples of measurement methods are aimed at MMF. However, some can be used for both types of fiber.

Basic Techniques in Optical Fiber Measurement

Refractive index is the most important parameter in multimode optical fibers. It determines the group delay characteristic and therefore the dispersion property of the fiber. To ensure good-quality production, continuous monitoring of the refractive index is a necessity while fabricating optical fibers. We describe several methods to cover this measurement.

The Reflection Method

The operating principle of the reflection method is based on the fact that the reflectivity of a dielectric surface depends on the difference of the refractive indices of the surrounding medium. To apply this principle to an optical fiber, a collimated light impinges at a right angle on the polished end face of a fiber, and the intensity of the reflected light is measured from the image.

The variations in the reflected light intensity can then be used to compute the refractive index.

An elaborate setup for the measurement of the refractive index by the reflection method is shown in Fig. 5.1 [1]. A highly focused laser beam is used as the light source. It is collimated in order to achieve high spatial resolution. The laser beam is first modulated by a chopper at a 1-kHz rate (to enhance the signal-to-noise ratio) and is purified by passing it through a polarizer and a λ/4 plate. The λ/4 plate converts the linearly polarized light to circular polarization. Reflected light, however, is polarized with the opposite sense of circular polarization. When it returns from the reflecting surface and passes the λ/4 plate and polarizer combination again, it is converted to a linear polarization at right angles to its original orientation and is then unable to pass the combination. This process decouples the impinging light from the reflected light and improves the stability of its operation.

The light source, after passing the λ/4 plate, is focused by a lens and forced through a spatial filter to purify its mode content. Actually, a tiny pinhole is used as a simple-mode filter. The filtered light

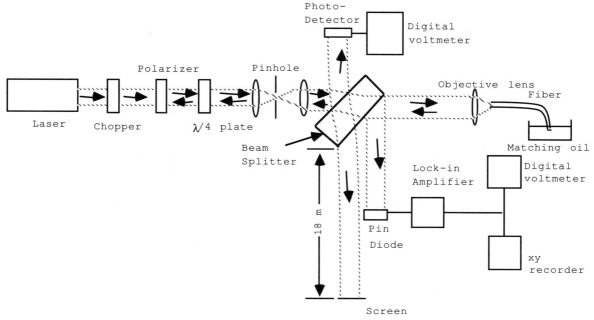

Figure 5.1 Optical setup for retractive-index measurement. (After W. Eickhoff and W. Weidel, p. 109, [1].)

(whose high order transverse modes have been removed) can be focused to a sharp spot. A second lens collimates the beam. A thick glass serves as a beam splitter. A portion of its light is focused on to the polished fiber end surface. The other end of the fiber is dipped into index matching fluid to reduce reflections. A small portion of the incoming light is directed to a solar cell monitoring system to test the stability of the light source. Light reflected from the end surface of the fiber is also split by the glass plate. One portion is sent to a screen where the spot size can be watched. The spot size can be minimized by adjusting the focusing lens. Minimum spot size indicates that sharp focusing on the fiber surface has been achieved. The second portion of the reflected light is detected by a PIN diode whose output is amplified and can be used to activate a recorder and a digital voltmeter.

The power reflection coefficient R is measured as a function of the lateral displacement of the fiber. It can be shown that [2]

$$R = \frac{P_r}{P_i} = \left(\frac{1 - n}{n + 1}\right)^2 \tag{5-1}$$

where P_r and P_i are the incident and reflected light power, respectively. Denoting the refractive index of the core as $n(r)$, and assuming the index difference between the core and cladding is much smaller than unity, we may let

$$n(r) = n_2 + \Delta n(r) \tag{5-2}$$

We may also write

$$R = R_2 + \Delta R \qquad (5\text{-}3)$$

where R_2 is the reflection coefficient of the cladding. Then we may write

$$\Delta n(r) \approx \frac{1}{4}(n_2^2 - 1)\frac{\Delta R}{R_2} \qquad (5\text{-}4)$$

Equation (5-4) suggests that the index variation Δn is approximately proportional to the reflection coefficient variation ΔR. If the lateral motion of the light source is geared to an x–y plotter (as in the setup in Fig. 5.1), the reflection coefficient and thus the index profile can be plotted directly (after calibration).

Although the concept involved in this method is very simple, the result of measurement obtained in this manner cannot be very accurate. First, the reflected light must be measured accurately because reflection from the core and cladding is almost the same. Second, the surface condition of the end face is very critical to the measurement [3].

The Near-Field Method

The near-field method is perhaps the easiest method to implement for measuring the refractive index profile of an optical fiber. However, its accuracy is not very good, primarily because of the effects of leaky modes.

The refractive-index distribution in this method is obtained by measuring the light power density at the far end of a short fiber that is uniformly illuminated by a Lambertian light source. Figure 5.2 shows the two versions of the setup for the implementation of this method. In Fig. 5.2a, an incoherent light source, such as a light-emitting diode, is used to illuminate the fiber end uniformly. A sharply focused microscope scans the near-field light power distribution at the other end. It can be shown that the light intensity distribution contributed by the guided modes is directly proportional to the index difference between core and cladding, $n(r) - n_2$. Figure 5.2b shows an alternative way of implementing the near-field method. A sharply focused beam of light scans the input end of the fiber, and a detector is used to intercept all the light arriving at the outer end. The angular range of the focused light source must be larger than the acceptance angle of the fiber such that the lightwaves are guiding and the power distribution is uniform over this range.

Gloge and Marcatili [4] derived an expression to show that the light-power density in the fiber core is proportional to the index difference

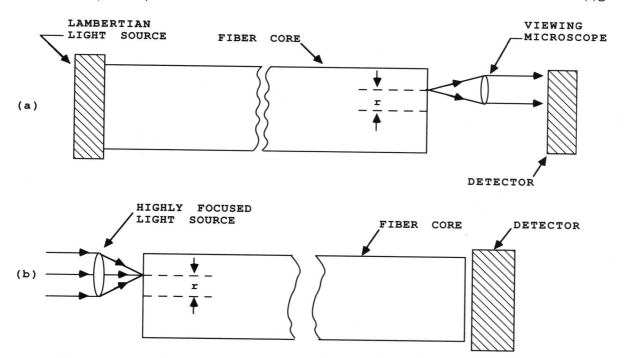

Figure 5.2 A setup for refractive-index measurement by the near-field method: (a) Lambertian light source with scanning microscope detection; after D. Marcuse, Principle of Optical fiber measurements, Academic Press, p. 70, 1981. (b) highly focused light source with uniform detector. [After D. Marcuse and H. M. Presby, *Proc. IEEE* **68**, 670 (1980) with permission © IEEE 1980.]

$n(r) - n_2$. If we indicate by n_1 the maximum value of $n(r)$, we can express the power density in terms of its maximum values

$$P(r) = \frac{n(r) - n_2}{n_1 - n_2} P_{max} \qquad (5\text{-}5)$$

The derivation of Eq. (5-5) is based on these four requirements: (1) illumination is uniform, (2) all propagation modes are excited uniformly, (3) the length of the fiber is sufficiently short so that the differential mode attenuation effects do not distort the profile, and (4) mode couplings do not exist.

Measurement of the near-field power distribution gives only the shape of the profile. It takes a second measurement to determine the index difference. Now by Snell's law

$$\sin \theta_m' = \frac{n_1}{n_0} \sin \theta_m = \frac{n_1}{n_0}\left(1 - \frac{n_2^2}{n_1^2}\right)^{1/2} = \frac{1}{n_0}\left(n_1^2 - n_2^2\right)^{1/2}$$

where θ_m' is the maximum trap angles modified by the index ratio n_1/n_0. Thus the index difference is $n_1^2 - n_2^2 = n_0^2 \sin^2 \theta_m'$ or

$$n_1 - n_2 \simeq \frac{n_0^2 \sin^2 \theta_m'}{2n_2} \qquad (5\text{-}6)$$

Equations (5-5) and (5-6) completely determine the refractive index profile $n(r) - n_2$.

The near-field method has a drawback related to the problem of leaky modes [5].

For better results, it is suggested that a short length of fiber should be used for measurements. Unfortunately, it is just for short lengths that the effect of leaky modes is greatest. The leaky mode problem can be avoided if the refracted near-field method [6] were adopted.

The Focusing Method

The focusing method utilizes the cylindrical lens action of the fiber core when illuminated by a collimated beam of light of uniform intensity at right angles to its axis. The focused image on an observation plane can be related to the refractive-index profile by the paraxial ray theory. This method has the advantage that core size becomes unimportant, making it equally applicable to fibers as well as to preform. The precision of the focusing method is better than that of the previously described methods. The principle of operation of the focusing method is schematically shown in Figure 5.3 [7]. The fiber used in the measurement is immersed in index-matching fluid. The core is illuminated with an incoherent filtered beam of light of uniform intensity perpendicular to its axis. A typical ray path is shown here for detailed discussion of its trajectory. The ray enters the fiber at a distance t from

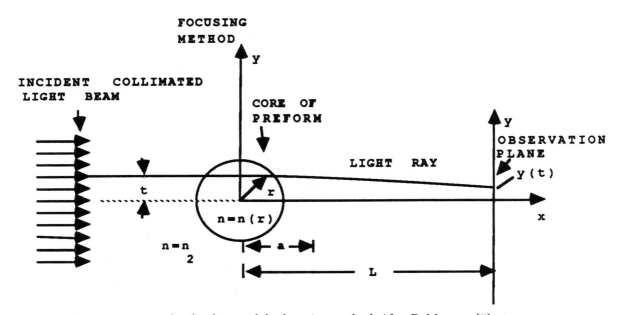

Figure 5.3 Ray trajectory for the theory of the focusing method. After D. Marcuse [7].

the optical axis. It experiences refraction in the core and leaves the core with a slightly different shape. At an observation plane located at a distance L away from the core center, the focused ray is at a distance y above its optional axis. The location of y is dependent on its entrance location t. By measuring the ray position t and $y(t)$, using the paraxial ray equation, we may write

$$n(r) - n_2 = \frac{n_2}{\pi L} \int_r^\infty \frac{t - y(t)}{\sqrt{t^2 - r^2}} \, dt \qquad (5\text{-}7)$$

or, instead of measuring the ray position t and $y(t)$, as in the ray-tracing method, we may measure the light-power distribution at the observation plane and find the desired function $t - y$ by integration. This can be done because the principle of conservation of power applies. Denoting P_i as the incident power density at a distance t from the core axis and $P(y)$ as that at y on the observation plane, we have

$$t(y) - y = \int_0^y \left(\frac{P(y')}{P_i} - 1 \right) dy' \qquad (5\text{-}8)$$

The refractive-index profile can now be obtained by using Eq. (5-8) in Eq. (5-7).

A computer program may be written to numerically integrate the equation and plot out the profile directly.

For this method to work, the observation plane must not be inside the fiber core. It must be placed so far away that rays have already crossed over after leaving the core. Good results can be obtained when the observation plane is placed just outside the core.

Interferometric Method

The operating principle of an interferometric method [8, 9] in optics resembles that of a Wheatstone bridge in electrical engineering. A Wheatstone bridge is used to detect small changes in electrical current in a balanced circuit. An interferometer compares the phase shift in wavefronts of two optical paths. If there is a phase difference between these two optical paths, the resultant interference pattern shows alternate bright and dark lines because the two wavefronts either enhance or cancel each other, forming the so-called interference fringes. A lateral shift of the interference fringe is a measure of the phase shift of light passing through it. The distance between two adjacent fringes corresponds to the relative phase difference of 2π.

The interferometric method can be applied to measure the refractive-index profile in optical fibers in either of two ways: the transmitted light or slab method and the transverse interferometric method.

The Interferometric Slab Method: The Transmitted Light Method

The slab method is conceptually the simpler and more accurate of the two, but it is destructive to the fiber being tested and requires elaborate and demanding sample preparation. For example, a thin sample (200–300 μm thick) must be cut from the fiber. The two faces must be cut parallel to a high degree of accuracy and polished. Because of the high accuracy and reliable results obtainable from this method, many laboratories and factories use this method to gauge their products.

Figure 5.4 is a schematic view of the slab interferometric method. The incident light is split into a dual beam by a semitransparent mirror and mirrors combination. One beam leads through the fiber slab to a microscope and combines with the other beam through a reference slab and microscope to yield the output on a screen.

Light passing through the slab parallel to its axis suffers a phase

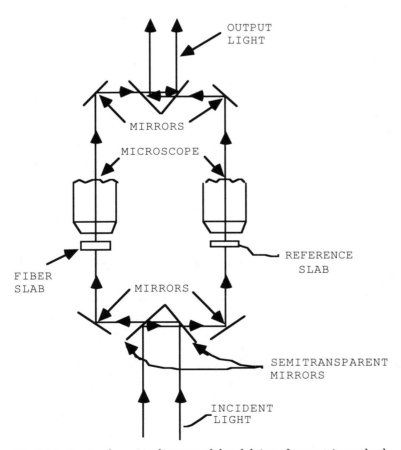

Figure 5.4 A schematic diagram of the slab interferometric method. [After H. M. Presby *et al.*, *Appl. Opt.* **14**, 2209 (1978), with permission.]

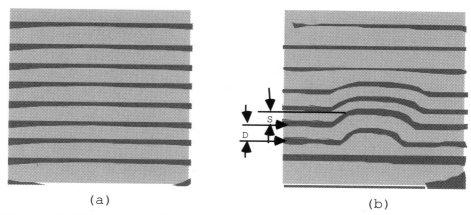

(a)   (b)

Figure 5.5 Field of view of interference microscope: (a) without graded-index slab sample; (b) with graded-index slab sample. [After F. T. Stone *et al.*, *Appl. Opt.* **14**, 151 (1975), with permission).]

retardation whose amount depends on the optical path length $L = n(r)d$, where d is the slab thickness. The phase retardation ϕ of the lightwave is

$$\phi = n(r)kd \qquad (5\text{-}9)$$

where $k = 2\pi/\lambda$. The reference arm contains a slab with refractive index n_2. A slight tilt is usually introduced in the mirrors to enhance the interference pattern. A field view of these patterns is shown in Fig. 5.5.

Figure 5.5a is the fringe pattern of the interferometer after introduction of a slight tilt without the fiber slab, which consists of parallel lines of alternating bright and dark bands distributed uniformly. The dark bands are regions where the phase front interferes destructively. Figure 5.5b shows the pattern with a graded-index fiber sample. Bulged bands are clearly visible. By measuring the fringe shift $S(r)$, which depends on its position r and the fringe position D, one can formulate

$$\frac{2\pi}{D} = \frac{k[n(r) - n_2]d}{S(r)} \qquad (5\text{-}10)$$

where $k[n(r) - n_2]d$ is the relative phase retardation between the fiber slab and the reference slab, or

$$n(r) - n_2 = \frac{\lambda S(r)}{Dd} \qquad (5\text{-}11)$$

The refractive index profile can thus be measured by knowing $S(r)$ and D from the interference pattern.

A computer program can be written to evaluate $n(r) - n_2$ by feeding into it the digitalized data of $S(r)$ and other constants. Typical data of the profile are shown in Fig. 5.6.

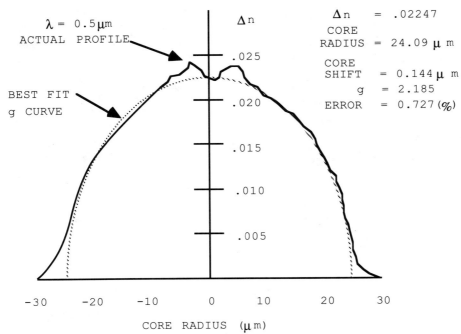

Figure 5.6   Refractive-index profile of graded-index fiber. The dotted curve is the best computer fit at g = 1.185. [After D. Marcuse and H. M. Presby, *Proc. IEEE*, **68**, 677, 1980 © IEEE, 1980.]

Transverse Interferometric Method

The destructive nature of the slab method can be avoided if the transverse interferometric method [10, 11] is used. The slab shown in Fig. 5.4 is replaced by a container filled with index-matching fluid into which a stripped fiber is immersed such that the light illuminates the fiber at a right angle to its axis. Details of the geometry of the setup are shown in Fig. 5.7. Notice that the light ray passes through regions of varying refractive index and the total path length must be expressed as an integral expression. The refractive-index profile of the fiber can thus be computed from the fringe shift observations obtained from the interference microscope.

Other methods

Scattering Methods

The refractive index profile of an optical fiber can be determined from an analysis of the scattering pattern generated when a laser beam is incident at a right angle to the fiber axis. In one approach, the backscattering fringe pattern obtained by using a laser beam incident on an unclad fiber is used to determine the refractive index and also the

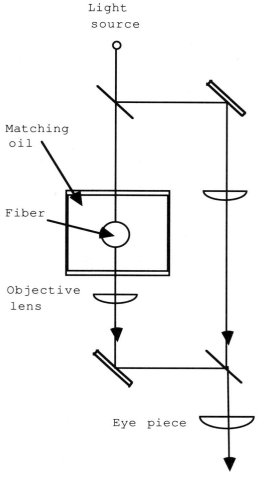

Figure 5.7 Setup for the transverse interferometric method. After D. Marcuse [7].

radius [12]. The ray that is internally reflected at the surface of an
unclad fiber travels backward. A scattering angle ϕ_m exists for which
the scattered light intensity becomes maximum. Figure 5.8a shows an
experimental setup, Fig. 5.8b shows the backscattered pattern, and
Fig. 5.8c shows the geometry of the scattering paths from which the
following equation can be derived:

$$
\begin{aligned}
\phi_m = \ & 4 \sin^{-1}\left\{\frac{2}{n\sqrt{3}}\left(1 - \frac{1}{4}n^2\right)^{1/2}\right\} \\
& - 2 \sin^{-1}\left\{\frac{2}{\sqrt{3}}\left(1 - \frac{1}{4}n^2\right)^{1/2}\right\}
\end{aligned}
\tag{5-12}
$$

This angle ϕ_m is determined solely by the refractive index of the fiber
n; hence, the index can be determined by measuring ϕ_m. The backscattering fringe pattern shows the presence of a ripple attributable to

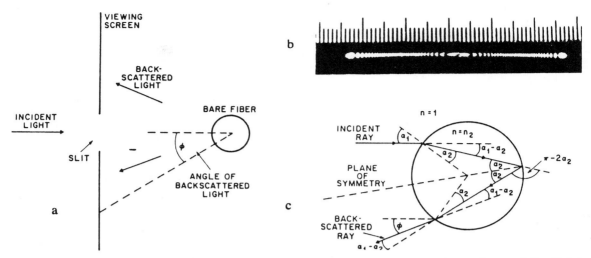

Figure 5.8 Scattering method for refractive-index measurement: (*a*) an experimental setup (after D. Marcuse, Principles of Optical Fiber Measurement, p. 7, Academic Press, 1981); (*b*) the backscattered pattern (after H. M. Presby, p. 280, [12]); (*c*) the geometry of the scattering paths, (after D. Marcuse, Principles of Optical Fiber Measurement, p. 7, Academic Press, 1981).

the interference between two rays incident on the fiber surface. The fiber radius can be determined from the period of this ripple. This method has been demonstrated to be applicable to measuring index distributions of step-index as well as graded-index profiles.

In another approach, the fiber is immersed in index-matching oil and the forward far-field scattering pattern is detected as a function of the scattering angle [13]. The method is limited to cases where the core radius–maximum index difference product is relatively small. A 5% error is introduced if this product is about 0.04 μm. Also, a large number of sampling points must be taken to present a reasonable result.

The Refraction Method

The refraction method monitors the amount of light power escaping sideways from the core to the cladding and beyond for determining the refractive index of the fiber. A light beam is focused on a spot at a distance r from the fiber axis end face (submerged in matching fluid) with a convergence angle that is larger than that which can be trapped by the fiber core. After entering, the fiber part of the ray remains trapped in the core while part of the ray contributes to the leaky modes. We wish to measure the escaped rays only. To accomplish this, the detector is shielded away from the trapped ray so that a narrow cone of light reaches the detector. This is shown in Fig. 5.9. Stewart [6] derived an expression for determining $n(r) - n_2$ from the following measured data: (1) the refractive index of the cladding n_2,

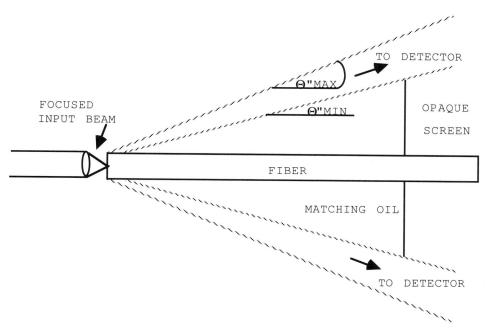

Figure 5.9 The refraction method. [After D. Marcuse and H. M. Presby, *Proc. IEEE* **68,** 674 (1980), © IEEE 1980.]

(2) the minimum angle θ''_{min} defined by the circular screen placed in front of the detector, (3) the maximum angle θ'_{max} of the incident core of light defined by either the focusing arrangement or additional apertures in the input beam, and (4) the light power measured as a function of the position of the input beam $P(r)$ and $P(a)$ when $r = a$. Then

$$n(r) - n_2 = n_2 \cos \phi_m (\cos \theta''_{min} - \cos \theta''_{max}) \frac{P(a) - P(r)}{P(a)} \quad (5\text{-}13)$$

An accuracy of 1 part in 10^4 in detectable index change has been reported.

X-Ray and Scanning Electron Beam Microscopic Method

To achieve high spatial resolution in measuring the refractive index of a graded-index fiber, an X-ray microanalyzer can be used. This is because with X-ray, the beam can be focused on a very small spot. It has been confirmed experimentally that the refractive-index variation is expressed approximately as the linear combination of the dopant [14], and the X-ray microanalyzer can be used effectively to measure the impurity densities. The specific X-ray generated by an impurity material is proportional to the density of that impurity. Thus we can use the X-ray microanalyzer to determine the index profile.

Scanning electron microscopes have also been used to measure the index profile of the fiber. The end face of a fiber is chemically etched to produce a three-dimensional picture, and the refractive-index profile can be obtained from a photograph taken by this microscope.

Some Comments on Fiber Index Measurement

1. Most of the methods described in previous sections are aimed primarily at measuring the index profile of MMF. The index profile for SMF is relatively harder to measure because of the small core size and low index difference of these fibers.
2. For measuring the index profile of SMF, the focusing method is perhaps one of the best. Besides its achievable accuracy of a few parts in 10^5, resolution better than 1 μm has been demonstrated.
3. For most accurate measurements, the transmission-type interferometric method should be used. However, it is destructive and very demanding in sample preparation.
4. The least accurate method is perhaps the reflection method. The reflected power distribution does not indicate the refractive index totally and correctly. A correction factor is needed to adjust the result. Nevertheless, it is the simplest method to implement.
5. In using the forward-scattering method to determine the refractive index, many points should be sampled. This could improve the resolution (0.2 μm has been reported).
6. The transverse interferometric method (TIM) is nondestructive. It is capable of resolving large refractive-index fluctuations that are sharply localized. Therefore, this method is more suitable for measurement of fiber indices with relatively high Δn values. Otherwise, large on-axis error may cause noticeable fluctuations in results. In order to increase the accuracy of the measurement, an expensive interference microscope should be used in place of an ordinary microscope.

Fiber Loss Measurements

In Chapter 1 we introduced the subject of fiber losses briefly. In Chapter 3 the loss mechanisms involving absorption, scattering, and radiation were discussed in some detail. In this chapter we describe the methods to measure these losses in optical fibers. Some preliminary remarks are in order before we discuss the methods of measurement.

Introductory Remarks

The Loss Parameter

To gauge fiber loss, we need a unit. It is customary to use a parameter such as 2α, the loss parameter, to describe the fiber loss in decibels per unit length. This poses no problem in measuring SMF losses as the loss per unit length of the fiber can be well defined. This is, however, not true in MMF. As a result of mode coupling, the power distribution along the fiber length of MMF often changes so much that it is difficult to establish a meaningful definition unless the operating condition of the fiber has reached a steady-state situation. Only under the steady-state condition can the power loss along the fiber be expressed by the exponential decay we discussed in Chapter 3. When we speak of steady-state power distribution, we have to use the ensemble averages over a length of the fiber or over many similar fibers to express the power distribution.

In MMF, one is interested in measuring the fiber's spectral attenuation rather than its attenuation at specific wavelengths. Most of the methods described in this section are limited to this effect.

Effect of Mode Coupling

Even with SMF in which the fiber loss can be described in terms of loss per kilometer, the total loss of an actual SMF may still depend on the coupling to radiation or cladding modes. Thus the measured fiber loss is subject to significant change before and after cabling or any other physical changes of the state of the fiber, such as bending or twisting.

Some fibers may have very little coupling between modes so that no significant exchange of power takes place even after traveling for 1 km or more. In this case, only the loss of its core material can be used to characterize the fiber losses. These losses can be determined by observing the fiber losses at low-order modes.

The Light Source

To measure the fiber loss, we need to launch the light into the fiber. Light of narrow spectral width and with tunable wavelengths is preferred. If a laser or an incoherent source of high radiance is used, then the effect of the spectral width of the source light needs to be discussed.

The light from the source must be focused on the fiber core by means of a microscopic objective lens. The spot size and its numerical aperture NA must match that of the fiber closely. The portions of the incident light beam that fall outside the fiber core are lost. Also lost is the light that impinges on the core at angles whose NA is larger than the fiber NA at the same core radius. Recall that the fiber NA is defined as

$$NA = [n^2(r) - n_2^2]^{1/2}$$

For a step-index fiber where $n(r) = n_1$ is constant, it is easy to match the source NA to that of the fiber. In MMF where the core index is graded, it will be very difficult to match these NA values. In this case, a compromise must be used.

It is not necessarily desirable to match the source NA to the fiber NA for the purpose of exciting the largest number of modes to satisfy the steady-state power distribution. Mode scrambles can be used for this purpose. Usually an incident light beam with a small NA value and a small spot size can be used to obtain adequate results.

Mode Scramble and Cladding Mode Strippers

Mode scrambles are used for the purpose of redistributing the power in the fiber in a desired manner. A long length of fiber itself is an ideal mode scrambler. Other schemes include a spliced fiber consisting of a section of graded-index fiber set in between two sections of step-index fibers [15], a piece of bent fiber, and other combinations.

Cladding mode strippers consist of an S-shaped fiber section immersed in index-matching fluid. This S-section allows cladding modes to leak out. It also removes the leaky modes as well as modes of higher orders.

Total Loss Measurement or Component Loss Measurement

Many types of loss mechanisms can inhibit the performance of an optical fiber, and one should be able to measure them either separately and/or in total. If one is interested only in finding out the total loss for a fiber length, one should choose a method that will do just that. For purposes of fiber research and development, however, it is often necessary to measure each component loss separately. This helps us to investigate the different loss mechanisms and suggests ways to minimize a particular kind of fiber loss. In the following descriptions, we will emphasize whether the specific method is applicable to measure component or total losses.

The Calorimetric Method

Many loss mechanisms dissipate their energy in heating up the surroundings. This is particularly true regarding the loss due to absorption, which one can measure with a temperature-sensitive device that detects changes in ambient temperature. For a good fiber, the fiber loss due to absorption is usually very small. Thus the temperature sensitivity of the measuring apparatus must be very high.

Because of the nature of this method, time is needed to establish an equilibrium for each change in temperature, and its accuracy is only marginal. But this is the only method we know that can measure the absorption losses. Moreover, a slight alteration in the setup enables us to measure the total loss, including the absorption loss, the radiation loss, and the scattered loss.

The principle of operation of the calorimetric method depends on the measurement of a change of the ambient temperature. The light source on the fiber provides the source of heat due to absorption. If Q is the amount of heat generated per second per unit length, which is proportional to the light power P carried in the fiber, we can write

$$Q = 2\alpha_a P \qquad (5\text{-}14)$$

where $2\alpha_a$ is the power loss coefficient due to absorption. The amount of heat flow out per second per unit length, on the other hand, is proportional to the temperature difference $T - T_0$, where T_0 is the initial temperature, or $q = C_1(T - T_0)$, where C_1 is a constant. Then the rate of change of temperature is

$$\frac{dT}{dt} = \frac{Q - q}{K} \qquad (5\text{-}15)$$

where K is the heat capacity of the fiber per unit length. At $t = 0$, that is, when the light source on the fiber is being turned on, $T = T_0$. Thus

$$2\alpha_a = \frac{K}{\rho}\left(\frac{dT}{dt}\right)_0 \qquad (5\text{-}16)$$

Thus, by observing the initial temperature rise, we can determine the loss coefficient provided K and P are known. We can measure P, the source light power, separately. To measure K, the heat capacity of the fiber, we can use an external heat source. Usually an electric heat source of a known quantity is used to heat the fiber up to the same temperature as before to evaluate K.

In carrying out this experiment, if we wish to measure only the absorption loss, then precautions must be taken to avoid the inclusion of strong light absorption other than the absorption mechanism we wish to measure. As the temperature difference to be measured is usually very small, a bridge balance technique could be used. Figure 5.10 is a schematic diagram for such a measurement [16, 17]. Two identical capillary tubes, one containing the fiber whose loss is to be measured, and the other, a reference, are both wound with resistance wire on the outside of the capillary walls to provide the two arms of a temperature bridge. The capillary tubes must be at least 20 cm long and the fiber sealed airtight in one of the capillary tubes. Light is launched onto the fiber through a mode stripper and scrambler, and the light output is monitored by a meter, also through a mode stripper.

The resistance wires on the capillary tubes are fed by an ac voltage source and the detecting amplifiers complete the bridge circuit. The amplifiers measure and record voltage difference due to the temperature changes. In the experiment, the thermal changes are recorded as a function of time, from which the initial slope of the curve is determined. With calibrated K and P values, the loss coefficient can be determined. If the apparatus is used to measure the total loss, then the outer surface of the capillary tubes should be covered with black

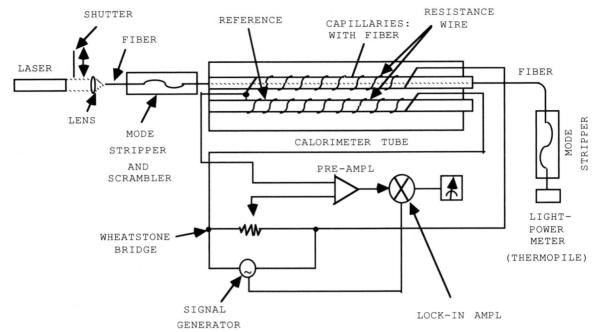

Figure 5.10 A calorimetric method for the measurement of fiber loss. After R. L. Cohen [16].

paint to ensure the absorption of the scattering and leaky mode energies.

With a 20-mW laser light-power injection onto the fiber, a loss co-efficient of 1 dB/km can be measured.

The Scattering Loss Measurement

In this section we describe a method that measures the scattering component of the fiber loss. Just as is true of the calorimetric method, this method does not offer high accuracy. But its principle of measurement is simple. All that is needed is a means for ensuring that the scattered light leaving the cladding reaches a detector. The loss coefficient $2\alpha_s$ can be obtained from an expression

$$2\alpha_s = \frac{\Delta P(z)}{LP(z)} \tag{5-17}$$

where L is the fiber length, $P(z)$ is the total light power at z, and $\Delta P(z)/L$ is the change in light power due to scattering as the measuring device is moving along the fiber length. Equation (5-17) is valid only when the criterion of exponential-power delay loss applies.

Scattering losses can be subdivided into Rayleigh scattering and scattering caused by fiber imperfections. They possess different characteristics that make them easily separable. For example, Rayleigh scattering is highly wavelength dependent. Its intensity changes as the inverse fourth power of the wavelength (or $1/\lambda^4$), while it does not vary with the direction from which the data are taken. On the other hand, scattering loss due to fiber imperfection does not depend on wavelength but is dependent on the scattering angle distribution.

The total amount of scattered light can be measured using an integrated detector device. A typical arrangement is an integrated detector cube, consisting of six faces of uniform square solar cells. Tynes [18] described such a cube in his experiment to measure the scattering loss. Small square cells having an area of 1 cm² on each side are carefully selected for uniform light sensitivity and electric resistance. The fiber whose scattering loss is to be measured passes through the centers of a pair of opposite sides. The cell cube is filled with index-matching fluid. The index of the matching fluid is purposely taken to be slightly higher than the cladding index n_2 in order to avoid the trapping effect of light in the cladding. The small dimension of the cell allows a higher resolution of the scattering loss measured as a function of position along the fiber length. An experimental setup is sketched in Fig. 5.11.

The integrated cube collects all the scattered light around the fiber regardless of the angle at which it is escaping from the fiber. The output of this integrated cube is a measure of photovoltaic voltages. A total loss detector is placed at the end of the fiber. By sliding the cube detector along the fiber, one can measure light scattered from a full length of the fiber and, using Eq. (5-17), determine the loss coefficient. The two detectors should be calibrated against a common source power to determine their absolute power relations.

The same setup can be used to separate the scattering loss into their component losses simply by using a variable-wavelength light source

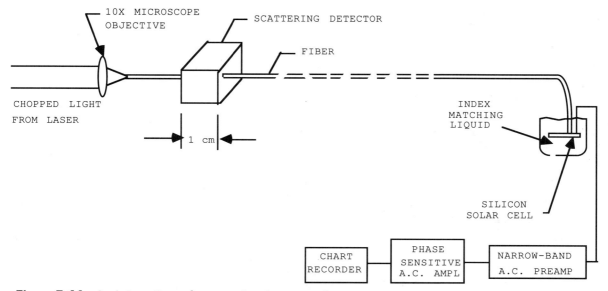

Figure 5.11 An integrating cube scattering detector. After A. R. Tynes [18].

and measuring the scattering losses in several wavelengths. It can be shown [19] that the loss coefficient can be expressed as

$$2\alpha_s = \frac{A}{\lambda^4} + B \tag{5-18}$$

where A is the slope of the $1/\lambda^4$ line and B is the intercept on the axis. Figure 5.12 shows such a measurement. The measured points fall on a straight line when $2\alpha_s$ is plotted against the $1/\lambda^4$ axis. This indicates that the wavelength-dependent portion of the scattering loss is due solely to Rayleigh scattering. The intercept shows the small amount of the scattering loss due to fiber imperfection.

In another measurement, the scattering loss is measured as a function of the angle. A short section of the fiber is illuminated while a sensitive detector travels along a semicircular path around the fiber to collect the forward-scattered light. It is found that the Rayleigh scattering intensity distribution of unpolarized light changes only slightly with the angle. Large peaks found at small angles may indicate the presence of tunneling modes.

The Cutback Method

A more accurate method to measure the total transmission loss in fibers is known as the *cutback method*. Relatively good accuracy on

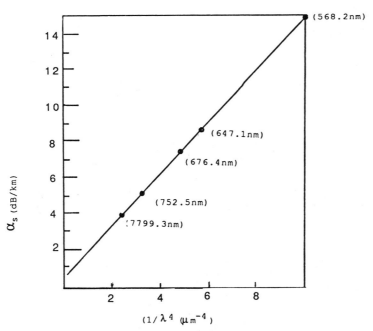

Figure 5.12 Experimental data of the total scattering loss plotted as a function of $1/\lambda^4$. After D. L. Philen and F. T. Stone [19].

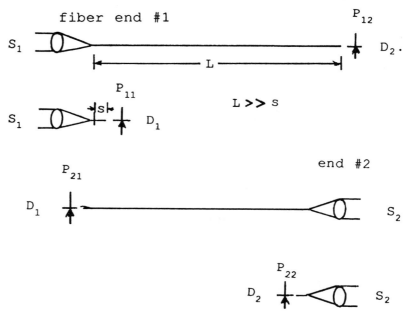

Figure 5.13 The cutback method.

the order of 0.1 dB has been achieved even for low-loss fibers. Unfortunately, the cutback method is destructive.

The principle of measurement of this method is described by Midwinter [20]. It requires having a source and detector pair at each fiber end. As shown in Fig. 5.13, S_1 is the power input to the fiber end 1, the power output at the remote end of the fiber is measured by a detector whose parameter is D_2. If K_1 represents the launching parameter of S_1, the measured output power at end 2 becomes

$$P_{12} = S_1 K_1 D_2 A \qquad (5\text{-}19)$$

where A is the total transmission loss of the fiber of length L. Next, the fiber is cut at the source end 1, and its output power is measured by a detector whose parameter is D_1. This gives

$$P_{11} = S_1 K_1 D_1 \qquad (5\text{-}20)$$

provided the launching condition remains the same and the loss of fiber at the cut is considered negligibly small.

Now, attach source S_2 to the end 2, and measure the power at end 1 using detector D_1. This gives

$$P_{21} = S_2 K_2 D_1 A \qquad (5\text{-}21)$$

where K_2 is the launching parameter of S_2.

The last step is to cut the fiber at end 2 and measure the power at end 2 using detector D_2. This gives

$$P_{22} = S_2 K_2 D_2 \qquad (5\text{-}22)$$

This fiber loss A can be obtained from Eqs. (5-19)–(5-22):

$$A = \left(\frac{P_{12}P_{21}}{P_{11}P_{22}}\right)^{1/2} \qquad (5\text{-}23)$$

Notice that all source and detector parameters drop out.

One important aspect of this method is that the source–detector pair at each fiber end must involve exactly the same launching condition during the measurement. Also, it is assumed that the fiber loss remains unchanged as the light transmission changes its direction. Although this method is classified as a destructive method, the destruction is not severe because the cuts are made at the ends of the fibers. In applying this method to measure fiber losses in a multimode fiber, one might question whether the launching conditions at both ends can be made to be the same with certainty.

Dispersion Measurements

In this section we discuss the methods for measuring dispersion characteristics of fibers. In general, the methods can be divided into two groups:

1. *Time-domain measurement.* Optical pulses are launched at the transmitting end, and the distorted output pulse is recorded at the receiving end of the fiber. The time delay or the broadening of short pulses is observed. The data are used to calculate the dispersion property of the fiber.
2. *Frequency-domain measurement.* The output of a light source is amplitude-modulated by a sweep frequency sinusoidal signal and is launched at the transmitting end of the fiber. At the receiving end, the baseband frequency response, that is, the response in terms of the modulation frequencies, is recorded. The dispersion is then expressed in terms of the bandwidth of the fiber.

Time-Domain Measurements

In time-domain measurements, monochromatic light pulses are injected at the transmitting end of the fiber and the output is recorded in real time at the receiving end. Either the time delay or the pulse broadening can be used to evaluate the dispersion. Chromatic dispersion can be obtained by observing the time-delay differences of pulses carried by light of different wavelengths. Alternatively, the broadened impulse response can be analyzed by Fourier transform technique to calculate the rms pulse width and the fiber bandwidth. We give some examples of using different techniques to achieve these objectives.

Dispersion Measurements with Raman Lasers

A convenient way to supply a light source with different wavelengths for the time domain measurements is to use a "fiber Raman laser." The "Raman effect" refers to the effect that when a light of high intensity passes through a liquid or solid material, the emerging light beam may contain new wavelength components in addition to those present in the incident beam. These new wavelength components appear as the Stoke lines, which are the longer wavelength components, and the anti-Stoke lines, the shorter wavelength components of the generated wavelengths. The Stoke lines are usually more intense and are therefore more important in this application. If a section of fiber is used for the solid, the laser is called the *fiber Raman laser*. Cohen and Lin [21] used this laser for their experiments to measure the dispersion of a fiber. Figure 5.14 is the schematic diagram of their experimental setup. The fiber Raman laser is pumped by a mode-locked and Q-switched Nd:YAG laser operating at a 1.06-μm wavelength. It supplies approximately 1 kW of pulse power to a single mode, low-loss "Raman fiber" 176 μm in length. The Raman fiber has a 6-μm silica core with borosilicate cladding. The loss is less than 6 dB/km for

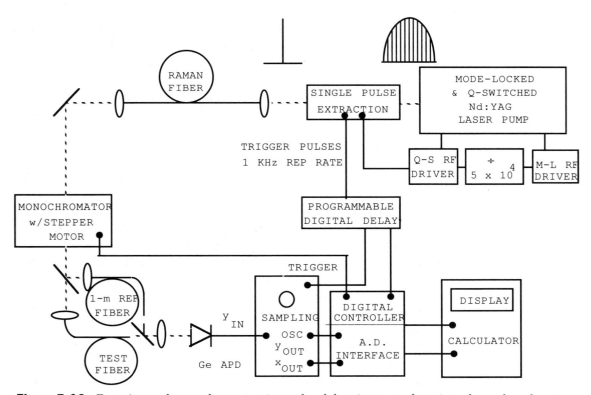

Figure 5.14 Experimental setup for measuring pulse delay times as a function of wavelength using a fiber Raman light source. [After L. G. Cohen and C. Lin, © IEEE 1978.]

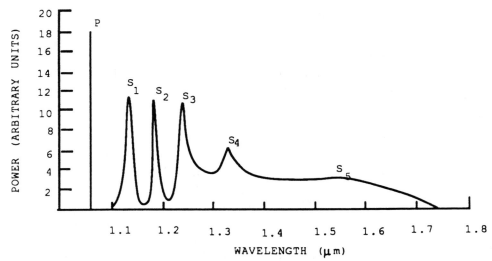

Figure 5.15 Raman spectrum of a single-mode silica fiber pumped at 1.065 μm. After L. G. Cohen and C. Lin [21].

wavelengths between 1.2 and 1.28 μm. The first Raman line at 1.12 μm is observed when the pump laser reaches a peak power of 70 W. The emission spectrum of the fiber Raman laser is as shown in Fig. 5.15. The five peaks corresponding to 1.12-, 1.18-, 1.24-, 1.31-, and 1.44-μm wavelengths are clearly visible. The light output from the fiber Raman laser is directed to the transmitting end of the fiber through a monochromator that is moving stepwise via a stepping motor such that at each step a monochromatic source of a certain wavelength is impinging on the fiber end. The signal is split into two paths, one through the test fiber and the other through a 1-m reference fiber. The recombined signal is then detected and displayed on a sampling scope. In order to trigger the sampling scope, a single pulse is extracted from the pump source and passed through a programmable digital delay to synchronize with the pulse.

The time delays measured for several single and multimode fibers relative to a pulse at 1.064 μm are shown in Fig. 5.16 as a function of wavelength. Each curve corresponds to one type of fiber. For example, fiber B-3 has a graded borosilicate core. Other fibers tested include B-4, B-5, and B-6, which have a small amount of B_2O_3 and a larger graded GeO_2 of different concentrations. All curves appear to have a minimum near 1.3 μm. This indicates the vanishing chromatic dispersion for these fibers.

Dispersion curves can be derived for each by taking the wavelength derivatives of the respective curves and plotting them as shown in Fig. 5.17. Notice that each fiber crosses the zero-dispersion point at slightly different wavelengths. For a larger dopant concentration, the zero crossing point shifts to longer wavelengths. This setup has an automatic digital control and is programmed such that for each pulse

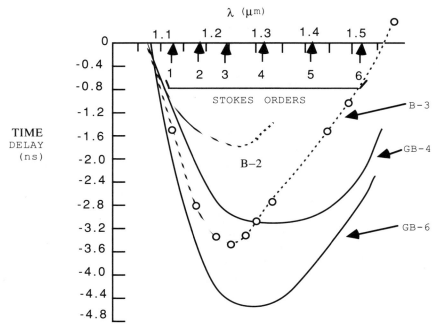

Figure 5.16 Transmission delay relative to a pulse at 1.064 μm versus λ. [After L. G. Cohen and C. Lin, [21] 855 (1978) © IEEE 1978.]

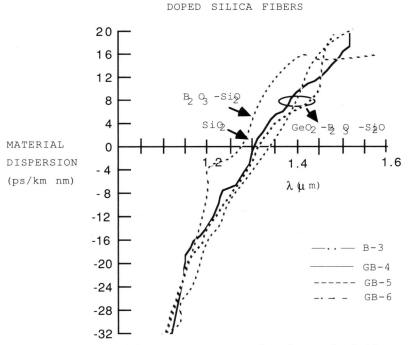

Figure 5.17 Material dispersion versus wavelength in multiple fibers. Fiber B-3, graded borosilicate core. Other fibers have a small amount of B_2O_3 and a larger graded GeO_2 concentration in increasing proportion from GB-4 to GB-6. [After L. G. Cohen and C. Lin, [21] 855 (1978) © IEEE 1978.]

wavelength, the time delay and its wavelength derivative are calculated and displayed automatically while the stepping motor changes
its monochromatic setting.

Measuring Pulse Broadening

Although it is desirable to observe the pulse shape in real time, with
a time scale on the order of a few picoseconds, most detectors are
inadequate for resolution of the waveform. As an alternative, measurements can be made by observing the broadening of the pulse as
it makes a single pass through the test fiber. An apparatus for measuring pulse broadening in MMF used by Gloge et al. [22] is shown in
Fig. 5.18. Outputs of two GaAs lasers oscillating at $\lambda_1 = 0.86$ μm and
$\lambda_2 = 0.9$ μm are launched on the test fiber (ca. 1 km in length) through
a half-mirror. The input and output waveforms are displayed on the
sampling scope, which is triggered by one of the signals. The results
of measurement are shown in Fig. 5.19. Figure 5.19a shows the transmitted pulses at two wavelengths. The left pulse corresponds to $\lambda_2 =$
0.9 μm, while the right pulse corresponds to $\lambda_1 = 0.86$ μm, which
was triggered 4.8 μm later. Both pulses have a half-power bandwidth
of 200 ps.

The corresponding received pulses are shown in Fig. 5-19b. The
individual pulse broadenings are not noticeable, but the pulse separation has been increased by about 3.6 ns. This indicates that the time
delay is more wavelength dependent, suggesting that material dispersion becomes a dominating factor in the total rms pulse width σ_t,
and $\sigma_t^2 = \sigma_{mat}^2 + \sigma_{mode}^2$. Calculation of the material dispersion delay

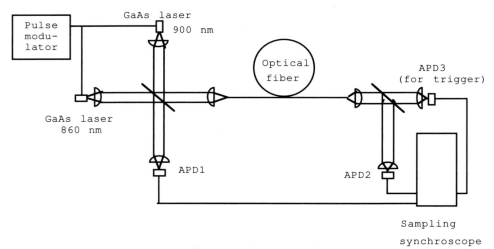

Figure 5.18 The use of two lasers for the separable dispersion measurement. After
D. Gloge et al. [22].

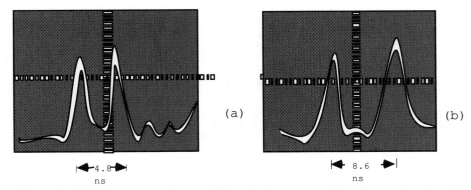

(a)

(b)

|◄—4.8—►|
ns

|◄ 8.6 ►|
ns

Figure 5.19 The measured waveforms of this system: (a) transmitted pulse; (b) received pulses—notice the increase in pulse separation. Right rules correspond to $\lambda_2 = 0.90 \ \mu m$; left pulses correspond to $\lambda_1 = 0.86 \ \mu m$. Time scale 2 ns/km per division. After D. Gloge et al. [22].

time for this experimental setup proved this is true (calculated τ_{mat} is 3.4 ns). Intermodal dispersion is almost negligible. This is possible for a long MMF because the effect of mode coupling tends to smooth out the modal dispersion.

This method has an advantage that the measured data can easily be interpreted as separable dispersion measurements in which the chromatic dispersion can be separated from the modal dispersion. Another method of measurement is described by Tenifugi and Ikeda [23].

Frequency-Domain Measurements

The time-domain techniques require a means for producing and detecting short pulses. If the impulse response or its Fourier transform is to be calculated, use of a computer is also required. In principle, the frequency-domain measurement is much easier to implement. It requires a signal generator that is tunable over a wide frequency range to modulate the light source for illuminating the transmitting end. At the receiving end, the signal is detected and is displayed on a spectrum analyzer. The dispersion measurement requires observing the change of the detected modulation as the signal frequency and the bandwidth of the light source are changed. Usually, the baseband frequency response, that is, the response in terms of the modulation frequency, is recorded. The frequency bandwidth of the fiber is the frequency difference at which the response is halved (the $- 3 - dB$ points). The block diagram of an experimental setup to measure the dispersion in a MMF in frequency domain is shown in Fig. 5.20. An incoherent light source (an LED) is used, but its spectral width must be filtered by a monochromator. The monochromatic source thus generated is first passed through an electrooptic modulator. The modu-

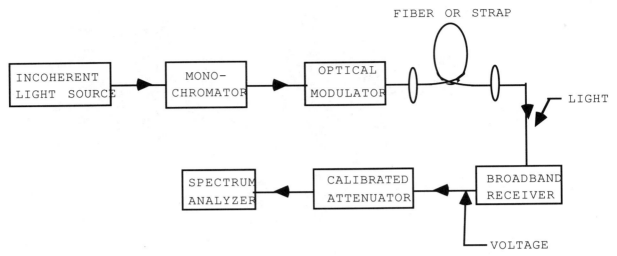

Figure 5.20 A block diagram showing apparatus for the measurement of frequency response of a fiber.

lated output is focused onto the fiber end or to a strap. The light output from the fiber or the strap is captured by a broadband receiver whose output is fed into a spectrum analyzer through a calibrated attenuator. As the modulation frequency is varied over a wide range, the spectrum analyzer displays the received output as a function of the modulation frequency. This is the baseband frequency response curve. By dividing the two sets of measurements obtained for the fiber and the strap, the transfer function of the fiber is obtained. This procedure is necessary because by comparison of the detected signal from the test fiber with the result from a much shorter strap, instrumental and launching effects can be eliminated.

In MMF where intermodal dispersion is of primary importance, another type of dispersion called "profile dispersion" might be experienced. If the source light has a broad-wavelength spectrum, and as the dopant concentration of the fiber core varies with the radius of a graded-index fiber, chromatic dispersion appears in addition to intermodal dispersion. This gives rise to the profile dispersion.

The broad-wavelength range of the modulated light helps to bring out the chromatic dispersion effect due to the graded-index profile. A fiber that has a broad signal bandwidth optimized at one wavelength may perform very poorly at another wavelength. Information on profile dispersion thus becomes very important in system design.

Gloge *et al.* [24] used yet another approach to obtain the frequency response. They used a free-running laser for generating a large number of sequential baseband frequencies spaced at uniform intervals. This is to be contrasted to the sinusoidally modulated intensity of a normal light source. A free-running krypton laser operating in the longitudinal cavity modes can generate many modes simultaneously. The frequency spacing between these modes is $\Delta f = v/2L_c$, where v

is the phase velocity of the light and L_c is the cavity length. Figure 5.21a shows the test frequency spectrum of such a laser operating at a 0.647-μm wavelength.

The light from such a laser is focused onto a MMF under test. Each fiber mode carries all the laser frequencies. At the receiving end, a detector is used to detect current at the beat frequency of two laser wavelengths. Because of the orthogonality of the modes, each mode can beat only with itself. This is identical to the case where the light signal had been intensity-modulated. Because of the different transit times of different modes, destructive interference takes place. The baseband frequency response at the output of a MMF drops off as shown in Fig. 5.21b. The input spectrum is the same as the laser output shown in Fig. 5.21a. The ratio of the output to the input frequency

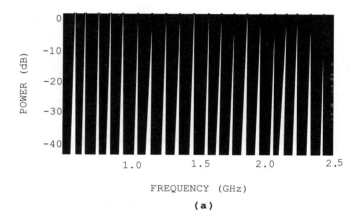

(a)

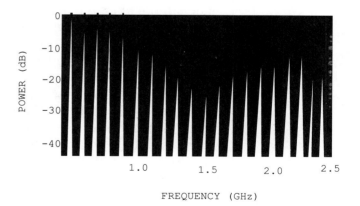

(b)

Figure 5.21 Beat frequency spectrum (a) at the input of MMF with a free-running krypton laser at λ = 0.647 μm; (b) at the end of a 30-m-long step-index fiber. After D. Gloge $et\ al.$ [24] **11**, 1534.

response is the frequency response of the fiber. By taking a lengthy exposure of the scope, the signal is averaged. This removes the problem of random fluctuations. The bandwidth of the fiber can easily be determined by taking the Δf at which the frequency response drops off to half.

The beat frequency method is simple to implement. The only disadvantage is that making intermodal dispersion measurements at many different wavelengths will require the use of a number of lasers, each being able to provide many closely spaced free-running wavelengths.

Guidelines for SMF Measurements

There are two measurements that are especially important for SMFs; the cutoff wavelength of the second order LP_{11} modes and the measurement of the dispersion properties other than the bandwidth scheme.

Cutoff Properties of the Second-Order LP_{11} Modes

The fundamental mode for a cylindrical optical fiber is designated as the LP_{01} mode. It has no cutoff wavelength. If the fiber is nonbirefringent, the two polarized modes can be assumed to propagate in the same way and pose very little problem in determination of the fiber property. But the cutoff properties of the LP_{11} mode become important because the attenuation, noise, and bandwidth of the fiber are all affected by the presence of the LP_{11} mode. Although the theoretical cutoff wavelength λ_c for the LP_{11} mode is at $V = 2.405$, the effective cutoff wavelength λ_{ce} can occur below λ_c as a result of the distribution of bends in the fiber, the presence of fluctuations in diameter, the refractive index, and the length. The existence of the LP_{11} mode below λ_c can cause bimode noise as well as a bandwidth decrease due to bimodal dispersion.

One technique for determining the effective cutoff of the LP_{11} mode is the scan of mode-field diameter (MFD) method [25]. The MFD is a measure of the width of the distribution of electric-field intensity, which can be used in predicting the cutoff wavelength. A variable wavelength light source is scanned through its spectral range and enters the fiber input end face through a lens and a mode stripper. The output near-field light pattern is observed. As wavelength increases from below the cutoff, the diameter of the LP_{01} mode and the larger LP_{11} mode increases at first (see Fig. 5.22). Then as cutoff is approached, it decreases sharply as the LP_{11} mode is eliminated. A further increase in wavelength increases the diameter again. Cutoff (effective) is determined by the straight lines fit to the decreasing and increasing portions of the curve.

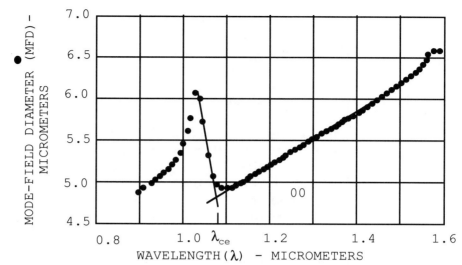

Figure 5.22 Scan of the mode-field diameter. After G. J. Grosso *et al.* [25].

Dispersion Measurement

Bandwidths for SMF are too high to measure directly. Moreover, in SMF, bandwidth depends on source spectral properties and can become complex when source fluctuation due to "chirp" effect is present. Chirp is the dynamic line-broadening effect of a laser under direct intensity modulation. It is more practical to measure the chromatic dispersion and derive the bandwidth therefrom. Besides the time-delay technique discussed in the section on time-domain measurements (above), many more techniques become available. Several dispersion measurements are based on phase-shift measurements. Usually the input light is sinusoidally modulated and the variation of the modulated phase shift over the fiber with wavelength is observed [26–28]. Another technique which is useful for short length fibers is the Mach–Zehnder interference technique for measuring dispersion [29]. New techniques are appearing almost every day. Readers are advised to consult a review book [30].

Conclusion

Measurement technique plays an important role in the advancement of optical fiber technology. Without the ability to measure the parameters involved in a fiber quantitatively, one has no means to gauge the improvement. Since the science of optical fibers is still rapidly developing, each new fiber may require new parameters with appropriate measurement techniques. Although some progress has already been made, the field still needs standardization. The techniques described in this chapter are just a few examples of what have been developed thus far. They only serve as introductory, basic techniques. We can expect much more to come.

Appendix: Commercial Test Equipment

In the past, major optical fiber manufacturers and research laboratories improvised their own measuring apparatus to control and characterize their products. Only very recently have standardized measuring instruments become commercially available. Two American companies, Hewlett-Packard (HP) and Tetronix, are among many who offer these instruments. A few are listed here:

Test Equipment

Hewlett-Packard	Tetronix
Optical Signal Sources	
HP8150A	Tetronix OT501/OT502/OT503
HP83401A	
MMF 50–125-μm fiber	
HP83400A λ: 1308 ± 10 nm	
SMF 9–125 μm	
	Power Meters
HP8151A pulse powermaker	S-42 optical sampling head
81511/81512 A optical heads	1000–1700 nm
HP8152A average powermaker	
450–1700-nm wire, two heads	
	Optical Time-Domain Reflectometer
HP8145A portable	OF150 for local-area network, 850 nm
	OF235 OTDR, 1300/1500 nm
	Cable Tester
HP8702A analyzer at a fixed	OF192 825, 850, 1300 nm,
wavelength (1300 nm)	40-dB range
	Measures bandwidth and loss in MMF
	Receivers
HP83410A P_{max} = 3 mW	OR 501/OR502
Responsivity 50 MHz: 20-dB PIN	700–1500 nm
Modulated bandwidth 300 kHz–3 GHz	Ge-APD
	Maximum limit input 20 dBm
	Frequency response 0.03–1500 MHz
	Optical Components
Lens	Opto Scope
81050A1 (450–1020 nm)	2815 for 850-nm system
NA = 0.2 fibers	Optical-Electrical Converter 7F10A
81050BL (900–1700 nm)	SMF/MMF optical input
NA = 0.2 fibers	700–1550 nm range
81010BL (900–1700 nm)	dc to 1-GHz frequency response
NA = 0.1 fiber	

Test Equipment (*continued*)

Hewlett-Packard	Tetronix	
Attenuating lens adapters	P6701	450−1050-nm range, dc to 700 MHz
HP81220FL	P6702	1000−1700-nm range, dc to 500 MHz
Fixed filter	P6751	spatial input head
HP81000FF		
Power splitter		
HP81000AS for 600−1200 nm		
HP81000BS for 1200−1600 nm		
Connectors		
011: HMS-10/HP		
012: FC/PC		

References

1. W. Eickhoff and E. Weidel, Measuring method for the refractive index profile of optical glass fibers. *Opt. Quantum Electron.* **7,** 109 (1975).
2. M. Ikeda and H. Yoshikiyo, Refractive index profile of a graded index fiber: Measurement by a refraction method. *Appl. Opt.* **14,** 814−815 (1975).
3. J. Stone and H. E. Earl, Surface effects and reflection refractometry of optical fibers. *Opt. Quantum Electron.* **8,** 459 (1978).
4. D. Gloge and E. A. J. Marcatili, Multimode theory of graded-core fibers. *Bell. Syst. Tech. J.* **52,** 1563 (1973).
5. M. J. Adams, D. N. Payne, and F. M. E. Sladen, Leaky rays on optical fibers of arbitrary (circularly symmetric) index profiles. *Electron. Lett.* **11,** 238−240 (1975).
6. W. J. Stewart, Advances in fiber-based components. *OFC/IOOC,* THB1, **191,** Reno (1987).
7. D. Marcuse, Refractive index determination by the focusing method. *Appl. Opt.* **18,** 9−13 (1979).
8. H. M. Presby, W. Mammel, and R. M. Derosiar, Refractive index profiling of graded index optical fibers. *Rev. Sci. Instrum.* **47,** 348−352 (1976).
9. B. C. Wonsiewicz, W. G. French, P. D. Lasay, and J. R. Simpson, Automatic analysis of interferograms: Optical waveguide refractive index profiles. *Appl. Phys. Lett.* **15,** 1048−1052 (1976).
10. M. E. Markic, P. S. Ho, and M. Epstein, Nondestructive refractive index profile measurement of clad optical fibers. *Appl. Phys. Lett.* **26,** 574 (1975).
11. L. Boggs, H. M. Presby, and D. Marcuse, Rapid automatic index profiling of whole-fiber samples. Parts I and II. *Bell Syst. Tech. J.* **58,** 867−882, 883−902 (1979).
12. H. M. Presby, Refractive index and diameter measurement of unclad optical fibers. *J. Opt. Soc. Am.* **64,** 280−284 (1974).
13. T. Okoshi and K. Hotate, Refractive-index profile of an optical fiber: Its measurement by the scattering-pattern method. *Appl. Opt.* **15,** 2756−2764 (1976).
14. O. C. Wells, *Scanning Electron Microscopy.* McGraw-Hill, New York, 1974.
15. W. F. Love, Novel mode scrambler for use in optical-fiber bandwidth measurement. *Conf. Opt. Fiber Commun.,* Washington, D.C., *1979,* Conf. Dig., p. 118, Paper 7hG2 (1979).
16. R. L. Cohen, Loss measurements in optical fibers. (1) Sensitivity limit of bolometric technique. *Appl. Opt.* **13,** 2518 (1974).
17. R. L. Cohen, K. W. West, P. D. Lazay, and J. Simpson, Loss measurements in

optical fibers. (2) Bolometric measuring instrumentation. *Appl. Opt.* **13**, 2522 (1974).

18. A. R. Tynes, Integrating cube scattering detector. *Appl. Opt.* **9**, 2706–2710 (1970).

19. D. L. Philen and F. T. Stone, Direct measurement of scattering losses in single-mode and multimode optical fibers. *Adv. Ceram.* **1** (1980).

20. J. E. Midwinter, *Optical Fibers for Transmission*. Wiley, New York, 1979.

21. L. G. Cohen and C. Lin, A universal fiber-optic (UFO) measurement system based on a near-IR fiber Raman laser. *IEEE J. Quantum Electron.* **QE-14**, 855–862 (1978).

22. D. Gloge, E. L. Chinnock, and T. P. Lee, GaAs twin-laser setup to measure mode and material dispersion in optical fibers. *Appl. Opt.* **13**(2), 261–263 (1974).

23. T. Tanifuji and M. Ikeda, Simple method for measuring material dispersion in optical fibers. *Electron. Lett.* **14**, 367 (1978).

24. D. Gloge, E. L. Chinnock, and D. M. Ring, Direct measurement of the base band frequency response of multimode fibers. *Appl. Opt.* **11**, 1534–1550 (1972).

25. G. J. Grosso and G. J. Cannell, A Novel Technique for cutoff wavelength measurement in single-mode fibers. *Proc. Eur. Conf. Opt. Commun., 8th*, Cannes, *1982*, p. 98 (1982).

26. B. Costa, M. Pulco, and E. Vezzoni, Phase-shift technique for the measurement of chromatic dispersion in fibers "using LeDs." *Electron. Lett.* **19**, 1074 (1983).

27. A. J. Barlow and I. MacKinzie, Direct measurements of chromatic dispersion by the differential phase technique. *Tech. Digest, OFC*, paper TUQ1, (1987).

28. R. Rao, Field dispersion measurement—A swept-frequency technique. *NBC Spec. Publ. (U.S.)* **683**, 135 (1984).

29. L. G. Cohen and J. Stone, Interferometric measurements of minimum dispersion spectra in short lengths of single mode fiber. *Electron. Lett.* **18**, 564 (1982).

30. G. Cancellieri and V. Ravaioli, *Measurement of Optical Fibers and Devices; Theory and Experiments*. Artech House, Dedham, Massachusetts, 1984.

Semiconductor Light Sources and Detectors

Part Two deals with light sources and photodetectors for optical fiber systems. Since the dimensions encountered in optical fibers are rather small, on the order of several micrometers to less than a hundredth of a millimeter in fiber diameter, light sources and detectors must be chosen with comparable size. Semiconductor devices become the most suitable choice. Besides the size compatibility, semiconductor light sources and detectors offer almost an unlimited choice of wavelengths in the visible to infrared range, reasonable power output, low power consumption, long life, low cost, and many other advantages. We devote two chapters, Chapters 6 and 7, to introduction of the semiconductor devices in optical fiber applications. Chapter 6 is concerned with light-emitting diodes and lasers and Chapter 7 with semiconductor detectors, including various versions of junction detectors and the photoconductors used for long-wavelength detection.

6

Semiconductor Light Sources: Light-Emitting Diodes and Injection Lasers

Introduction

Semiconductors are a group of elements or compounds having electrical conductivities intermediate between those of metals and insulators. The fact that the electrical conductivity of the semiconductor can be changed over several orders of magnitude by altering the temperature, excitation, and most importantly, the impurity content, makes this material the most valuable one in modern electronic industry. Optical fiber systems can use these materials advantageously.

Semiconductor Fundamentals

A glossary of technical terms in semiconductor fundamentals is prepared for quick reference.

Semiconductor materials. A semiconductor is an element or compound whose electrical resistivity ρ [in ohm-meters ($\Omega \cdot m$)] is within the range specified by this inequality

$$10^{-5} < \rho^{(\Omega \cdot m)} < 10^{10} \qquad (6-1)$$

The typical resistivity of germanium (Ge) is 0.83 $\Omega \cdot m$, silicon (Si) is 2.3 $\times$ 10^3 ($\Omega \cdot m$), and gallium arsenide (GaAs) is 10^6 ($\Omega \cdot m$), all measured at room temperature. In comparison, a good conductor such as copper (Cu), ρ_{Cu} is 1.67×10^{-8} ($\Omega \cdot$m) and a good insulator, diamond (C), has a resistivity $\rho_C = 10^{14}$ ($\Omega \cdot$m).

Intrinsic semiconductor. The spectral configurations of the semiconductor elements such as Si and Ge are respectively: $Si(14)$—$1S^2\, 2S^2\, 2p^6\, 3S^2\, 3p^2$, of which the outermost four electrons in the $3S^2$ and $3p^2$ subshells are the valance electrons; $Ge(32)$—$1S^2\, 2S^2\, 2p^6\, 3S^2\, 3p^6\, 3d^{10}\, 4S^2\, 4p^2$. Again Ge has four valance electrons in $4S^2\, 4p^2$ subshells. Both elements belong to group IV in the periodic table.

Intrinsic density n_i. One important property that characterizes a semiconductor is the intrinsic density n_i (in numbers per cubic centimeter). Each semiconductor element has its own unique intrinsic density and

$$n_i^2 = N_c\, N_v\, e^{-Eg/kT} \tag{6-2}$$

where

$$N_c = 2\left(\frac{2\pi\, m_e^*\, kT}{h^2}\right)^{3/2}, \qquad N_v = 2\left(\frac{2\pi\, m_h^*\, kT}{h^2}\right)^{3/2}$$

are the material constants with m_e^* and m_h^* the effective mass of electrons and holes, respectively, h is Planck's constant [$h = 6.63 \times 10^{-34}$ Joule-seconds (J-s)], k is Boltzman's constant [$k = 1.38 \times 10^{-23}$ joules/degree kelvin (J/K)], and T is the absolute temperature. The hole is an atom in which an electron has been stripped (lack of an electron).

The bandgap energy E_g. The other important semiconductor property is the bandgap energy E_g. Within this bandgap, no energy levels are allowed by quantum theory. Each semiconductor element has its own unique E_g, which exists between the conduction band edge E_c and the valence band edge E_v in the energy diagram. At 300 K, examples of n_i and E_g in semiconductors are:

	Si	Ge	GaAs
E_g (eV)	1.12	0.66	1.43
n_i (cm^{-3})	10^{10}	2×10^{13}	10^7

The energy-band diagram. Electrons in solids are not free to move as in a vacuum. The crystal forces among the atoms force the motion to obey the energy momentum law governed by quantum mechanics. Electrons can possess energies in certain momentum spaces and void in others. In semiconductors, the void energy gap is between the conduction band edge E_c and the valance band edge E_v such that $E_g = E_c - E_v$. At 0 K, all valence electrons fill up to the valence band edge E_v. There are no electrons allowed within the bandgap E_g nor above the conduction band edge E_c. At tempera-

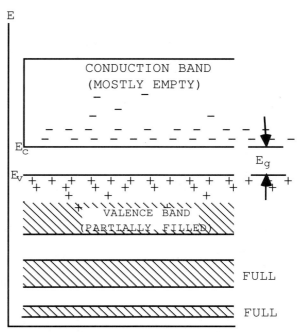

Figure 6.1 A simplified version of the energy-band picture of semiconductors.

tures above 0 K, there may be electrons above E_c, leaving holes in the valence band below E_v. A simplified energy-band diagram of a semiconductor is shown in Fig. 6.1, where the minus ($-$) signs signify electrons while the plus ($+$) signs indicate holes.

Charge carriers in semiconductors. Both electrons and holes are charge carriers. The equilibrium densities of electrons n and holes p are related to n_i and the energy levels as follows:

$$n = n_i e^{(E_F - E_i)/kT} \tag{6-3}$$

and

$$p = n_i e^{(E_i - E_F)/kT} \tag{6-4}$$

where E_F is the Fermi energy defined as an energy level such that at 0 K all energy levels below E_F are occupied by electrons and those above remain empty. The intrinsic energy level E_i equals $(\frac{1}{2})E_g$, and $E_i = E_F$ at 0 K.

Extrinsic semiconductor. If either a group III or a group V element is substituting some of the atoms in otherwise pure group IV elements, a process called *doping*, the resultant alloy becomes extrinsic. Because of the similarity of the crystal structures, these elements can substitute freely. Table 6.1 (which is a part of the periodic table) and Table 6.2 show some common semiconductor

Table 6-1 Common Semiconductor Elements

Group	II	III	IV	V	VI
		B	C		
		Al	Si	P	S
	Zn	Ga	Ge	As	Se
	Cd	In	Sn	Sb	Te

Table 6-2 Some Semiconductor Compounds

Group III–V Compounds		Group II–VI Compounds
Binary		
AlP	InP	ZnS
AlAs	InAs	AnSe
AlSb	InSb	AnTe
GaP		CdS
GaAs		CdSe
GaSb		CdTe
Ternary		
GaAlAs		
InGaAs		
GaAsSb		
GaAsP		
InAsP		
Quaternary		
GaInAsP, etc.		

elements and compounds. Impurity doping can be applied to both semiconductor elements and compounds.

p-Type semiconductor. If the doping impurity is taken from group III (e.g., Al or Ga) that has one less valence electron than the host element (e.g., Si), the alloy becomes a p-type semiconductor.

n-Type semiconductor. If the doping impurity is taken from a group V element (e.g., P, As) that has one more valence electron than the host element (e.g., Si), the result is an n-type semiconductor.

Impurity energy levels. Doping a group IV pure semiconductor element with impurity introduces an impurity energy level (or band) from which electrons may be acted on to increase the number densities of electrons or holes in the host material. For the n-type, an impurity level E_d is introduced in the bandgap closer to E_c so that at room temperature (300 K) nearly all electrons in E_d become ionized and excited into the conduction band to contribute to the electronic conduction. The number density of the n-type semiconductor can be calculated from the charge balance relation as

$$n = p + N_d^+ \qquad (6\text{-}5)$$

where p is the intrinsic hole density, N_d^+ is the ionized donor density in level E_d, and N_d is the doping density. For the p-type, an impurity level E_A is introduced in the band diagram closer to E_v such that at 300 K, electrons in the upper valence band may easily be excited and move up to occupying the impurity sites. The increase in p number can again be calculated by the charge balance relation as

$$p = n + N_A^- \qquad (6\text{-}6)$$

where N_A^- is the ionized acceptor of the doping density N_A.

Thus, the equilibrium density of semiconductor can be increased by the process of impurity doping.

Excess carriers in semiconductor. Electrons and holes can be created in excess of the equilibrium densities by applying external sources of energy such as optical energy and electron bombardment. The generated carriers appear in pairs and

$$\Delta n = \Delta p \qquad (6\text{-}7)$$

The increase in the net carrier densities is

$$n = n_0 + \Delta n = n_i e^{(F_n - E_i)/kT} \qquad (6\text{-}8)$$

$$p = p_0 + \Delta p = n_i e^{(E_i - F_p)/kT} \qquad (6\text{-}9)$$

where n_0 and p_0 are the electron and hole densities in equilibrium and F_n and F_p are the quasi-Fermi levels of electrons and holes, respectively.

Recombination of carriers. When the external energy is removed, electrons and holes tend to reunite. This is termed *recombination*. The time taken to complete a recombination process is called the *recombination time* τ_r.

Direct bandgap material. The energy-momentum diagram of a semiconductor material shows a bandgap energy E_g relative to E_c and E_v. For a simplified picture, the relations are shown in straight lines as in Fig. 6.1.

In an actual case, E_c and E_v are curves having both minimum and maximum. If the minimum of E_c and the maximum of E_v occur at the same momentum space, the material is said to be a *direct bandgap* one. In this case electrons and holes recombine without change in momentum. The excess energy is released in terms of radiation at a frequency proportional to the difference in the original energy levels, or $h\nu = E_2 - E_1$, where h is Planck's constant and ν is the frequency of radiation. This is shown in Fig. 6.2a. Gallium arsenide (GaAs) and InSb are the direct bandgap semiconductors.

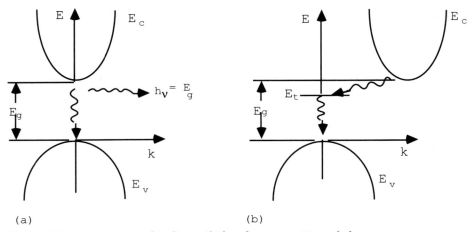

Figure 6.2   Direct (*a*) and indirect (*b*) bandgap transition of electrons.

Indirect bandgap material. If the minimum of the conduction band
edge E_c does not coincide with the maximum valance band edge E_v
in momentum space, we have an indirect bandgap material; Si and
Ge are these materials. The band diagram is shown in Fig. 6.2*b*. The
recombination takes place indirectly through an intermediate step
(or a trap level) in momentum. The released energy is usually in the
form of heat production.

Carrier motion in semiconductors. Motion of carriers in semiconductors is characterized by two types, the drift motion in response to
external field and the diffusion motion due to difference in carrier
density gradient in space.

Drift. Carrier motions in semiconductors in response to the external
field are not free as in vacuum. They suffer multiple collisions and
may change direction and momentum at all times. The net motion
in the direction of the external field resulted in an average drift
current density J_x as

$$J_x = q(n\mu_n + p\mu_p)E_x = \sigma E_x \qquad (6\text{-}10)$$

where E_x is the x-directed external field, J_x is the x-directed current
density, μ_n and μ_p are the mobilities of the electrons and holes, respectively, and $\sigma = q(n\mu_n + p\mu_p)$ is the conductivity.

Diffusion. Because of the difference in carrier concentration in space
created by the generation of excess carriers, carriers are diffused out
in space in the direction of the density gradient. The diffusion current density can be expressed as

$$J_{n(\text{diff})} = qD_n \frac{dn(x)}{dx} - qD_p \frac{dp(x)}{dx} \qquad (6\text{-}11)$$

where D_n and D_p are the diffusion constants of electrons and holes,
respectively, assuming a one-dimensional case for simplicity. Mo-

Table 6-3 Diffusion and Mobility Constants
of Semiconductors at 300 K

	D_n (cm²/s)	D_p (cm²/s)	μ_n (cm²/V·s)	μ_p (cm²/V·s)
Ge	100	50	3900	1900
Si	35	12.5	1350	480
GaAs	220	10	8500	400

bility and diffusion constants are related by Einstein's relation (eq.
6-12). These constants of some semiconductors at 300 K are listed
in Table 6.3.

$$\frac{D}{\mu} = \frac{kT}{q} \tag{6-12}$$

Semiconductor Junctions

A semiconductor junction is a metallurgically formed single-crystal
alloy of two semiconductor materials. Modern techniques of forming
junctions included the use of ion implantation, liquid-phase epitaxy
(LPE), vapor-phase epitaxy (VPE), and molecular-beam epitaxy (MBE).
Semiconductor junction is a building block in modern electronic devices. The most important problem encountered in forming junctions
is lattice matching. It affects the life and performance of the device
made with the junction.

p–n Homojunction

A junction formed with n- and p-types of the same host material, and
thereby the same E_g, is called a *homojunction*. The energy-band diagram is shown in Fig. 6.3.

On forming a step junction, at equilibrium, the Fermi energy levels
of both types should align ($E_{Fp} = E_{Fn}$); this causes the band edges to
tilt. A built-in potential ϕ_0 results such that $q\phi_0 = E_{cn} - E_{pn}$ and

$$\phi_0 = \frac{kT}{q} \ln \frac{N_a N_d}{n_i^2} \tag{6-13}$$

p–n Heterojunction

If the host materials of both types of semiconductor are different (i.e.,
$E_{g1} \neq E_{g2}$), the resultant alloy is a heterojunction. Figure 6.4 shows
the development of the complex band diagram, an n-type germanium
and a p-type GaAs. Notice that the total built-in voltage ϕ_0 (= V_{bi})

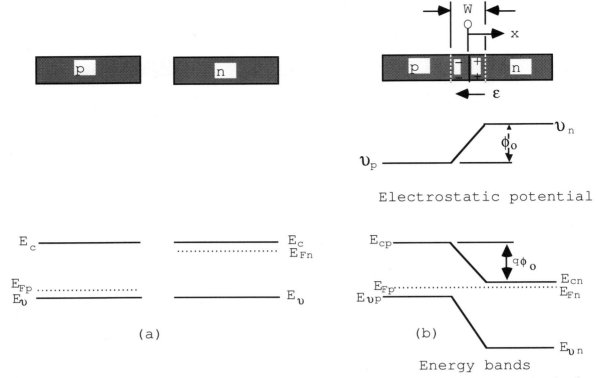

Figure 6.3 Energy-band diagram of a step $p-n$ homojunction: (a) before junction was made; (b) after junction.

equals the sum of the partial built-in potentials V_{01} $(= V_{b1})$ and V_{02} $(= V_{b2})$, and that the built-in potentials are different for electrons and holes. Thus the hole current from GaAs (p) to Ge (n) is expected to dominate because of the low potential barrier ϕ_{02} (or V_{b2}). The discontinuity in the band diagram favors the injection of majority carriers from the larger bandgap material regardless of the doping level.

$n-n$ Heterojunction and $p-p$ Heterojunction

Two n-type semiconductors or two p-type semiconductors of different bandgap energies can be joined to form a $n-n$-type or $p-p$-type heterojunction. The energy-band diagrams are shown in Fig. 6.5a and Fig. 6.5b, respectively. Notice that the effective potential barrier can be increased to favor carrier concentration. A Ge–Si $n-n$ heterojunction is used for illustration. It is observed that both sides of the junction are depleted. The barrier here acts like a metallic semiconductor contact.

The potential barriers added to the original steps in the $n-n$ and $p-p$ heterojunctions act well to confine holes or electrons in the structure.

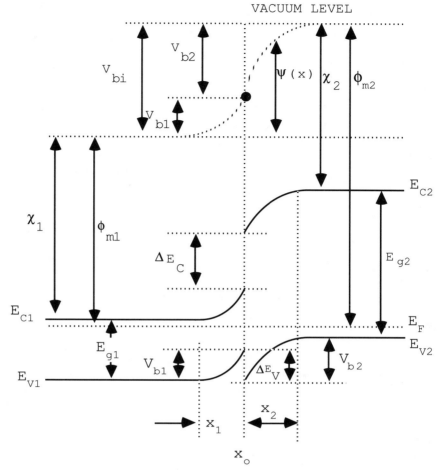

Figure 6.4 A $p-n$ heterojunction in equilibrium: ϕ_0 (for electrons) $= V_{b1} + V_{b2} + \Delta E_C$; ϕ_0 (for holes) $= V_{b2}$.

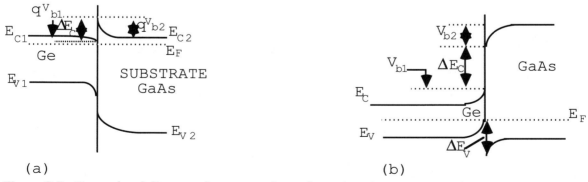

Figure 6.5 Energy-band diagram of an $n-n$ and $p-p$ heterojunction: (a) an $n-n$ heterojunction at equilibrium; (b) a $p-p$ heterojunction at equilibrium.

Double Heterojunction

If two heterojunctions are prepared on both ends of a semiconductor device, such as a laser, the structure is called a *double-heterojunction (DH) structure* (see subsection on laser structure).

Forward-Biased $p–n$ Junction

When external potential is applied to the junction whose polarity is to reduce the effect of the built-in potential ϕ_0, the result is a reduction in the barrier potential and allows easier drift of electrons and holes. Thus, carriers are said to have been injected across the junctions.

Reverse-Biased $p–n$ Junction

When the polarity of the applied potential is reversed, that is, to add to the built-in potential, a large increase in the barrier potential results. This will reduce the drift current, causing the current through the junction to be saturated, that is, practically independent of the voltage. But the junction is sensitive to excess carriers generated within the depletion region—a property useful for the detection of light-generated carriers. Reverse bias also increases the depletion width of the junction.

General Considerations of Semiconductor Light Sources for Fiber Optic Applications

Light sources for optical fiber systems should have sizes and configurations compatible with the fiber lightguide. A survey of available sources reveals that semiconductor light sources are probably the best candidates not only for their size compatibility but for many other considerations as well. Semiconductor light sources are capable of giving adequate power output. They are efficient, reliable, and easy to modulate even at high bit rates and have low noise figure, long life, and low cost.

Semiconductor light-emitting diodes (LEDs) and laser diodes (LDs) will be discussed. General requirements of a good light source are as follows.

Power Output

Although large power output from a source is usually desirable to overcome transmission losses in communication systems, care must be exercised in deciding on the output power of a light source for optical fiber applications. Optical fibers should never be overloaded with a power density that exceeds the material nonlinearity limit of the fiber in order to avoid the uncontrollable, nonlinear effects. Nonlinearity causes harmonic generation and increases noise production. At present, a power density of about 1 MW/cm² is considered as a limit. This converts into about 100 mW for a desirable semiconductor light source.

Coupling Efficiency

A light source must be coupled to the fiber efficiently; otherwise, light power will be wasted. Besides the size compatibility, a light source for optical fibers should have a radiation pattern (beam divergence diagram) compatible with the numerical aperture *NA* of the fiber. Means to ensure reflectionless engagement with the fiber are usually improvised. For example, a graded-index fiber has a light-intensity distribution assuming a Gaussian shape. It can be coupled efficiently with a light source with a similar intensity distribution. A SMF has a core diameter of only several micrometers. A light source must have an emission area similar to the fiber cross section. The light-acceptance angle of the fiber is limited by its numerical aperture. Thus, light sources with a broad radiation pattern (large divergence) will not be coupled efficiently to the fiber.

Source Efficiency

For a *p–n* injection-type light source such as LEDs and lasers, the source efficiency is a product of the carrier injection efficiency, the radiative efficiency, and the external quantum efficiency. We shall see how the selection of semiconductor materials, proper processing, and design consideration can affect the overall efficiency.

Carrier-Injection Efficiency

Most semiconductor light sources work on the principle of carrier injection across a *p–n* junction that recombines radiatively to produce light as described in the section on stripe-geometry DH lasers (below). But not all injected carriers are able to participate in the radiative recombination process. Some are lost as a result of trapping. Some may not actively participate in the process because of low mobility.

Thus, the carrier-injection efficiency can be expressed as

$$\eta_j = \frac{I_n}{I_n + I_p + I_{rec}} \tag{6-14}$$

where I_n and I_p are the electron and hole diffusion currents in the neutral p and n regions, respectively, after injection and I_{rec} is the recombination current within the depletion region due to the presence of some deep impurity levels within the bandgap. For good materials with careful processing, I_{rec} is usually small. Also, I_p is usually small compared to I_n in direct bandgap semiconductors as a result of the large mobility ratio. For example, in GaAs, the mobility ratio μ_n/μ_p is about 30. Thus, η_j can be made high by proper material selection and careful processing.

Radiative Efficiency and Internal Quantum Efficiency

Not all injected carriers participating in the recombination process can produce photons. Others may become nonradiative. The radiative efficiency can be expressed as

$$\eta_r = \frac{1}{1 + \tau_r/\tau_{nr}} = \frac{\tau}{\tau_r} \tag{6-15}$$

where τ_r and τ_{nr} are the minority carrier lifetimes that participate in the radiative and nonradiative recombination processes, respectively, and τ is the effective lifetime of the injected carriers defined as

$$\frac{1}{\tau} = \frac{1}{\tau_n} + \frac{1}{\tau_{nr}} \tag{6-16}$$

Shorter lifetime, in general, is desirable for high-speed operations. Lifetime is used as a criterion for selecting materials for light sources. For GaAs, the lifetime is on the order of 10^{-8} s, which is about four orders of magnitude shorter than Si. The lifetimes of AlGaAs and InGaAsP are similar to that of GaAs. These are the principal semiconductor materials of interest in producing light sources and detectors.

The internal quantum efficiency is

$$\eta_i = \eta_j\eta_r \tag{6-17}$$

External Quantum Efficiency

Not all photons produced by radiative recombination processes can escape from the crystal and reach the surface to serve as a light source. Some may have been refracted or reflected inside the crystal and at the surface due to differences in the refractive indices and the emission angle. Thus external quantum efficiency depends on the geometry of the structure and the interface conditions between the light source and the fiber.

The total source efficiency is then the product of the internal and external quantum efficiencies.

Spectral Emission and Spectral Width

The emission characteristics of a light source are required to match the spectral properties of the optical fiber. According to the loss characteristics of optical fibers discussed in Chapter 3 (Fig. 3.13), three wavelength regions are of particular interest for optical fiber communication systems. The first region is the shorter wavelength region (0.8−1.0 μm). The fiber loss within this wavelength range (the first-generation fiber) is low enough (a few decibels per kilometer) to allow for profitable design and installation of attractive short links for inter- and intraoffice data and signal transmission. For these systems, the most suitable light sources are the GaAs LEDs and lasers because they are reliable and inexpensive. To take advantage of the ultralow loss, zero-dispersion fiber properties at longer wavelengths (1.2−1.65 μm), the so-called second- and third-generation fibers, more sophisticated light sources must be chosen. For these second and third wavelength regions, ternary and quaternary semiconductor compounds are chosen to match the spectral emission and to satisfy the requirement of lattice matching.

Light-emitting diodes have a relatively wide spectral width, typically several hundred angstrom units. For MMFs used in short links, LEDs are suitable. However, light sources with wide spectral width may affect the material dispersion and profile dispersion of the fibers as discussed in Chapter 3. The material dispersion caused by wide source spectral width could be on the same order of magnitude as that produced by different modes in a graded-index fiber. Light-emitting diodes are therefore not suitable for long-haul systems where only SMF can be used. Lasers become the only source for SMF. Lasers with a narrow spectrum width (≥ 2 Å) are now available.

Source spectral width may also become a limiting factor in achieving high-density wavelength multiplexing in optical fiber systems. Wider source spectral width causes cross-talk between channels.

Modulation

Information can be imposed on optical carriers through the process of modulation. Intensity modulation can be used successfully for light sources such as the LEDs and the lasers. Intensity modulation is achieved by changing the driving current for the LEDs and lasers. It can be applied in either analog or digital forms. Frequency- and phase-modulation techniques are rarely used for modulation of these optical sources, however, because an optical light source is seldom a coherent source such as a radio-frequency oscillator. Frequency and phase modulation of an incoherent source introduces undesirable noises.

The rate of modulation is limited by the speed of the driving circuit and the response-time constant of the source. A fast response circuit enables a wider bandwidth signal to be handled with little distortion. Modulation linearity is a limiting factor for analog signals. The second and third harmonic discrimination margin signals for analog signals should be 40 and 50 dB, respectively. This is not so for a digital signal. Linearity plays a less important role in determining the signal quality in digital systems.

Operating Temperature and Device Life

Device life is defined as the mean time before failure. For light sources in optical fiber systems, a life span of 10^5 h can be considered as adequate. For long-haul systems, however, particularly those involving undersea cables, the demand is for 10^6 h or longer.

Life of an optical source may be very sensitive to the operating temperature. While the ambient temperature may not seem high, the junction temperature could be higher. An adequate heat sink should always be incorporated in the system to keep the device temperature low.

High temperature not only causes semiconductor light source to emit less light but may also severely shorten semiconductor life. Temperature change also affects emission wavelength. Laser wavelength may be affected by temperature on the order of 0.1 Å/°C.

Light-Emitting Diodes

Light-emitting diodes are the simplest light sources successfully applied to fiber optical systems thus far. The light-emitting area of an LED can be designed to be as small as the fiber's dimensions. LEDs can supply adequate power (0.05–2 mW) and are easy to modulate, simple in structure, long in life (10^6 hs), and low in cost. The disadvantage is that they do not couple efficiently to the fibers. In general, their coupling efficiency is from only a fraction of a percent to a few percent.

The light emitted from an LED is generated by spontaneous radiative recombination when a p–n junction made with direct bandgap semiconductor materials is forward-biased. Generally, not all light emitted at the junction area is readily accessible. For an LED to be more useful for optical fiber applications, it has to be specially designed. This includes (1) selecting proper semiconductor materials to emit the specific light spectrum required for the fiber, (2) tailoring the structure to confine the injection current, (3) arranging the layout to minimize the internal absorption, and (4) clearly defining the emission area for easy coupling. In this section, we describe some successful LED designs.

Semiconductor Materials for LEDs

The emitted light from a semiconductor can have a wavelength $\lambda = 1.24/E_g$, where λ is in micrometers and E_g is in electron volts. Thus, semiconductor material with a suitable E_g is the choice. Gallium arsenic, which has $E_g = 1.43$ eV, gives a wavelength of 0.867 μm. To add to the flexibility of changing the spectral emission, Al of variable composition x, as in $Al_xGa_{1-x}As$ ternary compounds, is used to cover the desired wavelength range. Both GaAs and AlGaAs can have a closely matched lattice structure, which accounts for the long life of LEDs (10^6 h is not unusual).

Structure Tailoring

Although a simple p–n junction constructed from direct bandgap semiconductor materials can produce radiative emission suitable for light sources, this simple structure is hardly efficient enough for optical fiber applications. We need to (1) shape the emission spot to a size compatible to the optical fiber, (2) confine the carriers so that more carriers can participate in radiative recombination, and (3) direct more light at the desired radiation angle for better coupling efficiency to fibers.

We shall see some structural designs in LEDs so as to fulfill these mentioned requirements. Generally, heterostructures and double heterostructures are used to shape and confine the carriers as typified in the forthcoming examples.

LED Structures

Two of the most successful LED structures for optical fiber applications are the surface emitter and edge emitter. These are actually modified laser structures specially designed for MMF. The basic design is a multilayer epitaxial structure of GaAs and AlGaAs compounds for the 0.8–0.9-μm wavelength range and InGaAs and InGaAsP compounds for the 1.3–1.65-μm wavelength range applications.

Surface Emitter (Burrus Type)

Figure 6.6 shows a cross-sectional view of a surface emitting-type LED developed by Burrus and Miller [1]. It has a multilayer GaAs p–n

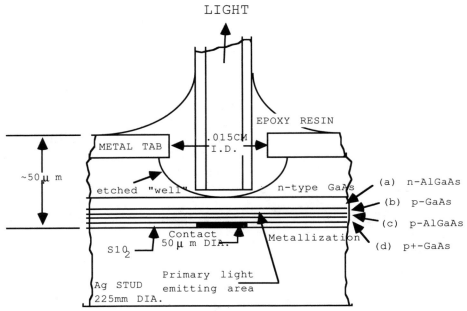

Figure 6.6 Cross-sectional view of Burrus-type surface-emitting LED. [After Burrus and Dawson (1970), *Appl. Phys. Lett.* 24, 97.]

heterojunction structure with a thin p-type GaAs active layer (10 μm) embedded between a double heterostructure (DH) of AlGaAs. The DH structure introduces potential barriers in the band structure as discussed in the Introduction (this chapter), which helps to confine the carriers in the active region and thus reduce the injection current. The emission area is about 2×10^{-5} cm^2. To bring the fiber end closer to the light-emitting area, a well is etched through the n-type GaAs substrate up to just above the active layer. The well also holds down the fiber whose polished end is butted against the emitting area and secured in place by index-matching epoxy resin. The thin active layer and the nearby butt joint to the fiber end reduce the absorption loss of optical power at the fiber interface. The emission from this type of structure is essentially isotropic (i.e., Lambertian), with an angular distribution of 120° beam width. The back metallic contact is about 50 μm in diameter. It is surrounded by an SiO$_2$ insulator in order to limit the contact area and bring up the current density. The side of the diode is heat-sunk.

The operating direct current is about 150 mA. It gives about 2 mW of emitted light power. The half-power spectral width has been measured to be between 350 to 400 Å. The emission response time is found to be about 1–2 ns, and the external quantum efficiency is observed to be 2–3%.

For a Lambertian source having a diameter smaller than the fiber core, the coupling efficiency to fibers is found to be proportional to the square of the numerical aperture [i.e., $(NA)^2$]. Thus, for a fiber with $NA = 0.14$, the coupling efficiency is approximately 2%. Coupling efficiency can be improved by using various types of lens arrangements. For example, if a hemispherical lens is included in the structure [2], the light extraction efficiency can be increased by a factor of 26 over that of a planar device. Other lens schemes offer further improvements [3].

Another factor in the LED source design is the diode frequency response. This is especially important when diodes of larger bandwidth (50–200 MHz) are sought. Generally, there is a trade-off between the bandwidth and the output power of a surface emitter.

If $p(\omega)$ is the power output and ω is the modulation frequency, then $p(\omega)$ and ω can be related by [4]

$$\frac{p(\omega)}{p_o} = \frac{1}{[1 + (\omega\tau)^2]^{1/2}} \qquad (6\text{-}18)$$

where p_o is the dc power output of the device and τ is the effective lifetime of the *injected carriers* defined in Eq. (6-16).

Equation (6-18) suggests that the inherent modulation speed is limited by the minority carrier lifetime τ. The effective carrier lifetime and thus the modulation bandwidth can be increased by increasing (1) the doping level of the recombination regions or (2) the carrier density. Increasing the doping level increases the nonradiative centers, which is accompanied by a decrease in quantum efficiency. For a doping greater than $10^{18}/cm^3$, the power bandwidth product becomes virtually constant as a result of sharp deterioration of the external efficiency. Both methods have their limitations. The τ value can also be reduced by decreasing w, the width of the active region.

The modulation bandwidth of high injection levels is proportional to $(JB/q\omega)^{1/2}$, where J is the injected current density and B is the bandwidth. The active-layer width is limited by the material growth techniques, although active layers as thin as 0.05 μm have been fabricated [5]. A more common range of layer thickness is 0.5–1.0 μm. Besides, as w is made smaller, higher barrier heights are required to confine the carriers in the active region.

To extend the use of LEDs to longer wavelengths, new materials are sought. So far, InGaAsP diodes grown on InP substrates achieve the best results. This is because an almost perfect lattice match between these compounds can be achieved, which, in turn, increases the reliability.

The bandwidth of InGaAsP surface LEDs ($\lambda = 1.25$ μm) is two to three times larger than that of AlGaAs diodes ($\lambda \approx 0.85$ μm) of the same doping density [6]. This is attributed to the fact that the carrier lifetime of an InGaAsP compound is shorter than that of an AlGaAs compound.

The Edge Emitter

The surface-emitting LED suffers from the broad emitting beam width, which makes coupling to a small-*NA* fiber very inefficient. Edge-emitting LEDs are designed to reduce the beam width. A high-radiance edge-emitting LED is shown in Fig. 6.7 [7]. The structure is an AlGaAs double heterostructure with separate optical and carrier confinement layers to provide a semidirectional light power output. Schematic diagrams of the refractive index and the aluminum concentration of the various layers are shown on the right-hand side of the figure. The active layer has a higher *n* value. The upper and lower layers with reduced refractive indices act as a partial internal wave guide for the spontaneous radiation. This arrangement helps to funnel the emitted power into narrow angles in the direction perpendicular to the junction planes. A separate pair of carrier confinement layers next to the active layer can also be seen. Further concentration of current flow is achieved by the SiO₂ layers, which open a width of 0.5 μm for the metallic contact. Edge-emitting LEDs offer higher coupling

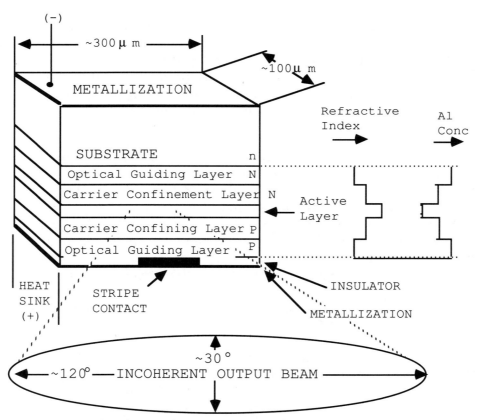

Figure 6.7   High-radiance edge-emitting LED. After M. Ettenberg *et al.* [7] © IEEE 1976.

efficiencies to low-NA fibers than surface-emitting LEDs. Its coupling efficiency is typically about four times better than that for a surface-emitting LED. The width of the emitting region is adjusted to suit the fiber dimension, typically 50–80 μm.

For a very thin active layer (500 Å, usually undoped), the emitter radiation bandwidth is about 30°. A thin active layer also reduces the carrier lifetime, thus improving the bandwidth.

Although the emitting area of an edge-emitting LED is normally less than that of the surface emitter, actually more absolute power can be coupled into the fiber with NA ≤ 0.3. If the fiber NA is greater than 0.3, however, the opposite is true.

Superluminescent Diode

A third device geometry that offers (1) increased light power, (2) well-directed output beam, and (3) reduced spectral bandwidth is an elongated version of the stripe-geometry edge emitter. It is also known as a *superluminescent diode* (SLD). It has an AlGaAs DH structure with a long stripe (1000 μm). Its radiation pattern from the edge is well directed to within 10°.

Several times in this book, we have mentioned the limited use of LEDs for long-wavelength transmissions. New technology development may have changed all this. During the last few years, in response to the increased demands to find a low-cost, high-speed, and reliable light source for the LANs and subscriber-loop system, research has been focused on LEDs again. Favorable results appeared frequently. Starting with the recognition that LEDs can be used with SMF successfully, even at high rates [8], both edge-emitting and surface-emitting LEDs that can handle power of several hundreds of microwatts at a speed of 400 mb/s have been reported [9, 10].

LEDs have also been used as a pumped energy source of Nd:YAG (neodymium–yttrium aluminum garnet) laser, to be discussed in Chapter 12.

Lasers

Laser, a coined acronym for "light amplification by stimulated emission of radiation," describes its principle of operation most vividly. It distinguishes itself from an LED in several respects. While light emission from an LED is from spontaneous emission of radiation, that from a laser is by stimulated emission. Spontaneous emission is a random process, resulting in incoherent emissions, thus yielding a wide spectrum. Stimulated emission is pulse-like, gives a highly coherent emission, and therefore has a narrower spectral width. A second difference concerns the light amplification. Light from LEDs is not amplified, whereas that from a laser is amplified. Thus, laser emission provides a much stronger light intensity than that from an LED.

To achieve light amplification in lasers, light emitted at the junction is made to pass many times within a cavity where for each passage, additional intensity is gained through added energy. We shall explain the mechanisms of laser operation in the following section.

Population Inversion and Stimulated Emission

If N_1 and N_2 are the instantaneous particle populations at energy levels E_1 and E_2, respectively, the equilibrium population usually follows Boltzmann's distribution law:

$$\frac{N_2}{N_1} = e^{-(E_2-E_1)/kT} = e^{-h\nu_{12}/kT} \tag{6-19}$$

provided the two levels contain an equal number of available states. Spontaneous emission occurs when electrons in E_2 fall to E_1 and emit a radiation of frequency ν_{12}. As $E_2 > E_1$, from Eq. (6-19), it can be concluded that $N_2 < N_1$. However, under certain conditions, say, by increasing the photon energy density, N_2 can be made to exceed N_1, producing the so-called population inversion. These electrons at higher or excited states need not wait for spontaneous emission to occur. They can be stimulated to fall to the lower level all at once and emit photons. Thus, the emission is coherent and occurs in a time much shorter than the mean spontaneous decay time. The stimulus is provided by the presence of photons of the proper wavelength.

In semiconductor junction devices, population inversion can be achieved by carrier injection over a p–n junction. If a p–n junction between degenerated materials is forward-biased, the band diagram is as shown in Fig. 6.8. Notice that the quasi-Fermi level F_n on the n-side now resides above the conduction band edge E_{cn} and F_p on the p-

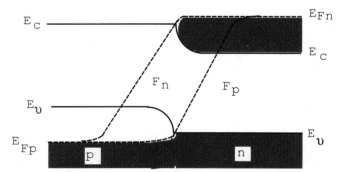

Figure 6.8 A degenerated forward-biased p–n junction.

side is found below the valence band edge E_{vp}. The transition region exposes a high concentration of electrons in the conduction band and a large concentration of holes in the valence band. These concentrations can be expressed as

$$N = n_i e^{(F_n - E_1)/kT} \tag{6-20a}$$

$$p = n_i e^{(E_i - F_p)/kT} \tag{6-20b}$$

In fact, Eq. (6-20a) or (6-20b) serves to determine F_n or F_F if n or p is known. It is also noticed that within the transition region, F_n varies from its value deep in the n-region ($F_n = E_{cn}$ deep in the n-region) to F_p in the p-region ($F_p = E_{vp}$ deep in the p-region). Similarly, the change for F_p is indicated by the dotted curves in Fig. 6.8. The difference in quasi-Fermi energy levels $F_n - F_p$ is indicative of the dynamic situation in the transition region. For a given transition energy corresponding to $h\nu$ in a semiconductor, population inversion exists when

$$F_n - F_p > h\nu \tag{6-21}$$

Now, the minimum requirement for population inversion to occur for photons in a band-to-band transition is $h\nu = E_c - E_v = E_g$; therefore

$$(F_n' - F_p) > E_g \tag{6-22}$$

Stimulated emission can dominate over the range of transitions from $h\nu = (F_n - F_p)$ to $h\nu = E_g$.

Increasing the forward bias tends to widen the transition region and to favor more population inversion and stimulated emission.

Resonant Cavity

To favor the condition for stimulated emission, the photon intensity must be raised. This is done by means of a resonant cavity. In p–n-junction devices, a resonant cavity can be formed by cleaving a stripe of a junction along a plane perpendicular to a crystal plane twice to form parallel and flat faces for a resonator. When the faces are partially coated, they serve as mirrors to reflect the emission back and forth between themselves to reinforce the light intensity required to sustain the stimulated emission. The length of the cavity must be such that

$$L = m \frac{\lambda}{2} \tag{6-23}$$

where m is an integer. But mechanical adjustment of the length to some particular $m\,\lambda/2$ is not necessary. Since $\lambda/2$ is so small for IR emission, many values of m and $\lambda/2$ can fulfill the resonant condition.

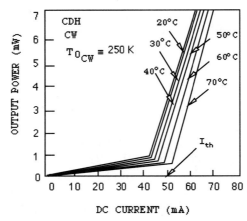

Figure 6.9 Optical power versus laser current for various temperatures.

The Threshold Current

The optical output power versus laser current resembles that shown in Fig. 6.9. For a definite operating temperature, below a threshold current I_{th}, the output of the laser is LED-like. Its efficiency is very low. But above I_{th}, the laser output increases very rapidly and almost linearly with further increase in current.

The differential external efficiency can approximate 100%. The threshold current can be defined as that current above which the injected carriers have reached the level of inverted population and that the optical field intensity is just enough to sustain stimulated emission. Above the threshold current, the optical power increases almost linearly with the laser current. Linear modulation can easily be achieved.

The threshold current is very temperature sensitive. It increases very rapidly with increasing temperature.

Laser Spectrum

The spectrum of the laser oscillation is very broad. Lasing at frequencies corresponding to all cavity modes is possible as shown in Fig. 6.10a. These modes correspond to the successive numbers of integral half-wavelengths fitted within the cavity as described by Eq. (6-23). With increased injection current, the cavity selects a most preferred mode or a set of modes to dominate the spectrum as shown in Fig. 6.10b. Theoretically, a pure monochromatic radiation is possible.

In most cases, however, even a single-mode laser still possesses a

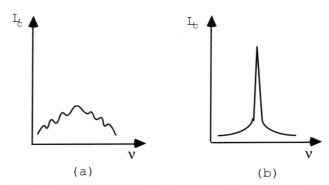

Figure 6.10 Light intensity versus photon energy $h\nu$ for a junction laser: (a) laser modes at threshold; (b) dominant laser mode above threshold.

spectrum width of approximately a few angstrom units. Increase in the injection current also shifts its wavelength toward the longer region due to heating (measured in joules) at the junction.

Laser Modes and Mode Stabilization

The active region of a laser source is usually an optical lightguide of rectangular cross section. The height of the rectangle is a fraction of a micrometer, while the width may vary with the particular design. Different kinds of modes can be developed in this lightguide: the transverse and longitudinal modes. It is often desirable to limit the modes to a single one. For a wider stripe of the lightguide, the emission is associated with a filamentary zone of highest gain where the refractive index is also the highest. Usually, a wider stripe tends to support more than one zone across itself. This causes fluctuations in the light-power output and makes coupling to the fiber very difficult. A single transverse-mode emission can be accomplished by making the width dimension very narrow.

In a single transverse-mode design, there exists several longitudinal modes determined by the length L of the parallel-plate Fabry–Perot cavity. By reducing this length, it is possible to obtain a single longitudinal-mode emission. This effectively reduces the gain of adjacent emission lines to the extent that only one line is favored. The emission from a single longitudinal mode of the laser is coherent.

Mode instability is the most urgent issue regarding semiconductor laser fabrication. Both transverse- and longitudinal-mode control are important. Various laser cavity structures have been successfully introduced to stabilize the laser modes. For transverse-mode stabilization, schemes such as index guiding and gain guiding have been used. For longitudinal-mode stabilization, various grating feedback schemes have been devised. We will describe these individual techniques later while exploring laser structures.

Laser Noises

The response of diode lasers to current pulses suffers stability problems in which the intensity of the output light fluctuates randomly. The most intrinsic mechanism of intensity fluctuations is the random nature of carrier recombination, which gives rise to quantum shot noise. McCumber [11] and Haug [12] showed that theoretically, the quantum noise intensity peaks at laser threshold current and decreases steadily for further increasing current.

When the transverse mode is unstable, a tremendous amount of intensity fluctuation can occur. At the same time, the output-current characteristic registers "kinks." Self-pulsation may also occur at the kink. For this reason, transverse-mode stabilization is essential for semiconductor lasers.

Self-pulsation may also be observed when a semiconductor laser is aged. In this case, nonradiative regions are generated near the facets or in the laser's active regions. When a laser is oscillating in multilongitudinal modes, the intensity of each mode is fluctuating significantly [13]. The fluctuation can again be traced to the random process of electron–hole recombination in the laser.

"Chirping" is a dynamic line-broadening effect experienced by a laser that oscillates in a single longitudinal mode under CW current injection. It is caused by the strong coupling of the free carrier density and index of refraction in the semiconductor structure and can be reduced by proper structural design.

Another source of noise comes from external reflections. Laser light can be reflected by discontinuities such as splicing and connectors. Optical feedback [14] causes laser oscillations to become unstable. In this case, the interference between the mode frequencies of the internal laser cavity and that of the external cavity can become constructive and destructive alternately.

Laser Structures

Modern laser design emphasizes the solutions to the problems outlined in the preceding section, subsections titled "Resonant Cavity" through "Laser Noises." The aim is to (1) reduce the threshold current by electrical current confinement schemes and potential barrier placement to increase the population inversion, (2) purify the laser spectrum by optical confinement schemes, (3) stabilize the lasering mode, and (4) reduce laser noise. In the following subsections, we shall present the progress achieved by different designs in attempting to solve these problems.

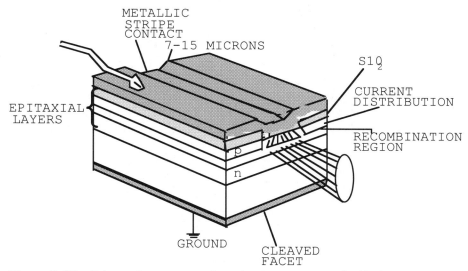

Figure 6.11 Schematic representation of stripe-geometry double-heterostructure laser. (After Botez and Herskowitz, *Proc. IEEE* **68**, 699. © IEEE 1980.)

The Stripe-Geometry Laser

The simplest type of practical laser is the stripe-geometry double-heterostructure laser developed by Kressel and Butler [15]. Figure 6.11 is a schematic representation of the laser structure. The substrate is a silicon-doped n-type GaAs. Multiepitaxial layers of compound semiconductor material are deposited successively. A layer of $Al_xGa_{1-x}As$ with Sn doping and $x = 0.3-0.4$ forms the lower n-wall layer that has a refractive index lower than the index n of the active layer.

The active layer is an undoped $Al_yGa_{1-y}As$ ($y = 0.05-0.1$). Its thickness is $0.05-0.2$ μm. On top of this comes a p-wall layer $Al_xGa_{1-x}As$ with the same x and n_2 values. The top p^+-cap is a Ge-doped GaAs p^+-layer for contact metallization. For achieving the purpose of current confinement, an insulating layer of SiO_2 is placed on top of the p^+-well layer. It has a narrow opening in the center to allow metallization to contact the p^+-well layer, thus preventing the spread of current across the width direction. The double heterostructure contains the p–p and n–n junctions, which help to confine the injected carriers into the active region to build up population inversion.

The n- and p-potential walls form a waveguide for the emission from the active layer to guide along the length. By cleaving the faces perpendicular to the length direction, a proper cavity length is established.

The normalized guide thickness D can be expressed as

$$D = \frac{2\pi d}{\lambda} \sqrt{n_1^2 - n_2^2} \tag{6-24}$$

The parameter D for planar guides is equivalent to the quantity V for fiber guides (see discussion in previous chapters). It is desirable to obtain $D \leq \pi$ in order to limit the number of transverse modes. For example, for $\lambda = 0.85$ μm, if the active-layer thickness is $d = 0.35$ μm, then $\Delta n = 0.2$ and $D \cong \pi$. The guide can support only one transverse mode. Mode confinement in the lateral direction can be further achieved by constricting the current flow to a narrow stripe by using proton-bombarded stripes on both sides of the active stripe.

The emission pattern of this laser is elliptical as shown in Fig. 6.11, with the transverse and lateral beamwidth, at half-power, of 40° and 10°, respectively. Narrow transverse beamwidths (15–25°) have been realized by growing very thin active layers ($d \cong 400-600$ Å).

The driving current of a stripe-geometry laser with a 15-μm-wide stripe is typically about 100 mA, with a quantum efficiency of 40–60%. In many applications, diodes that lase at a lower overall current are desired. This can be achieved by shortening the cavity length and improving current confinement. Low-threshold lasers ($I_{th} = 35-50$ mA) have been fabricated using shortened cavity length ($\leq 100-500$ μm) and two 15-μm oxide stripes to confine current [16].

The temperature-dependence of I_{th} is of the form $I_{th} \alpha \exp(T/T_0)$, where T_0 is a parameter characterizing the diode. For AlGaAs diodes, in the absence of carrier leakage, $T_0 \approx 120°$. This means that the threshold current at 70°C will be 1.5 times larger than that at 20°C. A heat sink must be provided to reduce the temperature rise. In case there is carrier leakage, T_0 will be lower. Thus I_{th} could be even larger. Carrier leakage can be reduced by providing adequate potential barriers due to double heterojunctions.

Stripe-geometry lasers, as a rule, have a spatially unstable optical mode when the driving current is raised above the threshold current. The modulation characteristics can be strongly affected by this instability. It is observed that the kink in the light-power output versus driving current characteristic is always associated with the nonlinearity. Kinks can also cause mode profile deformation, lateral wavelength shift, mode beam shifts, excess noise, and enhanced relaxation vacillations. Other laser structures are being tested in search of more stable operations.

Mode-Stabilized Laser Structures

Two problems, namely, nonlinear kinks in the L-I characteristics and the mode instabilities, have plagued the DH injection lasers. These problems can be solved by two approaches: (1) pushing the kinks to higher power levels than the one needed for optical communication (e.g., to 3–10 mW) and (2) developing "kinkless" laser structures using mode stabilization in two dimensions. The first approach is easier to implement. It can be done by confining light current to narrow stripes at the expense of increasing threshold current. This, the reader may recall, is done in stripe-geometry lasers as discussed in the last section. The second approach involves the creation of a lateral wave-confining structure and complex growth techniques. Again, there are two ways to accomplish mode stability in these structures: (1) by creating an index-guiding structure and (2) by using the gain-guiding structure. The index-guiding structure employs a transverse refractive index difference, and the gain-guiding structure has a transverse gain difference. We shall describe a few examples in each category to illustrate the arrangement.

Gain-Guiding Lasers

In planar lasers, if the optical mode distribution along the junction is determined by the optical gain, the lasers are identified as gain-guiding lasers. It has been recognized that transverse-mode instability of semiconductor lasers takes place by the deformation of the gain profile due to nonuniform carrier consumption through stimulated recombination. The spatial deformation of the gain profile induces deformation of the field profile and also changes the modal gain or output power. The most successful way to improve the stability of this kind is by narrowing the stripe width in conventional stripe-geometry lasers [17]. The power level at which transverse-mode instability shows up is a function of the stripe width w in stripe-geometry lasers. For example, Kobayashi [17] reported that for $w \sim 13 \ \mu$m, the mode becomes unstable at $p_{out} \leq 3$ mW. For $w \sim 6 \ \mu$m, the transverse mode is stable up to $p_{out} \geq 5$ mW. V-Grooved lasers also showed promise in transverse-mode control [18]. The structure and lasing spectrum of a V-grooved laser are shown in Fig. 6.12. The current is limited in a V-groove, resulting in better carrier confinement in the horizontal direction of the active layer. The room-temperature threshold current was ~ 50 mA for a 200-μm-long laser cavity (for $\lambda \sim 82 \ \mu$m). The spectrum exhibits several longitudinal modes and has a half-width of $\sim 2 \ \mu$m. The gain-guide laser is an antiwaveguide laser. Its far-field pattern exhibits two peaks for $w \leq 4 \ \mu$m.

No satisfactory gain-guiding structure has been demonstrated for long-wavelength operation ($\lambda \sim 1.0-1.7 \ \mu$m) using InGaAsP/InP. In a stripe-geometry laser designed for this wavelength range, the threshold current becomes very high and has difficulty operating above 50°C.

Gain-guiding lasers with a narrow stripe width ($w \leq 5 \ \mu$m) lase in multilongitudinal modes. The number of lasing longitudinal modes is typically 10–15 at $I \sim 1.5 I_{th}$. The large contribution of spontaneous emission to the lasing mode may be the origin of the multimode operation.

The undesirable characteristics of gain-guiding lasers are high threshold current, low differential quantum efficiency, and kink occur-

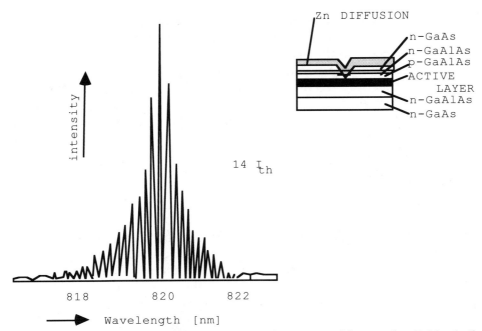

Figure 6.12 Structure and lasing spectrum of a V-grooved laser. After P. Marshall *et al.* [18].

rences at higher currents. These effects are caused mostly by carrier-induced index reduction, which leads to movement of the optical modes along the junction plane.

Index-Guiding Lasers

In an index-guiding laser, the active layer of the laser is a very narrow region that has a relatively higher index in the junction plane. The lasing mode is well confined in this region. If the index difference between the cladding is large, say, $\Delta n > 0.2$, it is called a "strong" index-guiding system. Otherwise, if Δn is between 0.01 and 0.02, it is called a "weak" index-guiding system. Usually, the laser structure is modified so as to induce an effective index step larger than that due to carrier-induced reduction. Hence the optical mode is essentially determined by the device's structure.

Buried Heterostructure Laser (BH laser)
A buried heterostructure laser [19, 20] is a strong index-guiding laser. Figure 6.13 shows a schematic diagram of a BH laser. It has a filamentary GaAlAs active region of 1 μm width embedded in GaAlAs with a higher AlAs content. The mole fraction of AlAs is chosen to be 0.05 for the active region and 0.4 for the cladding regions. There exist index steps in the vertical and horizontal directions to serve as index-guiding waveguides. Threshold currents as low as 4.5 mA have been

reported. Since the lasing spot width is limited to 1 μm, the useful CW powers are low (ca. 1 mW per facet), and the beam width in the junction plane is relatively large (40–50°). However, this low-power laser has been efficiently coupled into a 0.13-NA MMF. The mean lifetime at elevated temperature (70°C) is about 2000 h.

A more recent version of the BH laser has been described by Chinone et al. [21]. It has an additional layer of an index intermediate between the n-AlGaAs and the n-GaAs layers, sandwiched between the two. This additional layer ($\simeq$1 μm) acts as a guide layer and gives a large optical cavity structure so that its CW output power can reach 20 mW with a larger spot size. The threshold current is 20–40 mA. Linearity is good. However, carrier leakage to the guiding layer is found to be high.

The Transverse-Junction Stripe Laser (TJS laser)

Figure 6.14 shows the schematics of a transverse-junction stripe laser (TJS) [22]. Here the mode shape is controlled by lateral carrier injection across a transverse GaAs p–n homojunction, created by deep Zn diffusion. Zinc is selectively diffused by a two-step diffusion process. As a result, carrier density has a dip around the p–n junction in the active layer. This creates an optical confinement. In the deep Zn-diffused layer, the Zn is diffused down to the active layer with a diffusion width of 4–8 μm. The change in conduction type and carrier density results in a positive index step in the direction parallel to the

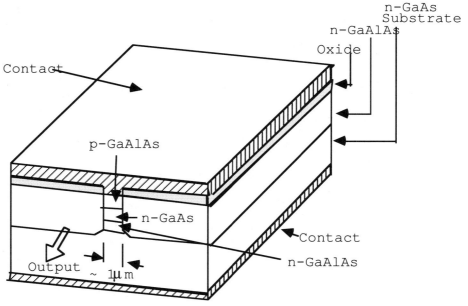

Figure 6.13 Schematic representation of a buried heterostructure laser. After T. Tasukada [19].

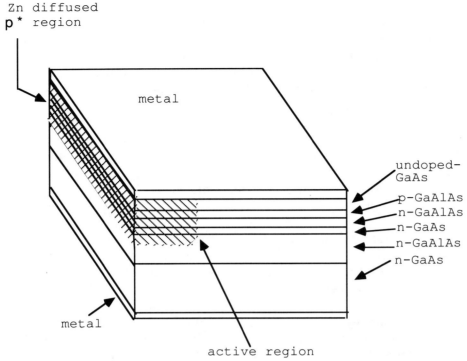

Zn diffused
p* region

metal

metal

undoped-GaAs
p-GaAlAs
n-GaAlAs
n-GaAs
n-GaAlAs
n-GaAs

active region

Figure 6.14 The TJS laser. [After H. Namizaka [22] © IEEE 1975.]

junction. This configuration replaces the metallic stripe contact of conventional stripe lasers in which carriers are injected vertically through the AlGaAs–GaAs heterojunctions. A lateral waveguide, for better mode stabilization, is achieved by two consecutive Zn diffusions. The mode is laterally bound by p^+–p and p–n GaAs junctions. Both the cathode and the anode can be placed on the top surface.

To prevent the rapid increase in the threshold current at elevated temperature due to the presence of a shunt path of current flow outside the active region, a self-reverse-biased p–n junction is incorporated into the TJS laser to concentrate the current path into the narrow active region. This reduces the I_{th} to as low as 20 mA. The lifetime of this laser at low temperature is estimated to be 12,000 h. By further incorporating a semi-insulating substrate to confine current to the active region, a TJS laser can operate at a temperature over 100°C with good efficiency. The transverse beam width is relatively large (~50°). Because of heating problems, the device is usually limited to operate below 4 mW per facet in CW operation.

Other Structures

Various types of long-wavelength GaInAsP/InP narrow stripe lasers have been experimented with worldwide; a few are shown in Fig. 6.15. Figure 6.15 shows transverse cross sections of a (a) mesa-substrate

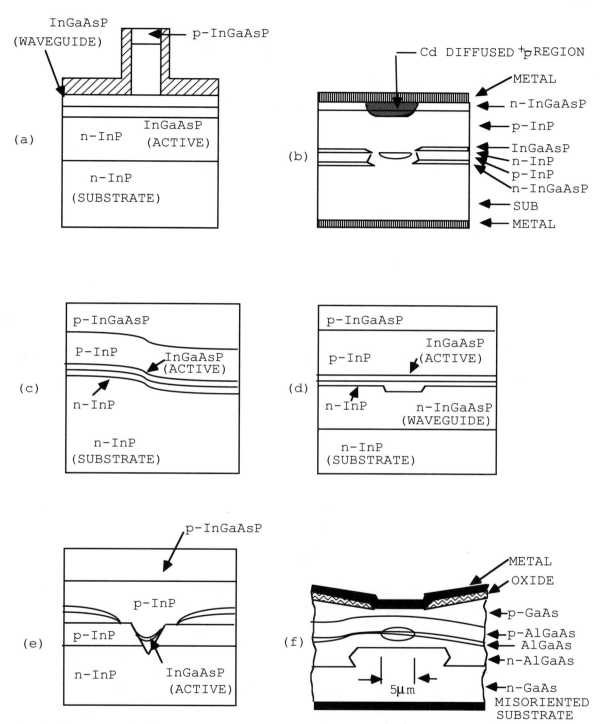

Figure 6.15 A number of index-guiding single-mode laser structures: (a) MSB [20]; (b) BC [21]; (c) TS [22]; (d) PCW [23]; (e) CSP [24]; (f) CDH [25]. After Y. Suematsu, Proc. IEEE, 71, 700, 1983 © IEEE; and D. Botez and G. J. Hershowitz, Proc, IEEE 68, 699, 1980. © IEEE 1980.

buried heterostructure (MSB) [23], (b) buried crescent (BC) [24], (c) terraced-substrate heterostructure (TS) [25], (d) planoconvex waveguide (PCW) [26], (e) channeled-substrate planar (CSP) [27], and (f) constricted double-heterostructure (CDH) [28].

All these structures are designed to control carrier flow and to stabilize the oscillation by the double heterostructure with index guiding configurations. Readers should consult the original papers for detailed construction and operation data. All these are experimental models for long-wavelength operations. No commercial products are available.

Distributed-Feedback Lasers

The narrow stripe-geometry index-guiding design of lasers effectively limits the transverse lasing mode to one, thus improving laser stability, but it still suffers some instability due to multilongitudinal modes. For long-haul transmission where single-mode operation is demanded, the suppression of multilongitudinal modes is important. From optics we know that control of longitudinal modes is possible by grating feedback. The design of distributed-feedback lasers is based on this principle. In terms of fabrication, long-wavelength grating is more attractive to implement because of the longer grating period.

In a distributed-feedback (DFB) laser, a periodic corrugated structure or grooves is spatially etched on the surface of a passive or active waveguide, which in this case lies between heterojunction layers. This is shown schematically in Fig. 6.16. Both the front and the back of the cavity are covered by a corrugated structure, which serves as distributed Bragg reflectors to the lightwaves (forward-scattered and backscattered waves). The pitch of the groove and the effective refractive index of the guide determine the lasing wavelength. This structure offers a high degree of spectral selectivity by the phase grating. Such a laser is difficult to fabricate, however, and commercial production is not yet available.

In an experimental laser described by Kobayashi et al. [29], the light-power versus current plot showed no kinks and the spectrum showed no mode-hopping. Operating at a wavelength of $1.5-1.6$ μm, the grooves are at 0.24-μm intervals. The typical cross section of the output waveguide is 0.54 μm^2 and, only the TE mode can be guided. The temperature-dependence of the lasing wavelength was 0.1 nm/°C, about one-fifth of the conventional value. Under 1.8 GHz modulation with a 100% modulation depth, the lasing spectrum showed no change with modulation. The dynamic wavelength shift was about 0.27 nm, corresponding to a bandwidth distance product of 185

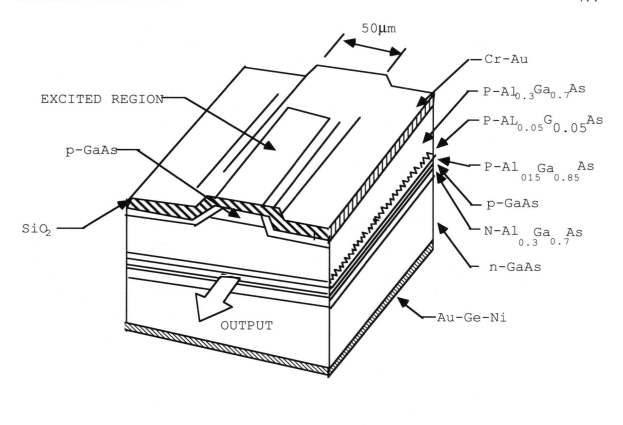

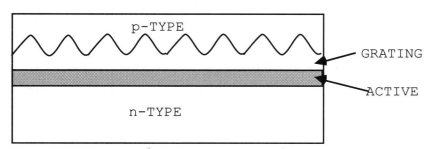

Figure 6.16 A distributed-feedback laser (DFB) and its enlarged grating structure. After Y. Suematsu *et al.* [30].

Gb/s·km, which is about 35 times better than a conventional laser. Lasing threshold current was about 100 mA [30].

Distributed-feedback lasers are useful in combating the problem of desparation for a system to operate at the fiber minimum attenuation wavelength (1.55 μm). This is because a DFB laser is stable over a wider range on temperature, mean power, and modulation, which often limit other types of laser structures, including the C[3] laser to be presented later. A new technique of growing the grating making it easy to fabricate and reproducible has been reported by Stockton [31]. A 380-μm-long device, operating at a low threshold current of 14 mA at 20°C has been reported. It can deliver light power of several milliwatts and has an estimated life expectancy well over 6000 h. This stable single longitudinal-mode laser operates over a wide range with side mode suppression better than -37 dB and modulated at 2 Gbit/s over an 80-kM link.

Cleaved Coupled-Cavity or C³ Lasers

A more recent development in semiconductor laser technology is the new electronically tunable, single-frequency laser source, known as the *cleaved coupled-cavity* (C³) laser. A schematic diagram of a C³ laser is shown in Fig. 6.17 [32]. Two standard Fabry–Perot cavities formed by separate stripe geometry having slightly different lengths are tightly coupled together optically through a separation of less

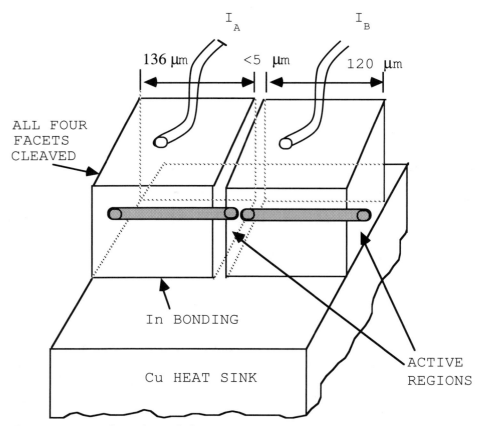

Figure 6.17   A cleaved coupled-cavity laser (C³). After W.-T. Tsang *et al.* [32].

than 5 μm. All the reflecting facets are formed by cleaving along crystallographic planes so that they are perfectly parallel. Each cavity is electronically controlled by independent dc current sources.

The electrodes are isolated and buried underneath the laser material, where they are perfectly aligned with respect to each other. To achieve these strict requirements, a heterojunction epitaxy layer is first cleaved at two ends to form a standard Fabry–Perot cavity of approximately 250 μm in length. This cavity is then covered with a thick ($\sim$5 μm) electroplated gold layer. The device is then recleaved near the middle to form two separate Fabry–Perot diodes. Because of the thick gold layer, these two cavities remain hinged together and are bonded with indium upside down on a copper heat sink.

Two stripe-geometry electrodes must be deposited on the epilayer before bonding. In the operation, in either tuned or untuned mode, the first cavity is operated above the threshold. The second cavity can be operated either above or below the threshold current. In a tuned mode, the laser wavelength can be controlled by varying the injection current level. A change in the injection current causes a change in carrier densities, which affects the refractive index of the medium and causes a shift in cavity modes in the second diode.

Single-frequency operation can be maintained under 2-Gb/s direct modulation with error rates of less than 10^{-10}. An experimental C³ laser has been used to transmit information at 420 Mbit/s through a 119 km unrepeated length of fiber with bit error rate $BER < 10^{-9}$. The chromatic dispersion at 1.55-μm wavelength was found to be 2.08 $\times$ 10^{-3} ps/km.

We are looking forward to a bright future with this laser.

Quantum-Well Lasers

Quantum Well

The idea of a quantum well was first introduced when a hydrogen atom was viewed from the quantum-mechanical point of view. It was revealed that energy can be expressed as $E_n = h^2 n^2 / 8 L^2 m^*$, where h is Planck's constant, n is an integer number corresponding to the levels, L is the width of the well, and m^* is the effective mass; L satisfies the deBroglie wavelength relationship ($\lambda = h/p \approx L$), where p is the momentum. In a double-heterostructure laser, if a not too thin active layer is sandwiched between two higher gap-cladding layers, it was found that discrete energy levels similar to the one-dimensional quantum well in hydrogen atoms exist [33]. The energy levels in an InGaAs well is as shown in Fig. 6.18. The confined particle energy levels E_g are denoted by E_{1c}, E_{2c}, and E_{3c} for electrons; E_{1hh}, E_{2hh}, and E_{3hh} for heavy holes; and E_{1lh}, E_{2lh}, and E_{3lh} for light holes. The electron–hole

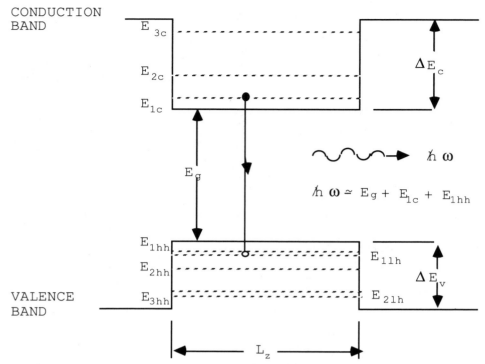

Figure 6.18 Energy levels in an InGaAs quantum well. [After N. Holonyak, *et al.* [33] IEEE 1980.]

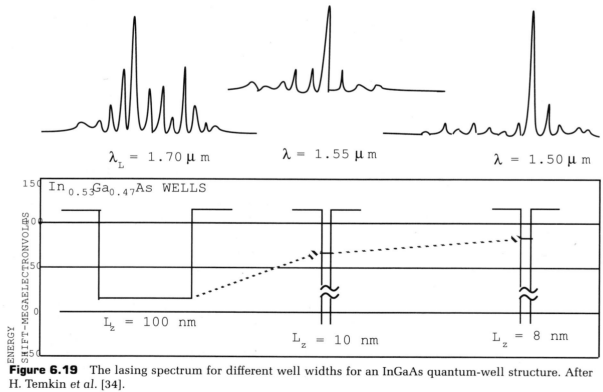

Figure 6.19 The lasing spectrum for different well widths for an InGaAs quantum-well structure. After H. Temkin *et al.* [34].

recombination in a quantum well follows the selection rule, and the radiative emission is given by

$$h\omega \approx E_g + E_{le} + E_{lhh}$$

$$= E_g + \left(\frac{h^2}{8L^2}\right)\left(\frac{1}{m_e^*} + \frac{1}{m_{hh}^*}\right)$$

(6-25)

Equation (6-25) shows that the energy of the emitted photon may be varied by varying the well width L. The lasing spectrum for different well widths for an InGaAs QW structure is shown in Fig. 6.19. For a narrow width of 10 nm, a distinct emission at 1.55 μm is observed [34].

Single-Quantum-Well and Multiquantum-Well Lasers

A series of quantum-well lasers have been fabricated. Those with one active region are called single-quantum-well (SQW) lasers (Fig. 6.20a), and those having multiple active regions are called multiquantum-well lasers (MQW) (Fig. 6.20b). In MQW, barrier layers are used to separate the active layers. An active schematic structure of an AlGaAs

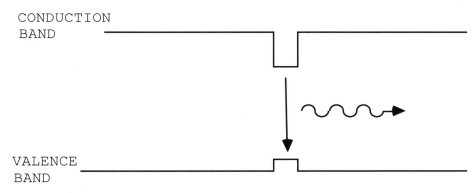

(a) Single quantum well.

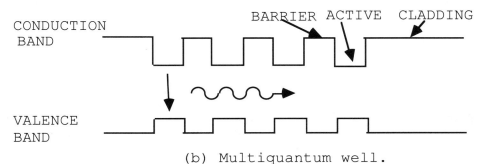

(b) Multiquantum well.

Figure 6.20 Energy-band diagram of single-quantum-well and multi-quantum-well lasers.

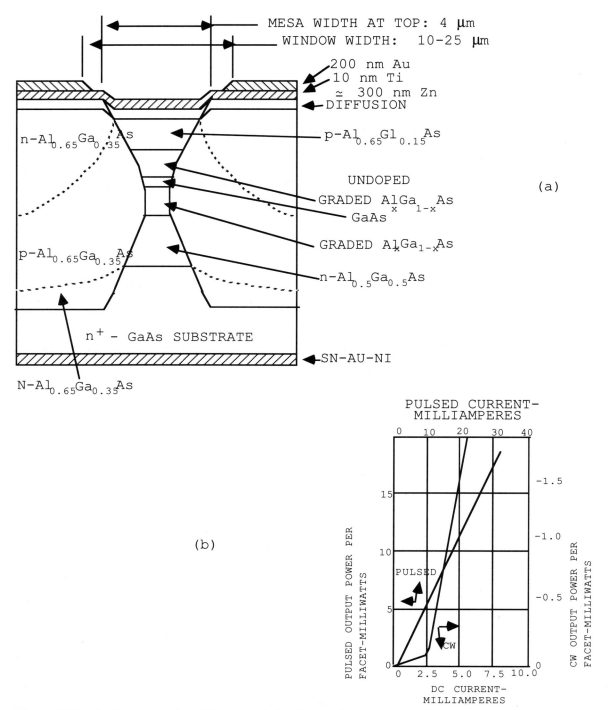

Figure 6.21 (*a*) Structure of an AlGaAs graded-index single-quantum-well laser; (*b*) power output versus current characteristics of the same laser. After S. D. Hersee *et al.* [35].

SQW laser is shown in Fig. 6.21a [35]. The output power versus the dc current characteristics of this laser are plotted in Fig. 6.21b. The results shown in these figures demonstrate the high performance of quantum-well lasers. For example, with this structure, the threshold current is about 200 A/cm². This compares with values in the range of 0.6–1.0 kA/cm² for regular DH lasers. Output power of about 2 W has been reported. Other advantages of quantum-well lasers over regular DH lasers are the possibility of lower threshold current, lower temperature-dependence of threshold, and lower chirp.

Materials for and Expitaxial Growth Techniques of LEDs and LDs

Recent interest in long-wavelength fiber applications promotes intensive material research for light sources and detectors working into these wavelengths. Possible quaternary semiconductor materials are listed in Table 6.4.

For LEDs, the well-studied material in long-wavelength ranges is the $Ga_xIn_{1-x}As_yP_{1-y}$/InP. The spectral widths of these LEDs are large. For example, at 1.5-μm wavelength a typical LED may have a spectral width of $\Delta\lambda_s = 120$ nm and at 1.6-μm wavelength, $\Delta\lambda_s = 140$ nm. The emission power of LED is limited by the total reflection at the emitting surface. Usually a few percent of the light power is coupled into the fiber.

Lasers have been fabricated with all those compound materials. The use of InP as a substrate has the advantage of low thermal impedance and is therefore preferred. Active research is progressing in this area.

Although we frequently mentioned the structures of the LEDs and LDs throughout this chapter, such as the p–n junctions, p–n hetero-

Table 6-4 Possible Quaternary Semiconductor Materials for Long-Wavelength Light Sources and Detectors

Substrate	Active Material	Energy Gap (eV)	Wavelength (μm)
InP	$Ga_xIn_{1-x}As_yP_{1-y}$	1.35–0.75	0.92–1.67
GaAs	$Ga_xIn_{1-x}As_yP_{1-y}$		0.65–0.9
InP	$Ga_xIn_{1-x}Sb_yP_{1-y}$		0.9–1.33
GaSb	$Ga_xIn_{1-x}Sb_yP_{1-y}$		2.0–3.0
InAs	$Ga_xIn_{1-x}As_ySb_{1-y}$		1.6–3.6
GaSb	$Ga_xIn_{1-x}As_ySb_{1-y}$		1.7–3.9
GaSb	$Al_xGa_{1-x}As_ySb_{1-y}$	1.24–0.73	1.0–1.7
InAs	$In_xAs_{1-x}Sb_yP_{1-y}$		1.8–3.6
GaAs	$Al_xGa_{1-x}As$		0.7–0.9

junctions, and double heterojunctions, we never discussed the techniques for making these junctions. A short note of the expitaxial growth techniques is added here.

In the early days of LED and LD application, LPE (liquid-phase-epitaxy) and VPE (vapor-phase-epitaxy) techniques were commonly used to make these devices. The deposition was slow, and only small pieces could be done at one time. Then came MBE (molecular-beam epitaxy) and MO-CVD (metalloorganic chemical-vapor deposition) (for a review of different epitaxy techniques, see Tsang [36]). These techniques are all well established in the industry. Detailed discussion of these techniques is beyond the scope of this book. However, we wish to mention a new addition to the epitaxy technique: CBE (chemical-beam epitaxy) [37]. This process combines the beam nature of MBE and the controlled use of an all vapor source as in MO-CVD. It may prove to be more advantageous than other techniques. For ex ample, it uses room-temperature vapor sources for both group III and V elements, thus obviating the demanding precision temperature control needed in MBE process. The beam nature of CBE also eliminated the flow pattern present in MO-CVD reactors, which often caused thickness and compositional nonuniformity.

Conclusion

We started this chapter by discussing the general considerations of the properties of light sources for fiber optical communications and introducing the light-emitting diodes and the injection lasers. A brief discussion, including material selection and examples of successful devices, were included in each category. Typical characteristics were cited. The purpose of this chapter is to introduce the up-to-date developments of LEDs and lasers without going into detailed analysis of the devices.

References

1. C. A. Burrus and B. I. Miller, Small-area double heterostructure aluminum-gallium arsenide electroluminescent diode sources for optical-fiber transmission lines. *Opt. Commun.* **4**, 307–309 (1971).
2. C. H. Gooch, *Injection Electroluminescent Devices.* Wiley, New York, 1973.
3. R. A. Abram, R. W. Allen, and R. C. Goodfellow, The coupling of light-emitting diodes to optical fibers using sphere lenses. *J. Appl. Phys.* **46**(8), 3468–3474 (1975).
4. W. B. Joyce, R. Z. Bachrach, R. W. Dixon, and D. A. Sealer, Geometrical properties of random particles and the extraction of photons from electroluminescent diodes. *J. Appl. Phys.* **45**(5), 2229–2253 (1974).
5. T. Yamoka, M. Abe, and O. Hasegawa, GaAlAs LED's for fiber-optical communication systems. *Fujitsu Sci. Tech. J.* **14**(1), 133–146 (1978).

6. I. Umebu, O. Hasegawa, and K. Akita, InGaAsP/InP DH LEDs for fiber-optical communications. *Electron. Lett.* **14**(16), 499–500 (1978).

7. M. Ettenberg, H. Kressel, and J. P. Wittke, Very high radiance edge-emitting LED. *IEEE J. Quantum Electron.* **QE-12**, 360–364 (1976).

8. L. Pophillat, Video transmission using a 1.3 μm LED and monomode fiber. *Proc. Eur. Conf. Opt. Commun., 10th*, Stuttgart, Paper No. 10B3 (1984).

9. D. M. Fye, R. Olshansky, J. LaCoursue, W. Powaznik and R. B. Lauer, Low current 1.3 μm edge-emitting LED for SMF subscriber loop applications. *Electron. Lett.* **22**, 2, 87 (1986).

10. T. Uji and J. Hayashi, High-power single-mode optical-fiber coupling to InGaAsP 1.3 μm mesa-structure surface-emitting LEDs. *Electron. Lett.* **21**, 418 (1985).

11. D. E. McCumber, Intensity fluctuations in the output of CW oscillators. *Phys. Rev.* **141**, 306–322 (1966).

12. H. Haug, Quantum mechanical rate equations for semiconductors. *Phys. Rev.* **148**, 338–348 (1969).

13. T. Ito, S. Machida, K. Hawata, and T. Ikegami, Intensity fluctuation in each longitudinal mode of a multimode AlGaAs laser. *IEEE J. Quantum Electron.* **QE-13**, 574–579 (1977).

14. R. Lang and K. Kobayashi, External optical feedback effects on semiconductor injection properties. *IEEE J. Quantum Electron.* **QE-16**, 347–355 (1980).

15. H. Kressel and J. K. Butler, *Semiconductor Lasers and Heterojunction LEDs*. Academic Press, New York, 1977.

16. M. Ettenberg and H. F. Lockwood, Low-threshold-current CW injection lasers. *Fiber Integr. Opt.* **2**(1), 47–63 (1979).

17. T. Kobayashi, H. Kawaguchi, and Y. Furukawa, Lasing characteristics of very narrow planar stripe lenses. *Jpn. J. Appl. Phys.* **16**, 601–607 (1977).

18. P. Marshall, E. Schlasser, and C. Wolk, A new type of diffused stripe geometry injection laser. *Proc. Eur. Conf. Opt. Commun., 4th*, pp. 94–97 (1978).

19. T. Tasukada, GaAs-Ga$_{1-x}$Al$_x$As buried heterostructure injection lasers. *J. Appl. Phys.* **45**(11), 4899–4906 (1974).

20. I. Mito, M. Kitamura, and K. Kobayashi, High efficiency and high power InGaAsP planar buried heterostructure laser diode with effective current confinement. *IEEE J. Lightwave Tech.* **LT-1**, 195–202 (1983).

21. N. Chinone, K. Saito, R. Ito, K. Aiki, and N. Shige, Highly efficient (GaAl)As buried-heterostructure lasers with buried optical guide. *Appl. Phys. Lett.* **35**(7), 513–516 (1979).

22. H. Namizaka, Single mode operation of GaAs-GaAsAlAs TJS laser diodes. *IEEE J. Quantum Electron.* **QE-11**, 427–431 (1975).

23. H. Namura, M. Sugimoto, and A. Suzuki, High power and high temperature operation laser diode with InGaAsP/InP buried heterostructure fabricated by single step LPE. *Electron. Lett.* **18**, 2(Jan. 1982).

24. T. Murotani, E. Oomura, H. Higuchi, H. Namizaki, and W. Susaki, InGaAsP/InP buried crescent laser emitting at 1.3 μm with very low threshold current. *Electron. Lett.* **16**, 566–568 (1980).

25. K. Moriki, K. Wakao, M. Kitamura, K. Iga, and Y. Suematsu, Single transverse mode operation of terraced substrate GaInAsP/InP lasers at 1.3 μm wavelength. *Jpn. J. Appl. Phys.*, **19**, 2191–2196 (1979).

26. H. Nishi, M. Yano, Y. Nishitani, Y. Akita, and M. Takusogawa, Self-aligned structure InGaAsP/InP DH Lasers. *Appl. Phys. Lett.* **35**, 232–234 (1979).

27. K. Aiki, M. Nakamura, T. Kuroda, J. Umeda, R. Ito, N. Chinone, and M. Maeda, Transverse mode stabilized Al$_x$Ga$_{1-x}$As injection lasers with channel-substrate planar structures. *IEEE J. Quantum Electron.* **QE-14**, 89–97 (1978).

28. D. Botez, Single-mode CW operation of "Double-Dovetail" constricted DIT (AlGa)As diode lasers. *Appl. Phys. Lett.* **33**(10), 872–874 (1978).

29. K. Kobayashi, K. Utaka, Y. Abe, and Y. Suematsu, CW operation of 1.5–1.6 μm

wavelength GaInAsP/InP buried-heterostructure integrated twinguide laser with distributed Bragg reflector. *Electron. Lett.* **17**(11), 366–368 (1981).

30. Y. Suematsu, S. Arai, and K. Kishino, Dynamic single-mode semiconductor laser with a distributed reflector. *IEEE J. Lightwave Technol.* **LT-1**(1), 161–176 (1983).

31. T. E. Stockton, A review of Fabry-Perot and DFB lasers in 1.55 micron fiber optic communications systems. *Proc. SPIE—Int. Soc. Opt. Eng.* **722**, 255–260 (1987).

32. W. T. Tsang, N. A. Olsson, and R. A. Logan, High-speed direct single-frequency modulation with large tuning rate in cleaved-coupled-cavity lasers. *Appl. Phys. Lett.* **42**, 650 (1983); Stable single-longitudinal-mode operation under high speed direct modulation in cleaved-coupled-cavity GaInAsP semiconductor lasers. *Electron. Lett.* **19**, 438 (1983).

33. N. Holonyak, Jr., R. M. Kolbas, R. D. Dupuis, and P. D. Dapkus, Quantum-well heterostructure lasers. *IEEE J. Quantum Electron.* **QE-16**, 170–186 (1980).

34. H. Temkin, K. Alari, W. R. Wagner, T. P. Pearsell, and A. Y. Cho, The lasing spectrum for different well widths for an InGaAs quantum-well structure. *Appl. Phys. Lett.* **42**, 845 (1983).

35. S. D. Hersee, B. DeCremoux, and J. P. Duchemin, Some characteristics of GaAs/GaAlAs graded-index separate-confinement heterostructure quantum-well laser. *Appl. Phys. Lett.* **44**, 476 (1984).

36. W. T. Tsang, ed., *Semiconductors and Semimetals*. Academic Press, Orlando, Florida, 1985.

37. W. T. Tsang, Chemical Beam Epitaxy. *J. Chem. Growth* **81**, 261 (1987); *IEEE Circuits Devices Mag.* **4**(5), 18–24 (1988).

7

Photodetectors

Introduction

Detection or demodulation is a process to recover the original signal carried by the modulated carrier wave in a communication system. In the case of a fiber optic system, the carrier is a lightwave in the infrared (IR) spectrum. Although there are many types of possible detector schemes, the only suitable detectors for any optical fiber system are those made with semiconductor materials. This is because only this type of detector can meet the size compatibility requirement for optical fibers. High quantum efficiency can also be incorporated in the design of semiconductor photodetectors.

For direct detection of either analog or digital signals in optical fiber communication, we shall discuss three types of detectors commonly used for this application: the $p–i–n$ photodiode (PIN), the avalanche photodiode (APD), and the photoconductor (PC). The photoconductor is a new addition that has aroused our interest for longer-wavelength applications only very recently.

In this chapter, we open our discussion with a brief comment on the requirements of a good detector. This is followed by individual device description and concluded with a comparison of various types of detectors for a variety of applications.

General Requirements of a Good Detector

A good detector must be sensitive and fast in response, low in noise, high in efficiency, compatible in size with the fiber, simple in design,

187

and low in cost. Not all requirements can be met in one design. Usually several requirements must be compromised in the design of detectors, and, therefore, one must choose carefully.

Quantum Efficiency

The quantum efficiency of a detector is a measure of the effectiveness of the generation of electron–hole pairs (EHPs) in response to incident photon energy falling on it. The quantum efficiency η_Q can be expressed [1] as

$$\eta_Q = (1 - R)[1 - e^{-\alpha d}] \qquad (7\text{-}1)$$

where R is the surface reflectivity, α is the absorption coefficient, and d is the width of the absorption layer. Equation (7-1) shows that high quantum efficiency can be achieved if (1) the surface reflectivity R is

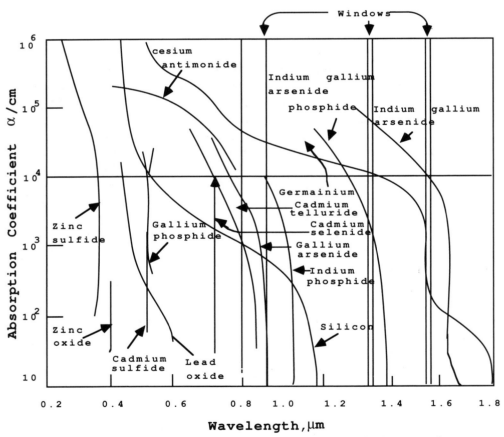

Figure 7.1 Absorption coefficient as a function of the wavelength of semiconductor materials. The vertical bands are the windows of maximum transparency in silicon fibers used in optical communication. (*Source: Laser Handbook.* North-Holland Publ., Amsterdam, 1972.)

kept low and (2) the product αd is large. In practical optical detectors, R can be reduced by antireflection coating on the detector surface. The absorption layer thickness d should be as wide as is practical, and α should be as large as possible. Figure 7.1 shows a plot of the absorption coefficient as a function of wavelength for several semiconductor materials. A line for $\alpha = 10^4$/cm is drawn on the graph to show the desired absorption coefficient for optical detector applications. Windows of maximum transparency in silica fibers are drawn between pairs of bars for each region: the shorter-wavelength region ($0.8-0.9$ μm), the 1.3-μm region, and the 1.55-μm region. These superimposed windows allow us to judge which material is best suited for a certain wavelength region. For example, in the $0.8-0.9$-μm region, silicon has an absorption coefficient of $500-1000$/cm. Although this value seems low, because of its highly developed technology, silicon is still considered the most suitable material for detectors in this wavelength range because the criterion $\alpha d \gg 1$ can be achieved readily by choosing a larger width d. An absorption layer width of $30-60$ μm can be easily obtained. It can also be seen from the graph that silicon becomes useless in longer wavelength ranges. Other materials, such as InGaAs and InGaAsP, are more suitable for these long-wavelength regions.

The quantum efficiency can also be affected by the speed of operation. Ideal quantum efficiency assumes full depletion of the absorption layer. If this is not true, a slower component of the signal, known as the *diffusion tail*, will appear, causing the device to have different low- and high-speed quantum efficiencies.

Sensitivity

The sensitivity of a detector can be expressed in many ways. Before the days of digital communication, the sensitivity of a detector was expressed in terms of the responsivity in amperes per watt. In optical fiber systems, we now express it as the amount of electrical current the detector can produce for a watt of input optical power, or

$$I_{\text{ph}} = \frac{\eta q \bar{p} \lambda}{hc} \qquad (7\text{-}2)$$

where η is the quantum efficiency (use η instead of η_Q as it can include the coupling efficiency), q is the electronic charge, $\bar{p}$ is the average optical power at a wavelength λ, and hc is the photon energy. The responsivity is defined as

$$r = \frac{I_{\text{ph}}}{\bar{P}} = \frac{\eta q \bar{p} \lambda}{hc} \qquad (7\text{-}3)$$

For an ideal detector, $\eta = 1$, and the responsivity is directly proportional to λ. This is shown as the dotted straight line in Fig. 7.2, where

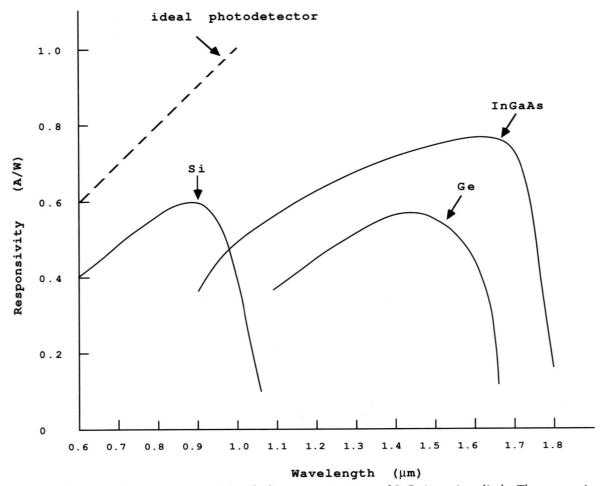

Figure 7.2 Typical current responsivity of silicon, germanium, and InGaAs $p-i-n$ diode. The responsivity of an ideal photodiode is shown as the dashed line.

the responsivity is plotted against the wavelength λ. The responsivities of a silicon detector, a Ge diode, and a InGaAs diode, are also shown as solid curves on Fig. 7.2.

Silicon, because its absorption coefficient decreases very rapidly at longer wavelengths, soon loses its responsivity above $\lambda > 1.00$ μm. At these wavelengths, the absorption layer width d can never be wide enough to satisfy the relation $\alpha d \gg 1$, and its quantum efficiency also drops drastically.

Equation (7-2) represents only one factor that determines detector sensitivity; the other factor is the noise level, which determines how weak a signal can be to be detected. Thus, the minimum detectable power as a function of the bandwidth would be a more complete description of detector sensitivity. In digital communication systems, a convenient way to express the sensitivity is the minimum detectable power for achieving a minimum bit error rate. We shall return to these descriptions when individual detector types are discussed.

For long-wavelength detection (1.0–1.7 μm), Ge and InGaAs p–i–n diodes have been developed. Their responsivity curves are very attractive, as can be seen from Fig. 7.2. But both are troubled by the large dark current due to low bandgap energy (see [3]). Other materials, such as InGaAsP/InP and AlGaAsSb, have been developed.

Noise in Detectors

Detector Noise [2] degrades signals and impairs system performance. In optical fiber systems, several sources of noise exist that are associated with the detecting and amplifying processes of signals. In a direct detection process where the output current of a photodetector is a linear function of the incident optical power, the noise is primarily shot noise. It consists of the quantum noise, the dark current noise, and the surface-leakage current noise. The quantum noise, which arises from intrinsic fluctuations of photon excitation of carriers, is fundamental in nature and sets the ultimate limit for detector sensitivity. Both the number of electron–hole pairs generated and the generation rate are statistically random quantities that satisfy Poisson statistics. The shot noise current due to the combination of the signal current and the dark current I_d is

$$\langle i_{\text{shot}}^2 \rangle = 2q[I_{\text{ph}} + I_d]B \tag{7-4}$$

where B is the bandwidth, I_{ph} is the total light-induced current, and $\langle i_{\text{shot}}^2 \rangle$ is the rms of current fluctuation due to shot noise. The term I_{ph} should be differentiated from I_{sig}, which is different from I_{ph} if modulation of less than 100% is involved. The dark current I_d is that part of the collected carriers without illumination, due to thermal generation of electron–hole pairs. Noise due to dark current is sensitive to temperature changes, since the intrinsic carrier density is proportional to $e^{-E_g/2kT}$. In silicon, for example, an increase of 50°C above room temperature results in an increase of the dark current by a factor of 20. The dark current must be kept as small as possible in detectors. In some semiconductor materials in which the bandgap energy is small, such as Ge and InGaAs, the tunneling effect may cause the dark current to increase. This may present fundamental limitations to device performance at long wavelengths [3]. Surface-leakage current can be eliminated by careful design and processing and is usually neglected in our analysis.

In the case where the detector output is followed by an amplifier, then the noise from the amplifier, which is dominated by Johnson noise, must be considered. The noise contribution due to thermal noise at the input of a preamplifier is

$$\langle i_{\text{th}}^2 \rangle = \frac{4kTB}{R_{\text{L}}} \tag{7-5}$$

where R_L is the load resistance of the detector or the resistance of the amplifier, k is Boltzmann's constant, and T is the temperature in degrees kelvin. The total noise contribution is now the sum of the shot noise and the thermal noise. Shot noise is a signal-dependent noise, whereas thermal noise is signal-independent. Thermal noise may be reduced by increasing R_L, but at the penalty of reducing the maximum available bandwidth B as $B = 1/(2\pi R_L C_d)$, where C_d is the detector capacitance.

In the case of an avalanche photodiode, where photocarriers are multiplied in the avalanche process, the expression for the shot noise is modified by two factors: the multiplication factor M and an excess noise factor F. This is

$$\langle i_{shot}^2 \rangle = 2qFM^2IB \tag{7-6}$$

where I is the total current. The avalanche multiplication factor M is

$$M = \frac{1}{(1 - V/V_{br})^2} \tag{7-7}$$

and

$$F(M) = M\left[1 - (1 - k)\left(\frac{M-1}{M}\right)^2\right] \tag{7-8}$$

Note that $k = \gamma_h/\gamma_n$, the ratio of the hole ionization coefficient to that of the electron [4].

The noise of a photoconductor comes from the thermal effect. The finite dark current, when the device is not illuminated, gives rise to a random fluctuation known as Johnson noise current, the mean square of which is

$$\langle (I_j^2) \rangle = \frac{4kT}{R_D} B \tag{7-9}$$

where R_D is the dark resistance of the photoconductor. The effect of Johnson noise in the accompanying amplifier should also be considered.

The effect of noise in detectors can usually be expressed in terms of the signal-to-noise ratio. It limits the usefulness of a detector. We will discuss this property in each of the detectors in the following sections.

The Speed of Response

Three main factors that limit the speed of response of photodetectors are (1) the diffusion time—the time for carriers generated within a diffusion length from the junction to diffuse into the depletion region, (2) the drift time of carriers through the depletion region, and (3) the effect of the junction capacitance.

The Diffusion Time

If d is the depth of photon absorption, the diffusion time can be expressed as

$$t_{\text{diff}} = \frac{d}{2D} \qquad (7\text{-}10)$$

where D is the diffusion constant of the minority carriers. Since the absorption coefficient is strongly dependent on the wavelength, so is d and the diffusion time. To ensure that as few carriers as possible are generated outside the depletion region, we require that at a desired wavelength, the depletion width $w(w \Rightarrow d)$ be larger than $1/\alpha(\lambda)$. If this criterion cannot be met, which may be the case with low reverse bias, diffusion time may become a limiting factor of the response time.

The Drift Time

Within the depletion width, carriers drift cross this region with a velocity known as the *saturated velocity*. For a narrow depletion width, even at moderately low reverse-bias voltage, the field within this region is sufficiently high that only saturated velocity is considered, so that

$$t_{\text{drift}} = \frac{w}{v_s} \qquad (7\text{-}11)$$

where w is the depletion region width and v_s is the saturated velocity. If $v_s = 10^7$ cm/s, the drift time $t_{\text{drift}} \sim 10$ ps/μm.

Junction Capacitance Effect

Junction capacitance of a p–n diode is expressed as

$$C_j \cong \frac{A}{2} \left[\frac{2q\epsilon}{\phi_0 + V} \frac{N_d N_a}{N_a + N_a} \right]^{1/2} \qquad (7\text{-}12)$$

where A is the junction area, ϕ_0 is the built-in voltage, V is the reverse-bias voltage, and N_a and N_d are the carrier densities of the p- and n-sides, respectively. Usually $V \gg \phi_0$, and for $N_a \gg N_d$,

$$C_j \cong \frac{A}{2} (2q\epsilon N_d)^{1/2} V^{-1/2} \qquad (7\text{-}13)$$

The junction capacitance shunts across an input resistance R_L of the amplifier. The time constant $R_L C_j$ gives rise to a bandwidth B such that

$$B = \frac{1}{2\pi R_L C_j} \qquad (7\text{-}14)$$

The combination of R_L and C_j acts as a lowpass filter to limit the frequency response of the receiver. Frequency response can, therefore, be improved by reducing C_j, which is accomplished by reducing the diode area A, reducing the carrier density N_d, or increasing the reverse-bias voltage V. However, the area should be at least as large as the fiber cross section. Reducing N_d will increase the series-resistance of the diode such that it may no longer be neglected in the above calculation. Increasing V increases the drift length, thus affecting the carrier drift time as the carrier velocity is saturated. Careful design should consider a compromise among these factors.

Signal-to-Noise Ratio and Probability of Error

Signal-to-Noise Ratio

The signal-to-noise ratio S/N [5] is defined as the ratio of the mean square of the signal current to the sum of the mean-square noise currents, as

$$\frac{S}{N} = \frac{\langle (i_{ph})^2 \rangle}{\langle (i_n)^2 \rangle} \tag{7-15}$$

where i_{ph} is the signal component of the optical current I_{ph} and $\langle (i_n)^2 \rangle$ is the sum of the noise-contributing fluctuation currents. These are

$$\langle (i_n)^2 \rangle = \langle (I_{ph})^2 \rangle + \langle (i_Q)^2 \rangle + \langle (i_d)^2 \rangle + \langle (i_L)^2 \rangle + \langle (i_{th})^2 \rangle$$

where $\langle (i_Q)^2 \rangle = 2qI_{ph}B$ $=$ quantum noise
$\langle (i_d)^2 \rangle = 2qI_0B$ $=$ dark-current noise
$\langle (i_L)^2 \rangle = 2qI_LB$ $=$ leakage-current noise
$\langle (i_{th})^2 \rangle = 4kTB/R_L$ $=$ thermal noise

If the signal current is expressed in terms of signal power as in Eq. (7-2), the signal-to-noise ratio can be written as

$$\frac{S}{N} = \frac{(P_0\, \eta q/h\nu)^2}{(2q^2 P_{ph}\, \eta/h\nu + 8\pi\, TBC_d)B} \tag{7-16}$$

In Eq. (7-16), the noise contribution due to dark current and leakage current has been neglected for simplicity since they are small in well-designed diodes. Power P_0 is the signal power included in the side bands of the total optical power P_{ph}.

The Bit Error Rate

In digital modulated systems that require only to identify the "1" or "0" state in each sampling interval, one is interested in knowing the probability of making a false identification.

The arrival of a pulse may be represented by a collection of photons. On the average, there will be 10–20 photons per pulse. Since photon shots are random, there is the probability that a pulse may fail to contain a single photon. This indicates a false identification or error. How often this kind of error occurs in a system actually limits the speed of operation of bit rate. If $p(n,\Omega)$ is the probability of detecting n photons per unit time interval, and Ω is the average rate of photons per pulse, then by Poisson statistics, one can write

$$p(n,\Omega) = \frac{(\Omega)^n}{n!} e^{-\Omega} \qquad (7\text{-}17)$$

If we let $\Omega = 20$ and $n = 0$, representing the case where 20 photons are registered per pulse, we call this a "1." If there is no photon in a pulse, we call this a "0."

Plugging in our numbers, the probability is

$$p(0,20) = \frac{(20)^0}{0!} e^{-20} = 2 \times 10^{-9}$$

If we assume a system containing equal numbers of zeros and ones, the above probability gives rise to a bit error rate of $\frac{1}{2}(2 \times 10^{-9}) = 10^{-9}$. This is a reasonable number and has been universally adopted as the standard bit error rate BER for digital systems. In fact, many authors derive the sensitivity of a detector using the minimum detectable power as a function of the bandwidth or bit rate for a specified BER (10^{-9}). Assuming a Gaussian approximation, the S/N ratio can be related to the bit error rate [6] as

$$BER = \frac{1}{2} \operatorname{erfc}\left(\frac{S/N}{\sqrt{2}}\right) = \frac{1}{2} \operatorname{erfc}(Q) \qquad (7\text{-}18)$$

where $Q = S/N/\sqrt{2}$ and erfc is the complementary error function defined as

$$\operatorname{erfc}(x) = 1 - \frac{2}{\sqrt{\pi}} \int_0^x \exp(-t^2)dt \qquad (7\text{-}19)$$

For an error rate of 10^9, $Q = 6$ and the corresponding $S/N = 6\sqrt{2}$, or about 9.3 dB.

The Minimum Detectable Power

The minimum power P_{min} required to obtain a reliable identification of the signal for a given bandwidth is obtained by assuming $S/N = 1$. A plot of P_{min} with bit rate gives useful information about the detector performance. One such plot is shown in Fig. 7.3. First, we calculate the power required to identify a signal for a bit error rate of $10(-9)$ for a finite number of photon energies. These are marked as lines corresponding to 10, 100, and 1000 photons per bit, respectively. The 10-

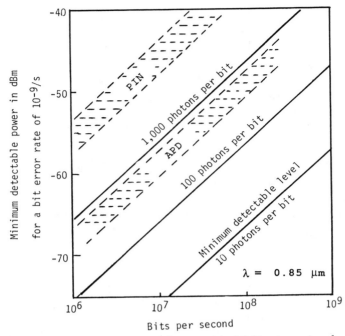

Figure 7.3 Minimum signal power required to give a signal-to-noise ratio of unity in a photodiode detection system, $\lambda = 0.85 \ \mu m$.

photons line is accepted as the limiting case for detection. Then the minimum detectable receiver power using PINs and APDs are estimated separately from the power equations. When these values are plotted on the graph, they appear as lines or bands for each type of detector as shown.

PIN Photodiode

PIN Photodiode Structures

In Chapter 6, section on power output, we mentioned that a reverse-biased $p-n$ junction can be used to detect light signals. The response time of the junction diode can be improved by confining the generation of electron–hole pairs in the depletion region exclusively. This suggests a wider depletion region. To achieve this purpose, an intrinsic layer is introduced at the junction, forming a $p-i-n$ or "PIN" photodiode. This has been the most commonly used photodetector structure for optical communication systems. Silicon PIN photodiodes can be used successfully up to 1-μm wavelengths, and germanium and other compound semiconductor PIN diodes are used for the longer-wavelength range.

The basic structure of a PIN photodiode is shown in Fig. 7.4. A thick i-layer is grown on an n-substrate. A thin p^+-layer is grown on top of that, forming the basic p–i–n diode. A metallic contact with an opening to admit light serves as one contact. The other contact is on the substrate. When a reverse bias is applied, the intrinsic region is fully depleted. The incident light penetrates the thin p^+-layer and reaches the i-layer. There, absorbed photon energy generates electron–hole pairs (EHPs). The high electric field in this region rapidly sweeps the electrons and holes to the n- and p-sides of the diode, respectively, causing current to flow in the external circuit in proportion to the illumination level. Of course, any photons absorbed by the top p^+-layer and the bottom n-layer also produce EHPs. But these carriers will have to diffuse back into the depleted region before they can be

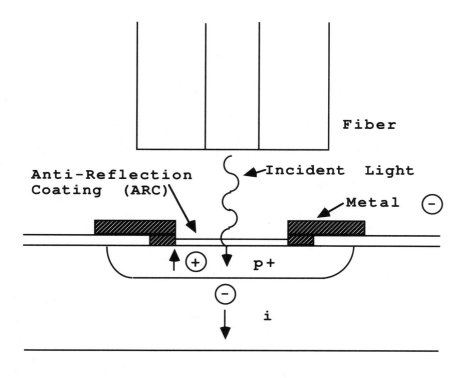

Figure 7.4 A p–i–n diode structure. (After Y. Suematsu, Proc. IEEE, 71, 692, 1983 © IEEE 1983.)

collected. Since the diffusion process takes time (10^{-9} s), it affects the response time of the diode. In comparison, the transit time of carriers in the depletion region is very short ($\approx 10^{-12}$ s). A good diode should, therefore, be designed to limit the absorption of photons only in the i-layer.

Another factor that limits the response time of a diode is the shunt capacitance associated with the junction. The junction capacitance can be kept small by (1) limiting the diode area and (2) reducing the doping density of the i-region. But the area of the diode should be compatible with the optical fiber. Too small a diode restricts power-handling capability. Reducing the doping density indicates that a careful purification process is necessary. Also, if the density is low, the depletion width increases. This could add transit time for the carriers thus spoiling the frequency response.

Several methods have been improvised to minimize the absorption of photons in the top p-region. One way is to use higher-bandgap p-type GaAlAs for the top layer, which allows light to pass through the p-layer without absorption because higher-bandgap GaAlAs is transparent to optical radiations [7].

Another way is to use side or bottom illumination. In a top-illuminated diode, the metallic contact on the p^+-layer usually has a hole to admit light to the p^+-layer. The large area necessary for this structure increases the diode area and thus the junction capacitance. By turning the diode upside down, that is, to admit light from the substrate, which can be made with transparent n-type semiconductor material, the whole substrate can be exposed to illumination. This increases the quantum efficiency of the diode without increasing junction capacitance [8]. Similar results can be achieved by using side illumination. In this case, illumination enters the i-layer from the sides [9].

p–i–n Photodetector Performance

For the 0.8–0.9-μm wavelength range, a silicon p–i–n photodiode can easily attain a quantum efficiency of 85% from dc to 1 Gbit/s. The noise source is primarily from shot noise as the dark current is sufficiently low and causes no additional noise. Generally, the signal output from a detector requires amplification. This makes the amplifier noise very important for the receiving system. Using Eqs. (7-4) and (7-5), we can express the signal-to-noise ratio of a p–i–n photodiode as

$$\frac{S}{N} = \frac{I_{ph}^2}{2q(I_{ph} + I_d)B + 4kTB/R_L} \tag{7-20}$$

Using Eq. (7-14) to eliminate R_L, we have

$$\frac{S}{N} = \frac{I_{ph}^2}{2q(I_{ph} + I_d)B + 8\pi C_d kTB} \tag{7-21}$$

Let us define a minimum signal current from which we can calculate the minimum detectable power. For $S/N = 1$, Eq. (7-21) yields a signal current

$$(I_{ph})_{min} = [2Bq(I_{ph} + I_d) + 8\pi C_d kTB^2]^{1/2} \tag{7-22}$$

and

$$P_{min} = \frac{2hc}{\eta q\lambda} [2Bq(I_{ph} + I_d) + 8\pi C_d kTB^2]^{1/2} \tag{7-23}$$

In case the receiver is amplifier-noise-limited, that is,

$$8\pi C_d kTB^2 > Bq(I_{ph} + I_d)$$

then

$$P_{min} = \frac{2hc}{\eta q\lambda} B(2\pi kTC_d)^{1/2} \tag{7-24}$$

The minimum detectable power is then directly proportional to the bandwidth.

Avalanche Photodiode

The Avalanche Mechanism

In the PIN-type photodiode discussed under "p–i–n Photodetector Performance" (above), at most one EHP is generated for each photon absorbed. The process, therefore, possesses no gain mechanism. However, for a wide depletion width operating at a high reverse bias, the field within the region may reach a value sufficiently high ($E > 10^5$ V/m) to produce avalanche breakdown. The mechanism of avalanche multiplication is schematically shown in Fig. 7.5. In this case, electrons and holes traversing under such a high field can acquire sufficient kinetic energy to produce additional EHPs through inelastic collisions. The avalanche is measured by a multiplication factor M defined as

$$M = \frac{1}{1 - \int_0^w \gamma(x)dx} \tag{7-25}$$

and the breakdown criterion is

$$\int_0^w \gamma(x)dx = 1 \tag{7-26}$$

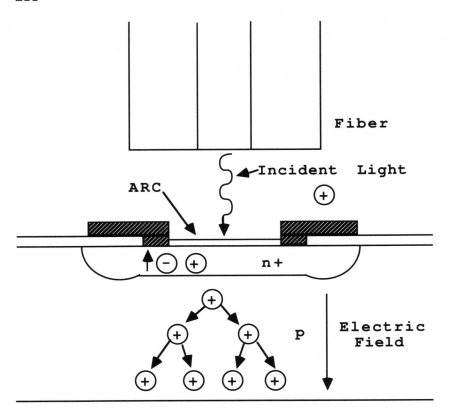

Figure 7.5 Structure of an avalanche photodiode showing the avalanche multiplication mechanism. (After Y. Suematsu. Proc. IEEE, 71, 692, 1983 © IEEE 1983.)

where $\gamma(x)$ is an ionization coefficient defined as $\gamma = Ae^{-B/E}$, A and B are material constants, and E is the electric-field intensity. The ionization coefficient for electrons and holes may be different. The multiplication M factor therefore includes a factor $k = \gamma_h/\gamma_n$, which is the ratio of the ionization coefficients of holes and electrons [10]. Thus

$$M = \frac{1 - k}{e^{-[(1-k)\delta]} - k} \qquad (7\text{-}27)$$

where

$$\delta = \int_0^w \gamma(x)\,dx$$

Responsivity and Signal-to-Noise Ratio in an APD

The responsivity of an APD is that of a simple photodetector Eq. (7-3) modified by a multiplication factor

$$r = \frac{I_{ph}}{p} = \frac{\eta q \lambda}{hc} M \qquad (7\text{-}28)$$

The noise is modified in a more complicated manner. Excess noise is generated by the avalanche process, and the excess noise factor F is as stated in Eq. (7-8).

Similarly, the shot noise and the dark-current noise are all modified by $M^2 F$ as

$$\langle (i_{sh})^2 \rangle = M^2 (2 q I_{ph} BF)$$

$$\langle (i_d)^2 \rangle = M^2 (2 q I_d BF)$$

and the signal-to-noise ratio, for $k > 1$, becomes

$$\frac{S}{N} = \frac{(P_a q \eta / h)^2 M^2}{(2 q i_d + 2 q P_{ph} q \eta / h) M^2 F + 4 (kT/R_L) B} \qquad (7\text{-}29)$$

A silicon APD has a structure very similar to the p–i–n diode shown in Fig. 7.4. The design criterion regarding optimization of the absorption region also applies. The principal difference lies in the bias. In APDs, a reverse-bias voltage as high as 300 V is used, whereas only 5–10 V is customarily used in a p–i–n diode. The multiplication factor increases with increasing bias voltage, but the multiplication process is slower, since it takes time to build up an avalanche. Thus, gain and bandwidth are traded off. The highest bandwidth obtainable is when the gain is unity.

Equation (7-29) shows that S/N changes with M. But the F factor is also a function of M in a complicated manner, as shown in Eq. (7-8). The response time of an APD also depends on the ionization coefficient ratio. The worst-case ratio occurs when $k = 1$. Either $k = 0$ or $k = \infty$ minimizes noise. This accounts for the success of silicon APDs ($k \ll 1$) at short wavelengths where low noise and high gain can be achieved. A GaAs APD where $k \sim 1$, on the other hand, is more noisy. The optimum gain of an APD should not be above 50–100 because stability and temperature effects will become difficult problems to handle. Stabilization schemes are usually required.

Germanium APDs are sensitive to wavelengths below 1.8 μm and may have a broadband quantum efficiency better than 40% but have an inherently high multiplication noise level [11–15].

Long-Wavelength APD

High-quality APDs for detecting long-wavelength radiation are considerably more difficult to design than those for short wavelengths. At long wavelengths, one needs semiconductor materials with narrow bandgaps, such as Ge and InGaAsP, to absorb low-energy photons. But low-bandgap materials have a large dark current at high reverse voltage as a result of the tunneling effect. Most recent research is focused on solving these problems.

One design is known as SAM–APD, an APD with a separated absorption-and-multiplication region [10, 16, 17]. This is shown schematically in Fig. 7.6.

The diode has a heterostructure. Photons are absorbed in a thin, active layer of InGaAs. The photogenerated holes are swept into a layer of InP, which has a significantly larger bandgap. In such a structure, the electric field never exceeds values that would induce significant leakage current from quantum-mechanical tunneling. The highest electric field occurring at the p–n junction is in the high-bandgap region. In indium phosphate, the holes are more highly ionizing than electrons; the device is thus optimized for the injection of holes by being constructed with n-type material. However, the detector efficiency and response speed are both low, rendering it useless in high-bit-rate applications. This is true because the bandgap energy

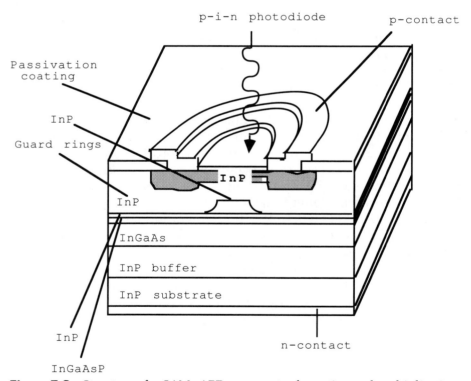

Figure 7.6 Structure of a SAM–APD, a separate absorption-and-multiplication avalanche photodetector. (After N. Susa *et al.*, IEEE *J. Quantum Electron*. QE-14, 864–870, © IEEE 1980.)

difference between InP and InGaAs introduces an energy step that traps holes generated in the InGaAs active layer and causes recombination (which reduces the efficiency) or simply delays their travel to the multiplication region (which slows down the response). Considerable research remains to be done before SAM–APDs made of long-wavelength InGaAs/InP can be extensively employed in practical optical fiber systems.

Photoconductors

Photoconductivity is a long-recognized fundamental phenomenon in semiconductor materials [12], but its application as a photodetector in optical fibers is relatively new. In fact, renewed investigations, particularly for long-wavelength optical fiber applications, showed that photoconductors may even be advantageous over the use of other detector schemes, such as PIN and APDs [12].

When a semiconductor bar is uniformly illuminated, if the photon energy is greater than the bandgap energy E_g of the semiconductor, EHPs will be generated. In response to an applied electric field, the conductivity of the semiconductor bar increases with the illumination intensity. If G_{op} is the optical generation rate, the equilibrium carrier density is determined by the difference between the generation and recombination rates. Let Δn and Δp be the excess carrier concentrations of the electrons and the holes, respectively. Then, the new carrier densities become $n = n_0 + \Delta n$ and $p = p_0 + \Delta p$, respectively. Since $\Delta n = \Delta p$ for the EHP, then the charge balance relation states that

$$G_T + G_{op} = \alpha_r np = \alpha_r[(n_0 + \Delta n)(p_0 + \Delta p)] \tag{7-30}$$
$$= \alpha_r[(n_0 + p_0)\Delta n + n_0 p_0 + (\Delta n)^2]$$

where G_T is the thermal generation rate and is equal to $\alpha_r n_0 p_0$ and α_r is the recombination coefficient.

Thus

$$G_{op} = \alpha_r[(n_0 + p_0)\Delta n + (\Delta n)^2] \tag{7-31}$$

If one defines $\tau = 1/(n_0 + p_0)\alpha_r$, where τ is the carrier lifetime, then for low excitation, one may neglect the second-order term Δn^2, leaving

$$G_{op} = \frac{\Delta n}{\tau} \tag{7-32}$$

The change in conductivity due to illumination can be written as

$$\Delta\sigma = qG_{op}(\tau_n\mu_n + \tau_p\mu_p) \approx qG_{op}\tau(\mu_p + \mu_n) \tag{7-33}$$

since for most simple recombination processes, $\tau_n = \tau_p = \tau$. Thus, when G_{op} is changed, the conductivity changes proportionally. The device using this phenomena can be used to detect weak optical signals. Cadmium sulfide light switches and silicon photoconductor detectors are examples of this application. However, the relatively low gain bandwidth product of a silicon photoconductor strongly limits its usefulness in applications requiring both high sensitivity and quick response [13]. For maximum photoconductive response, materials with high mobility and long lifetime are essential.

An InGaAs/InP Photoconductor Structure

A long-wavelength photoconductor of this type is shown in Fig. 7.7 [18]. A thin layer of indium gallium arsenide, either n-, p-, or i-type, that can absorb radiation having a wavelength as long as 1.65 mm, is grown on an i-type InP substrate. For good crystalline match, the composition has been chosen as $In_{53}Ga_{47}As$. Metallic electrodes are deposited in an interdigital pattern on the surface so that alternate electrodes can be connected to a power supply. In this manner, the carriers generated between the conducting layers have the shortest distance to travel before they are collected.

Even before this photoconductor is illuminated, thermally generated EHPs have been collected on the electrodes, electrons to the anode, and holes to the cathode. When short-circuited, a dark current flows in the external circuit. This dark current causes Johnson noise to become excessive in a photoconductor. When the layer is illumi-

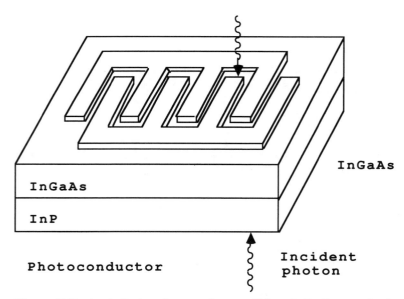

Figure 7.7 An InGaAs photoconductor. [After S. R. Forrest [78], *Spectrum* **23**, 76 (1986), with permission.]

nated, the absorbed photons generate additional EHPs, giving rise to a signal current due to illumination.

Interestingly, a photoconductive process may possess current gain. This is because an electron moves faster than a hole. The hole persists for a long time after the end of an optical pulse. The current gain in a photoconductor is simply the ratio of the hole lifetime to the electron transit time. Thus, the response time or bandwidth is inversely proportional to the hole lifetime. This means that the higher the gain, the lower the bandwidth. It is not possible to design a photoconductor with arbitrarily higher gain and also arbitrarily wide bandwidth. The gain and bandwidth are physically linked. They represent the trade-off between sensitivity and frequency response as in all photodetectors.

Noise in Photoconductors

Photoconductors have many sources of noise. The dominant noise arises from the dark current. It flows whenever an external voltage is applied. This finite dark conductivity generates a randomly fluctuating noise current known as *Johnson noise current*. The mean square of the current is

$$\langle (i_T)^2 \rangle = \frac{4kTB}{R_c} \tag{7-34}$$

where R_c is the dark resistance of the conducting layer.

The photoconductor signal current is equal to the gain multiplied by the primary photogenerated current, which is a function of the time-averaged power of the incident light.

$$S = mI_{ph} = \frac{m\eta qP\lambda}{hc} \tag{7-35}$$

Therefore

$$\frac{S}{N} = \frac{m\eta qp\lambda}{hc} \cdot \frac{R_c}{4\pi kTB} \tag{7-36}$$

Equation (7-36) shows that S/N increases with increasing conductive layer resistance R_c and the gain factor m. Unfortunately, increasing the gain to improve sensitivity decreases the frequency response as described in the preceding section.

In contrast to photodiodes, the photoconductive detector requires a photon energy of only the difference between the band edge and the impurity level, which is usually a few tenths of a millielectron volt. Consequently, photoconductors can be used as detectors well into the infrared, out to wavelengths of perhaps 30 μm where the photon energy is 40 meV. This is one major advantage of photoconductors over all types of other photodetectors.

Comparison between Photodetectors

It is informative to compare the three basic types of photodetectors. The basis for comparison is the relative sensitivity or the minimum detectable power (time averaged signal power) for a prescribed operating condition. For digital optical fiber communication, the sensitivity is usually expressed in terms of the minimum detectable signal power as a function of the bit rate for a *BER* of 10^{-9}. The bit error rate is a measure of how often a transmitted digit "1" will be mistakenly identified by the receiver as a "0" as described by the probability function in Eq. (7-17). The figure 10^{-9} is considered as sufficiently low for most applications. Figure 7.8 shows such a plot based on the calculated and experimental data for an InGaAs structure operating at 1.3-μm wavelength. The solid curves are the theoretical calculations, and the dots indicate reported experimental data. The minimum time-average power rather than peak power is specified to yield a value that is independent of the transmitted bit pattern. The unit of power is the number of decibels below one milliwatt of power (dBm). Figure 7.8 shows that the avalanche photodiode is the most sensitive, followed

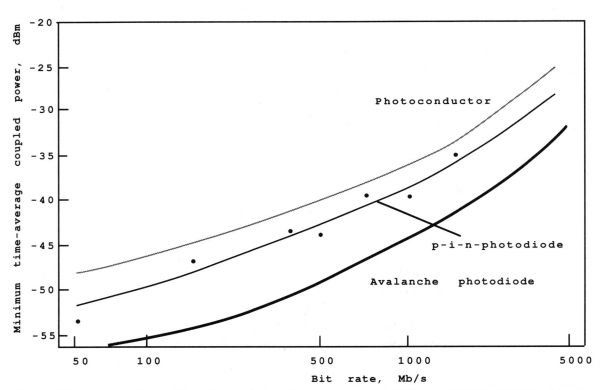

Figure 7.8 A comparison of three types of photodetectors. [After S. R. Forrest [12] *IEEE J.* **LT-3**, © IEEE, 1985.

by the p–i–n photodiode, which can be used at even higher bit rates. The photoconductor is the least sensitive. This figure also shows that the sensitivity of all three types of photodetectors decreases with increasing data rate. This is because the noise of all these detectors depends increasingly on bandwidth.

Although the p–i–n diode possesses no gain, the diode can be designed to have low background noise and high quantum efficiency. The bandwidth of an InGaAs/InP p–i–n diode may exceed 10 GHz. But it is usually limited by extrinsic effects such as the L–C time constant of the detector and the external amplifier circuit.

The sensitivity difference between the p–i–n diode and the photoconductor is not apparent except at low bit rates. There the high Johnson noise of a photoconductor makes it significantly less attractive than the p–i–n photodiode for applications demanding high sensitivity. At higher bit rates, the detectors appear to be nearly comparable.

Avalanche photodiodes have internal gain, but they also tend to be very "noisy." The excess noise indicated by the factor $F(M)$ [in Eq. (7-8)] increases with the gain except for very small $k(=\gamma_h/\gamma_n)$. For $k < 1$ (as in silicon), the statistical fluctuations in the collected current are considerably smaller than in materials with $k = 1$ (as in GaAs). At some high value of gain, the avalanche noise of an APD dominates all other noises, including the amplifier noise. A further increase in gain only reduces the signal-to-noise ratio. This trade-off occurs when the noise from an APD equals the noise from the amplifier. This gain is referred to as the *optimum avalanche gain*.

The highest bandwidth (when the gain is unity) of an APD is ultimately limited by the carrier transit time across the depletion region, just as in a p–i–n photodiode. With gain, the bandwidth is reduced, since it takes time to build up an avalanche as discussed earlier in this chapter in the section on long-wavelength APD.

The best combination of bandwidth and sensitivity can be achieved with a p–i–n photodiode at very low and very high bit rates, whereas the APD is useful for moderate bit rates between 100 Mbit/s and 4 Gbit/s.

Photoconductors may also have gain depending on the ratio of the hole lifetime to that of the electron. But the gain is usually insufficient to offset the increase of the dark-current noise to the extent that would render its performance superior to that of reverse-biased detectors such as the p–i–n and the APD. On the other hand, the simplicity of the photoconductor and its potential compatibility with electronic devices enhance its potential as a detector in future monolithic integrated optical electronic circuits. Long-wavelength applications of photoconductors have become increasingly interesting as no other diodes have been found to perform as well as they do above a wavelength $\lambda > 3.0$ μm.

Recent Developments in Photodetector Technology

New developments of photodetector technology have two main goals: (1) to develop photodiodes with improved sensitivity for operation at long wavelength and very high bandwidth and (2) to develop structures with improved performance at low cost.

In order to increase the sensitivity, avalanche photodiode design is favored, provided the increased noise current at long wavelength can be reduced. One approach is to use a multi-quantum-well structure. In this design, the avalanche region is a sandwiched structure, consisting of the narrow-bandgap layers such as GaAs, alternating with a layer of wide-bandgap material, such as AlGaAs. The narrow-bandgap layers function as quantum wells in confining carriers. The structure creates a boundary through which the holes and electrons can have markedly different ionization coefficients and thereby reduce the noise. This is because the edges of the valence and conduction bands have been offset at the interface, which enhances the ionization coefficient of the electrons dramatically. Unfortunately, this technique does not render an easy solution when InP material is used, where long-wavelength operation is desired. For the $0.8-0.9$-μm range, silicon works well and only simple structure is needed. Research is under way for new materials and techniques for long-wavelength application.

To develop a structure with good performance and lower cost, one comes to investigate the $p-i-n$ structure again. But this time with the help of integrated electronics technology, one can incorporate a $p-i-n$ diode with a metal semiconductor field-effect transistor (MESFET) or transistor [19]. This integrated approach has the potential for increased sensitivity at high bandwidths and at a lower cost. The replacement of a conventional discrete photodiode and field-effect transistor by its integrated counterpart reduces the capacitance of the photodiode, thus increasing the bandwidth. For high frequencies, however, the FET should be replaced by a bipolar transistor. Figure 7.9 shows the schematics of an integrated $p-i-n$ photodiode and a FET.

An array of $p-i-n$ photodiodes has been constructed to increase the bandwidth. A 12-photodiode array is sketched in Fig. 7.10 [20]. Each diode is addressed through an optical fiber. The array may be used in series or in parallel. The effective bandwidth with a 12-element array, each operating at 50 Mbit/s, would give a total bandwidth of 600 Mbit/s. Again, to apply this technology to long-wavelength materials, such as with indium phosphate-based materials, the technology is not yet mature.

The simplicity of using a photoconductor as a detector has aroused much interest recently. As discussed in earlier sections, the device can be integrated with a transistor easily. Further, for operations at

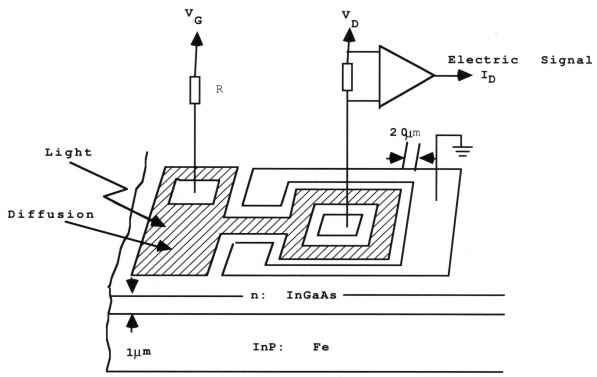

Figure 7.9 A $p-i-n$ and MESFET integrated photodetector assembly. After R. Leheny *et al.* [19].

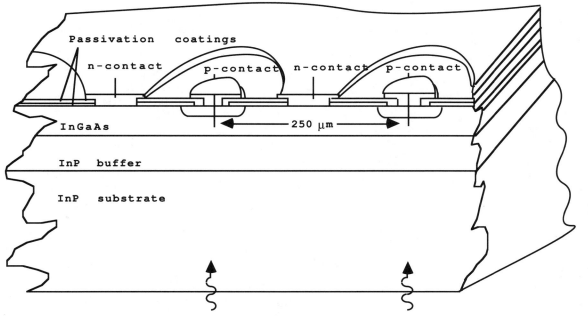

Figure 7.10 A linear array of $p-i-n$ diodes. [After Y. Ota and R. C. Miller, *Int. Soc. Opt. Eng.*, with permission.]

even longer wavelengths ($\lambda > 1.8$ μm), the photoconductor may well be the best candidate yet to date.

Some recent laboratory measurements of optical detectors are listed in Table 7.1 [21]. No commercial products of these devices are available as yet.

Encouraging experimental results on coherent detection instead of direct detection of optical fiber communication signals have been reported recently [21–23]. The essential component of a coherent detection system is a tunable single-frequency, single-polarization stabilized semiconductor laser with a narrow spectral width. It is used as a local oscillator to beat with the optical signal by either the heterodyne or homodyne technology to obtain a microwave-frequency electronic signal. At microwave frequencies, the signal can be amplified electronically, a technique that has been perfected since World War II. A sensitivity enhancement of 100 times that of direct detection can easily be achieved. Sharp electronic filters are readily available at microwave frequencies. They can be used to separate the signals if multiplexed. The use of sharp microwave filters instead of broad optical filters can increase the selectivity of the receiver by a factor of 1000.

Table 7-1 Optical Detector Sensitivity Levels for Various Photodetectors

Type	Material	Bit Rate (Mbit/s)	Wavelength (μm)	Normalized Detector Sensitivity (dB$_m$)	Transmission Distance (km)	References
Photo conductor	InGaAS	2000	1.51	−30.3	—	Chen (Bell Lab.), 1985
	InGaAs	1000	1.30	−35.9	—	Chen et al., (Bell Lab.), 1984 [6]
p−i−n diode	InGaAs	1200	1.53	−36.6	—	Singthe, 1984
				−36.6	113.7	British Telecon. Res. Lab.
		420	1.55	−35.6	119	Tsang et al., 1983 Bell Lab.
		565	1.3	−40.2	—	Smith et al., 1982
					—	British Telecon. Res. Lab.
		420	1.3	−35.0	84	Boenke et al., 1928 Bell Lab.
Avalanche photodiode	Ge	2000	1.55	−33.6	51.5	Yamada, 1982 Nippon Telephone Telegraph Corp.
		420	1.55	−40.4	108	Tsang et al., 1983 Bell Lab.
		4000	1.3	−22.0	–	Takano et al., 1985 Nippon Elect. Corp
		1200	1.3	−31.7	22.7	Yamaha et al., 1979 Nippon T & T Corp.
Integrated p−i−n MISFET	InGaAs/EnP	100	1.3	—36.0	—	Kasahara et al., 1984 Nippon Electric Corp.
	1 × 12 InGaAs p−i−n array	45	1.3	−40.5	—	Kaplan Farrest, 1956 Johnson Bell Lab.

This table was selected from the data list in Henry [21].

Thus, coherent detection becomes really attractive. The only hurdle before the technology can go commercial will be to develop truly coherent semiconductor lasers that can be tuned over a sizable part of a fiber's large bandwidth. The race is on.

Conclusion

In this chapter, we started by prescribing the requirements of a good photodetector for optical fiber applications. Three types of photodetectors, the p–i–n photodiode (PIN), the avalanche photodiode (APD), and the photoconductor (PD) were selected for discussion. The structure, principle of operation, sensitivity, speed of response, bandwidth, and noise characteristics of each type were discussed and compared. Materials suitable for making photodiodes in different wavelength regions were investigated. Recent and future trends in long-wavelength photodiode design and research were explained with comments. In particular, we wish to mention a review article by Linke and Henry [24] on coherent optical detection that summarizes promising techniques for detection of laser signals with increasing sensitivity. The use of homodyne detection of phase-modulated on–off keying gives sensitivity approaching the theoretical limit.

Table 7.1 lists more recent laboratory measurements of optical detector sensitivities for various long-wavelength photodetector types.

References

1. H. Melchior, Demodulation and photodetection techniques. In *Laser Handbook* (F. T. Arechi and E. O. Schulz-Dieber, eds.), pp. 725–835. North-Holland Publ., Amsterdam, 1972.
2. F. N. H. Robinson, *Noise in Electrical Circuit*. Oxford Univ. Press, London, 1962.
3. S. R. Forrest, M. DiDomanice, Jr., R. G. Smith, and H. J. Stocker, Evidence for tunneling in reverse-biased III-V photodetector diodes. *Appl. Phys. Lett.* **36**, 580–582 (1980).
4. R. J. McIntyre, Multiplication noise in uniform avalanche diodes. *IEEE Trans. Electron Devices* **ED-13,** 164–168 (1966).
5. R. G. Smith, Photodetectors for fiber transmission systems. *Proc. IEEE* **68**(10), 1247–1253 (1980).
6. S. D. Personick, Receiver design for digital fiber optic communication systems. Parts I and II. *Bell Syst. Tech. J.* **52**(6), 843–906 (1973).
7. H. Kressel, ed., *Semiconductor Devices for Optical Communication*. Springer-Verlag, Berlin, 1980.
8. T. P. Lee and T. Li, Photodetectors. In *Optical Fiber Telecommunications* (S. T. Miller and A. G. Chynoweth, eds.), pp. 593–626, Academic Press, New York, 1979.
9. O. Krumpholz and S. Maslowski, Schnelle Photodioden mit Wellen langenuablangigen Demodulationzeigenschafton. *Z. Angew. Phys.* **25**, 156–160 (1968).
10. R. J. McIntyre, The distribution of gains in uniformly multiplying avalanche photodiode theory. *IEEE Trans. Electron Devices* **ED-19,** 702–713 (1972).

11. M. C. Brain, Responsivity and noise characterization of Ge avalanche photodiode throughout wavelength range 1.1-1.7 μm. *Electron Lett.* **15**(25), 821–823 (1979).

12. S. R. Forrest, The sensitivity of photoconductor receivers for long wavelength optical communications. *IEEE J. Lightwave Technol.* **LT-3**(2), 347–360 (1985).

13. M. DiDomenico and O. Svelto, Solid-state photodetection, a comparison between photodiodes and photoconductors. *Proc. IEEE* **52**, 136 (1964).

14. G. E. Stillman, L. W. Cook, G. E. Bulman, N. Tabatabaie, R. Chin, and P. D. Dapkins, Long wavelength (1.3 to 1.6 μm) detectors for fiber-optical communications. *IEEE Trans. Electron Devices* **ED-29**, 1355 (1982).

15. H. Ando, H. Kaube, T. Kimura, T. Yamako, and T. Kaneda, Characteristics of germanium avalanche photodiodes in the wavelength region of 1-1.6 μm. *IEEE J. Quantum Electron.* **QE-14**, 804–809 (1978).

16. S. R. Forrest, Sensitivity of avalanche photodetector receivers for high-bit-rate long-wavelength optical communication systems. In *Semiconductor Semimetals* 329–385 (W.-T. Tsang, ed.), Vol. 22, Part D, Chapter 4. Academic Press, Orlando, Florida, 1985.

17. J. C. Campbell, A. G. Dentai, W. S. Holder and B. L. Kasper, High performance avalanche photodiode with separate absorption grading and multiplication regions. *Electron Lett.*, **19**, 818–820 (1983).

18. S. R. Forrest, Optical detectors: These contenders. *Spectrum* **23**(5), 76–84 (1986).

19. R. F. Leheny, R. Nahory, M. Pollack, A. Ballman, E. Beeke, J. DeWinter, and R. Marin, Integrated InGaAs p-i-n FET photoreceiver. *Electron. Lett.* **16**, 353–355 (1980).

20. Y. Ota and R. C. Miller, 12-channel PIN and LED arrays and their package for 1.3 μm applications. *Proc. SPIE—Int. Soc. Opt. Eng.* **839**, 143–147 (1987).

21. P. S. Henry, Lightwave primer. *IEEE J. Quantum Electron.* **QE-21**(12), (1985).

22. T. Okohsi, Heterodyne and coherent optical fiber communication recent progress. *IEEE Trans. Microwave Theory Tech.* **MTT-30**(8), 1138 (1982).

23. S. Machida and Y. Yomamoto, Quantum limited operation of balanced mixer homodyne and heterodyne receivers. *IEEE J. Quantum Electron.* **QE-22**(5), 617–624 (1986).

24. R. A. Linke and P. S. Henry, Coherent optical detection. *IEEE Spectrum* **24**, 52–57 (1987).

PART THREE

Optical Fiber Communication Systems

In Parts One and Two, Chapters 2–7 inclusive, we described the components of a fiber communications system: the optical fiber, the light source, the detector, and the auxiliary components. We are now ready to assemble these components to complete a system. This part contains two chapters (Chapters 8 and 9). In Chapter 8 we discuss the methods used to modulate the light source of a transmitter and to demodulate it in a receiver. Design considerations of an optical fiber system, including the choices of wavelength, multimode or single-mode fibers, design parameters, system design, and optimization, are also discussed. Chapter 9 presents the data of some systems currently in use around the world.

8

Optical Fiber Transmitter and Receiver

Introduction

The properties of some essential components of an optical fiber communication system, such as the fiber, the light source, and the detector, have been treated in previous chapters. Now we must complete the system by introducing the information onto the optical frequency carrier and reclaiming it at the receiving end. This includes the modulation and detection processes. Also included in this chapter are considerations of parameters important to a system design. Repeaters, multiplexing schemes, and coherent receptions are also discussed in the conclusion of this chapter.

The Choices of Optical Fibers and Operating Wavelengths

The design of a system usually starts with a list of specific desired requirements and constraints. Inputs for the design of a communication link are (1) the link length and the number of terminals, (2) the type of signal or data to be transmitted, and (3) the speed of transmission or the bandwidth demanded. Other constraints may include the fidelity of the system needed to satisfy a certain signal-to-noise ratio or bit error rate, and the fact that the system must be cost-effective compared to other links.

Long Wavelength versus Short Wavelength

The optical fiber loss characteristics discussed in Chapter 3 clearly show that optical fibers operate best at long wavelengths. This is because Rayleigh scattering loss decreases as λ^{-4}, and the chromatic dispersion shows the possibility of zero dispersion in the 1.3–1.6-μm wavelength range for silica fibers. For wavelengths beyond 1.7 μm, infrared (IR) absorption loss for silica fiber increases rapidly with wavelength, thus limiting its longest useful wavelength to below 1.7 μm. Within the wavelengths between 0.8 μm and 1.65 μm, for a silica fiber, there exist three windows that favor application to communication systems. The first window is from 0.8 to 0.9 μm; the so-called first-generation fiber operates within this wavelength range. The second and third windows are at 1.3 and 1.55 μm, respectively. Present technology shows that the losses of silica fiber at 1.3 and 1.55 μm are only 18% and 9%, respectively, that at 0.85 μm. It might, therefore, seem logical to choose the longest wavelength outright. However, the problem of choosing which window to operate rests with the light source and the detector. For short-wavelength operation in the 0.8–0.9-μm range the GaAs LED is a proven successful light source with good spectrum property, long life, and low cost. Silicon photodetector and Si avalanche photodetectors (APDs) provide adequate detection for the receivers. However, these devices become useless at long wavelengths. This is because at long wavelengths, we have to use semiconductor materials with narrow bandgap energies such as Ge and InGaAsP complex compounds for both the light source and the detector. But these materials have their share of trouble. Basically, materials with lower bandgap energies contribute to larger leakage current in the devices. They are also difficult to purify. This fact and the defect recombination process due to reduced material purity lower the quantum efficiency and yield lower launch power in a light source and exhibit more noise and larger dark current in a detector. Silicon is useless for wavelengths beyond 1.0 μm because of its low absorption coefficient. Even Ge and InGaAsP, which are used for detection at longer wavelengths, are less sensitive and noisy, and InGaAsP used for making lasers requires careful design to stabilize wavelength fluctuations. This device is also sensitive to temperature variations. Although stabilized InGaAsP/InP lasers are now available as light sources, the device is still in an experimental stage and is very expensive. The life span of this laser is very short. Thus, the price one has to pay for taking advantage of the low fiber loss and the zero dispersion at long wavelengths is, indeed, very high at the present time. However, this situation may change as the technology matures.

Single-Mode Fiber (SMF) versus Multimode Fiber (MMF)

Single-mode fibers have many advantages over multimode fibers:

1. Lower fiber attenuation due to lower refractive-index differences (0.3–0.5% vs. 0.8–2.0% for MMF). Thus, these fibers require less doping and thereby have lower Rayleigh scattering loss and provide larger resistance to nuclear radiation and to the effect of gaseous hydrogen.
2. Lower joint and coupling losses (0.1 dB for splices and 0.5 dB for a connector loss), which amounts to about half that for the MMF.
3. Simpler parameter-dependence because there are fewer to consider. For example, with MMF, the joint loss depends on the refractive-index differences and the index g of the graded-index profile. In SMF, simply knowing the near-field mode profile is sufficient. Also, SMF is more tolerant to small distortions during joint fusing. Couplers, such as taps, stars, and wavelength multiplexers, have lower losses because the power-transfer mechanisms are wave-like. In MMF, they are ray-like. They can also be more lossy.
4. Single-mode fibers have lower dispersion and therefore higher bandwidth. In MMF, the bandwidth seems to reach a saturation around 2GHz · km.
5. Single-mode fibers can have many applications other than for transmission. For instance, the controlled polarization effect in SMF can be used for sensors; SMF can easily be integrated with other devices to function in integrated optics and many other applications.

On the other hand, SMF has its concerns: (1) SMF has a much smaller fiber diameter (9–10 μm at 1.3-μm wavelength), making the joining of fibers more difficult; (2) less power can usually be coupled into a SMF; and (3) for wavelengths greater than 1.5 μm, it can lead to bend and microbend sensitivity, thus affecting cabling and environmental performance. Besides, although a SMF supports a single mode, it possesses two polarizations. The modal noise becomes large when both polarizations are present. This limits the distance between terminals with no repeaters.

One of the stingiest demands for operating a SMF is the requirement of a more sophisticated light source. It requires a single-frequency, single-polarization, and stabilized source in order to achieve all the benefits accredited to it. An MMF, on the other hand, can be powered by a much cheaper and more reliable LED source. Therefore, the choice between SMF and MMF depends on the application. Generally, if only short distance and moderate bit rate are involved, such as an interoffice communication system or data transmission system, an MMF system can be installed easily and cheaply. On the other hand, if extremely long distance is involved, as in an undersea cable where as few repeaters as possible will be tolerated, SMF becomes the only suitable choice.

Modulating a Transmitter

The functional block diagram showing a typical optical fiber telecommunication system was given in Fig. 1.1 in Chapter 1. In this section, we discuss the processor block in more detail. Processors may include modulators, multiplexers, and switching devices. But our discussion in this section will be limited to the modulation of the light source.

Modulation is a process that enables information to be impressed onto a light source, called the *carrier for efficient transmission*. In communication systems where the carrier is a single-frequency source, there are three ways to perform modulation: by amplitude, frequency, and phase modulation schemes [1]. If the carrier has a voltage wave in the form of

$$e(t) = A_c \sin (\omega_c t + \phi_c) \tag{8-1}$$

where A_c is the amplitude, ω_c the angular frequency, and ϕ_c the phase angle of the carrier, then any variation of either one of these parameters represents a modulation process. For example, if the amplitude A_c is varied in accordance with the modulation frequency ω_m, while keeping the other parameters constant, then

$$e(t) = A_c (1 + m_a \cos \omega_m t) \sin(\omega_c t + \phi_c)$$

$$= A_c \sin \omega_c t + \frac{m_a A_c}{2} \cos(\omega_c - \omega_m)t \tag{8-2}$$

$$- \frac{m_a A_c}{2} \cos(\omega_c + \omega_m)t$$

where we have omitted ϕ_c for simplicity; m_a is the modulation index defined as the ratio A_m/A_c, which is the amplitude ratio of the modulating signal to the carrier amplitudes. This is the amplitude-modulation process. The result of the modulation, the so-called modulation product, consists of the original carrier plus upper and lower side bands whose amplitudes are both equal $(m_a A_c)/2$ and whose frequencies are $\omega_c - \omega_m$ and $\omega_c + \omega_m$, respectively.

On the other hand, if the amplitude and the phase are kept constant, and the frequency is varied according to the modulating frequency, then we have, for frequency modulation:

$$e(t) = A_c \sin (\omega_c + m_f \sin \omega_m t)$$

$$= A_c \left[J_0(m_f) \sin \omega_c t \right]$$

$$+ J_1(m_f) \left[\sin(\omega_c + \omega_m)t - \sin(\omega_c - \omega_m)t \right] \qquad (8\text{-}3)$$

$$+ J_2(m_f) \left[\sin(\omega_c + 2\omega_m)t - \sin(\omega_c - 2\omega_m)t \right]$$

$$+ J_3(m_f) \left[\sin(\omega_c + 3\omega_m)t - \sin(\omega_c - 3\omega_m)t \right] + \ldots$$

where m_f, the modulation index, is now defined as $(\Delta\omega/\omega_m)$, and J_0, $J_1, \ldots$ are Bessel functions of order n, n = 0, 1, 2, . . . , n. The frequency spectrum acquired by the frequency-modulation scheme is infinitely larger unless m_f is small.

Similar expressions can be derived for a phase-modulation system as

$$e(t) = A_c \sin(\omega_c + m_p \sin \omega_m t) \qquad (8\text{-}4)$$

In Eq. (8-4), we have neglected θ_c for simplicity. Here only the modulation index $m_p[= (\Delta\theta/\theta_m)]$ is defined differently from the frequency-modulation case, and thus the frequency spectrum of a phase-modulated carrier is similar to that of a frequency-modulated case.

Not all these modulation schemes are applicable to optical communication systems. This is because the optical carriers are generated by LEDs or LDs, neither of which are coherent emitters (i.e., not single-frequency generators). The spectrum width of an LED is on the order of several hundred angstroms around a center wavelength. That of an LD is at least a few angstroms wide. Therefore, they do not qualify as a single-frequency source. For example, at 0.85-μm wavelength, the center frequency is 3.5×10^{14} Hz. A hundred-angstrom spectral width of an LED corresponds to a frequency spread of $\pm 10^{13}$ Hz, which is larger than the frequency deviation of an FM system. This means that direct frequency or phase modulation is inappropriate with this source. Even with LDs, the frequency spread can be misinterpreted as signal variations. Therefore, for optical communication with today's technology (except for the coherent system), amplitude or intensity modulation is the only sensible way to transfer information to carriers.

This statement, however, does not rule out the possibility of using frequency and phase modulation in optical fiber communication systems in an indirect way to combat noise and distortion. For example, the signal can first be frequency- or phase-modulated onto a subcarrier. The subcarrier with the information can then intensity-modulate the optical carrier. This method can improve the noise performance of an analog signal and avoid cross-modulation in frequency-division multiplex (FDM) schemes.

Recently, the stability of lasers has been improved such that a very narrow spectral width light source has become available. Coherent modulation becomes increasingly important in modern systems.

Direct Modulation

The method and approach to modulate an optical source depends highly on the type of light source. Amplitude modulation, also known as *intensity modulation,* can be applied directly to light sources such as LEDs and lasers. For some types of lasers, such as the Nd:YAG laser, only external modulation schemes can be used. In a direct modulation process, the signal is applied to the bias voltage or current that drives the light source (LEDs or LDs). If the light output is nearly directly proportional to the driving current (or voltage), a linear operation is possible.

Typical light output driving current characteristics of an LED and laser are shown in Fig. 8.1 [2, 3]. For LEDs, the characteristic is linear at low driving currents. In lasers, a linear portion of the characteristic appears after a threshold current I_{th} is reached. Both devices can be modulated directly for analog or digital signals. For lasers, a prebias (previously set bias) may be required to ensure linear operation. For a pulse-code-modulation (PCM) system, the modulation may simply involve switching between zero and some preset maximum value. A simple driving circuit of a pulsed LED operation is shown in Fig. 8.2*a* [4]. Here a voltage pulse is connected directly to the LED to supply the energy required for spontaneous emission that emits light. This scheme requires a large current and may cause problems of load fluctuation in the power supply and trouble in circuit design.

To ease the loading effect when the signal pulses are switched on and off in the previous circuit, the LED is prebiased at a constant voltage and uses a transistor gate logic to turn the injection current on and off. This circuit is shown in Fig. 8.2*b*. To achieve high modulation

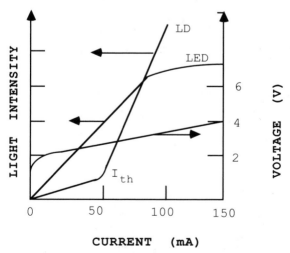

Figure 8.1 Light output as a function of the driving current for LED and laser.

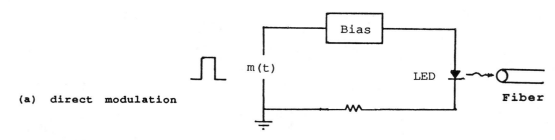

(a) direct modulation

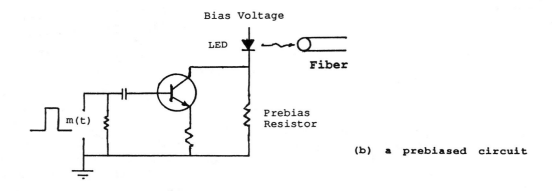

(b) a prebiased circuit

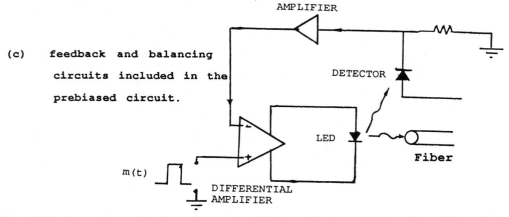

(c) feedback and balancing circuits included in the prebiased circuit.

Figure 8.2 Intensity-modulation schemes for LED: (*a*) direct modulation; (*b*) a prebiased circuit; (*c*) feedback and balancing circuits included in the prebiased circuit.

rates, one must use a very low impedance driving circuit to compensate for the LED capacitance. A prebias resistor is used to provide a dc bias for the LED. The primary limiting factor on the spread of response is the spontaneous recombination time.

For an analog modulation scheme, high linearity in the *L–I* characteristic is a necessity. Various feedback and equalization techniques must be employed. One such circuit is shown in Fig. 8.2*c*. Here part

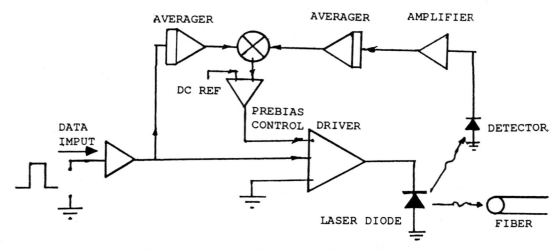

Figure 8.3 Prebias and compensated stabilization circuits for laser modulation.

of the LED output light is detected, amplified, and compared with the original output. The resultant output is used to regulate the LED injection current. Since it is important that the signals compared are of the same mode, the system is more suitable for single-mode lasers. To modulate a laser, the biasing circuit can be the same as in the LED except that the laser needs to prebias beyond its threshold current. A prebias circuit is advantageous since the switching current can be made much smaller than the bias current, thus avoiding the time delay necessary for building the current density up from zero. To protect a voltage surge from damaging the laser when the modulation signal is momentarily removed, another control is introduced to the modulation level so as to reduce the laser output current when the signal goes to zero. This circuit is shown in Fig. 8.3 [5]. One disadvantage of the circuit is that critical balancing and calibration of various effects are necessary in order to compensate for the difference in modulation signal levels and in local photodetector output waveforms. The bandwidth is also limited by the relaxation oscillations arising from external resonance effects usually present in the circuit.

External Modulation

If a light source cannot accept a direct intensity modulation, then external means of modulation must be provided. This involves more equipment and circuitry and is therefore more costly. But external modulation has many advantages over direct modulation [5], including (1) relieving the burden imposed on the power supply of the light source, thus improving its stability; (2) providing flexibility of introducing the signal in any form through suitable transducers; and (3) providing the possibility of integrating light sources and transducers to improve performance and reduce cost.

The performance characteristic of external modulation schemes is rated by the figure of merits, usually defined as the power required in milliwatts per megahertz $(P/\Delta f)$ to achieve a given degree of modulation at a wavelength λ. Other factors include optical insertion loss, bandwidth, extinction ratio, and temperature sensitivity. The physical compatibility of the modulator to the optical fiber system as well as the cost are additional important considerations.

External modulators can be of either reactive or absorptive type. They can be applied either in bulk form in which the optical beam is freely propagating, or in waveguide where the optical beam is confined in a guiding structure as in thin film stripes. By far, the reactive-type modulator in bulk form has received the most attention since many materials exist that can be used in this type of modulator. However, reactive-type modulation in thin-film form has generated increasing interest recently because it requires less power to modulate and it is compatible with the advent of integrated optics. The absorptive type of modulator has not yet been well developed.

Reactive-type modulators are based on the field-induced interactions between the optical field and the signal field through electrooptic, acoustooptic, and magnetooptic effects. The electrooptic effect has been the most widely exploited for high-speed light modulators. The two other field-induced effects, the acoustooptic and the magnetooptic effects, have received only limited attention because of their poor frequency response and difficulty in material selection.

Electrooptic Modulators

In many electrooptic materials, for example, potassium dihydrogen phosphate (KDP), lithium tantalate ($LiTaO_3$), and lithium niobate ($LiNbO_3$), an applied electric field can induce variation of the electrical polarizability, which, in turn, produces a change in the refractive index of the material. The light beam passing through this material can then be polarization-modulated if the crystal is properly orientated with the light beam. Many electrooptic modulators have been proposed, built, and evaluated, and the work in this area is reviewed by Kaminow and Turner [6].

The Electrooptic Effect
If an applied electric field on an optical crystal results in a change in its refractive index, the effect is generally referred to as the *electrooptic effect*. Kaminow [7] derived an expression to describe this effect as

$$\Delta\left(\frac{1}{n^2}\right) = \alpha_{ij}E + \beta_{ij}E^2 \tag{8-5}$$

where α_{ij} and β_{ij} are coefficients of the linear and quadratic field effect, respectively; i and j are reduced tensor indices that change with the crystal orientation in a complex manner. (In fact, α and β are coefficients in tensors.)

The linear field-effect term, $\alpha_{ij}E$, is called *Pockel's effect*. The quadratic field-effect term, $\beta_{ij}E^2$, is the *Kerr effect*. Both can be used to modulate optical transmissions. For fiber optic transmission, the linear Pockel effect is more useful. To this effect, we may choose an optical crystal whose $\beta_{ij} = 0$ so that only the linear effect exists.

To simplify the analysis, let us choose a crystal, such as KDP (potassium dihydrogen phosphate, KH_2PO_4), and orientate it in a direction so that the electric field is in the z-direction perpendicular to the crystal plane, x–y. In this orientation, all coefficients are reduced to zero except α. Using Eq. (8-5), we can write

$$n - n_0 = \frac{1}{2} \alpha n_0^3 E_z \qquad (8\text{-}6)$$

where n_0 is the refractive index before the electric field was applied. The components of n in the x- and y-directions are

$$n_x = n_0 + \frac{1}{2} \alpha n_0^3 E_z \qquad (8\text{-}7a)$$

$$n_y = n_0 - \frac{1}{2} \alpha n_0^3 E_z \qquad (8\text{-}7b)$$

Equations (8-7a) and (8-7b) suggest that since the refractive indices in the x- and y-directions are different, the velocity of propagation of a wave polarized along the x-axis differs from that of a wave polarized along the y-axis. After the waves travel through a length L of the crystal, there will be a phase difference between these two components. The phase shifts are, respectively

$$\phi_x = \frac{2\pi}{\lambda} n_x L = \frac{2\pi}{\lambda} L n_0 \left(1 + \frac{1}{2} \alpha n_0^3 E_z \right) = \phi_0 + \Delta\phi \qquad (8\text{-}8a)$$

and

$$\phi_y = \frac{2\pi}{\lambda} n_y L = \frac{2\pi}{\lambda} L n_0 \left(1 - \frac{1}{2} \alpha n_0^3 E_z \right) = \phi_0 - \Delta\phi \qquad (8\text{-}8b)$$

where

$$\Delta\phi = \frac{\pi}{\lambda} L \alpha n_0^3 E_z = \frac{\pi}{\lambda} \alpha n_0^3 V \qquad (8\text{-}9)$$

Since $E_z = V/L$, where V is the applied voltage, the emergent light-wave is elliptically polarized, and the degree of ellipticity changes with the applied voltage V.

When a KDP crystal is used as a modulator to a CW optical source, it is placed between a pair of crossed polarizers with its z-axis (L-

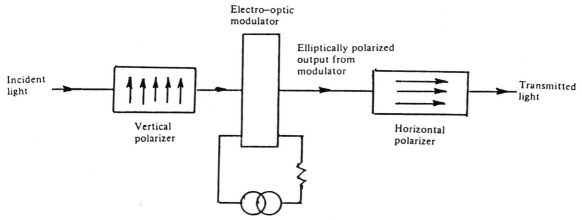

Figure 8.4 A KDP external modulator. [After F. S. Chen [5] © IEEE, 1970.]

dimension) along the direction of the lightwave propagation as shown in Fig. 8.4. A modulator source voltage is applied along the L-dimension in the z-direction. Notice that in this arrangement, the electrodes for implementing the electric field intercept the lightwave. Proper provision must be made to clear the path. This can be achieved by leaving an open hole in the electrode.

An unpolarized incident CW light is led to pass through a vertical polarizer and impinges on the crystal x–y-plane. The output wave can be resolved into two orthogonal components represented by $E_1 = E_0$ and $E_2 = E_0 e^{-i\phi}$, where ϕ is the phase difference indicated in Eq. (8-9) and $\phi = 2\Delta\phi$. The resultant of these two complex fields is

$$E = \frac{E_0}{\sqrt{2}} (e^{-i\phi} - 1) \qquad (8\text{-}10)$$

and the corresponding output intensity is

$$I_m \alpha EE^* = 2E_0^2 \sin^2 \frac{\phi}{2} = I_0 \sin^2 \frac{\phi}{2} \qquad (8\text{-}11)$$

where E^* indicates the complex conjugate of E. The fractional intensity transmitted through the modulator is

$$\frac{I_m}{I_0} = \sin^2 \frac{\phi}{2} = \sin^2 \left(\frac{\pi}{\lambda} \alpha n_0^3 V \right) = \sin^2 \left(\frac{\pi}{2} \frac{V}{V_m} \right) \qquad (8\text{-}12)$$

where $V_m = \lambda/(2\alpha n_0^3)$ is the voltage required for maximum transmission, or $I_0 = I_m$; V_m is often called the half-wave voltage, which is the voltage required to generate a phase of 180°. Thus, a beam of plane polarized light incident on the modulator would have its plane of polarization rotated by 90° when a voltage $V = V_m$ is applied to the modulator.

Equation (8-12) suggests that the modulation will not be linear at low modulating voltages. Therefore, it is advisable to bias the modu-

Table 8.1 Characteristics of Some Electrooptic Materials

| Material | λ_{min} | Linear Electro-optic Coefficient α^a (pm/V) | | Retractive Index | | |
		Ordinary	Extra-ordinary	n_0	n_e	ε_{12}
KDP (KH_2PO_4)	0.5	10.6		1.51	1.47	42
KD*P (KD_2PO_4)	0.5	26.4		1.51	1.47	50
$LiNbO_3$	0.5	30.8		2.29	2.20	18
$LiTaO_3$	0.5	30.3		2.175	2.180	43
GaAs	0.8	1.6		3.5	—	11.5

[a]Most materials have more than one linear electrooptic coefficient. We have quoted only the one used in Pockel's effect [8].

lator with a quarter-wave plate at the output. The $\lambda/4$ plate introduces a fixed phase shift of $\pi/2$, thus translating the operating point to the middle of the I–V curve (intensity vs. voltage) where the slope is steepest. The linearity of the operation can be improved. The power needed for the crystal to reach a peak phase retardation ϕ_m $[= (2\pi/\lambda)\alpha n_0^3 V_p$, where V_p is the peak modulating voltage $V_p = (E_z)_m L]$ can be found as

$$P = \frac{V_p^2}{2R_L} = \frac{\phi_m^2 \lambda^2 A \varepsilon \, \Delta f}{4\pi \alpha^2 n_0^6 L} \qquad (8\text{-}13)$$

where A is the area, $A\varepsilon/L$ is the capacitance C of the modulator, and $\Delta\phi = 1/(2\pi R_C)$. The maximum modulation bandwidth must be less than Δf for the modulated signal to be a faithful representation of the applied modulating voltage.

Materials useful for electrooptic effect are listed in Table 8.1 for reference.

The Acoustooptic Effect

The acoustooptic effect accounts for the change in the refractive index of a medium caused by the mechanical strains accompanying the passage of an acoustic wave (strain wave) through the medium. The refractive-index changes are caused by the photoelastic effect, which occurs in all materials on the application of a mechanical stress. Pinnow [8] showed that the change in refractive index is proportional to the square root of the total acoustic power. Consider the case of a monochromatic lightwave of wavelength λ incident on a medium in which an acoustic wave of wavelength Λ has produced sinusoidal variations in the refractive index. The interaction of the optic wavefront and the acoustic wavefront sets up a diffraction grating such that optical energy is diffracted out of the incident beam into various orders. One of these is the Bragg regime. This picture is illustrated in

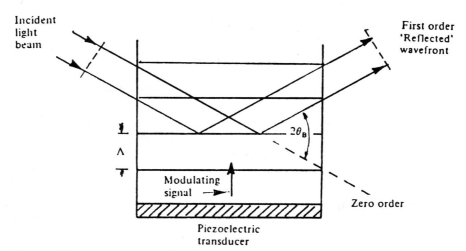

Figure 8.5 Acoustooptic effect by Bragg diffraction.

Fig. 8.5. The incident beam, on entering the acoustic field, is extensively rediffracted before leaving the acoustic field. The acoustic field acts like a "thick" diffraction grating. The situation is very much like that of Bragg diffraction of X-rays from planes of atoms in a crystal. Significant amounts of light will emerge only in those directions in which constructive interference occurs. This occurs when the Bragg diffraction condition is met. That is, $\sin \theta_i = \sin \theta_d = (m\lambda/2\Lambda)$, $m = 0, 1, 2$.

If only the first order is used, then $m = 1$, the equation for the Bragg angle q_B becomes $\sin \theta_B = \lambda/2\Lambda$. At the Bragg angle, the modulation depth $(I_0 - I)/I_0$ is given by $\sin^2 \phi/2$, where $\phi = 2\pi/\lambda(\Delta n \, L/\cos \theta_B)$, Δn is the amplitude of the refractive-index fluctuation, and L is the length of the modulator [9]. The movement of the acoustic waves through the medium gives the effect of a moving diffraction grating. As a consequence, the frequency of the mth-order diffracted beams are Doppler-shifted by $\pm mf_m$, where f_m is the frequency of the sound wave. The change in frequency can be used as the basis of a frequency modulator. In Bragg modulation, as only the first order is important, the optical frequency is shifted by $\pm f_m$ depending on the relative motion of the two sets of waves. Thus a frequency-modulated (FM) light beam can be obtained. The bandwidth of the modulator is limited by the transit time of the acoustic wave across the optical beam. A bandwidth of only 50 MHz with an LiNbO$_3$ crystal can be achieved. But the operating voltage is low in comparison with that required for an electrooptic modulator (a few thousand volts).

Magnetooptic Effect

The simplest magnetooptic effect of interest for optical modulation is the Faraday effect. When a beam of plane-polarized light passes through a substance subjected to a magnetic field, its plane of polarization is observed to rotate by an amount proportional to the magnetic-field component parallel to the direction of propagation. The rotation of polarization is given by $\theta = VBL$, where V is the Verdet constant, B is the magnetic flux parallel to the direction of propagation, and L is the path length in the material. The Faraday effect is small and wavelength-dependent. Its application to modulators is limited. However, it can be used to build an optical isolator and magnetic memories [10].

Modulation Format

Many methods are available for modulation of the optical carrier by intensity variations. Analog and digital signaling are the two most widely used schemes.

In analog modulation the original signal waveform is preserved in the modulated carrier, and the receiver must reproduce the signal as close as possible in waveform, frequency, and phase to the original input. The fidelity of the transmitted signal is dictated by the signal-to-noise ratio S/N of the system. The transmitter requires a high degree of spectral purity, and the receiver requires a high degree of linearity. Noise is usually a limiting factor in good reception.

In digital coding, a simple binary coding in which only "0" or "1," representing the "off" and "on" of a transmitter, respectively, are transmitted. The receiving end is only required to determine whether a signal is above or below a certain threshold. The fidelity is dictated by the nature of its error occurrence or the bit error rate BER. System noise does not create serious problems. But the original signal needs to be converted into a digital signal for transmission and then decoded at the receiving end. The timing for the coding and decoding is synchronized by a clocked digital arrangement.

Analog Modulation

Analog modulation needs no special coding. The intensity of the light source is varied in accordance with the amplitude of the modulating signal. The signal waveform resides in the side bands of the carrier as indicated in Eq. (8-2). At the receiving end, if direct detection is used, only the amplitude variation is reclaimed. To take the advantage of the extremely wide bandwidth available in an optical fiber system, frequency or phase modulation can be applied to a subcarrier before intensity modulation of the optical carrier is carried out. In this case, heterodyne detection is used to recover the signal. The output of a heterodyne receiver becomes

$$e(t) \propto A_c A_0 \cos[(\omega_c - \omega_0)t + (\phi_c - \phi_0)] \qquad (8\text{-}14)$$

where A_0 is the amplitude, ω_0 the frequency, and ϕ_0 the phase of the local oscillator. All these parameters can be used to carry information.

Heterodyne detection has other advantages. If the signal source is shot-noise-limited, it can be shown that the signal-to-noise ratio in heterodyne detection is a factor 2 greater than that in direct detection. However, the improvement can easily be offset by the necessity of stabilizing both the local oscillator and light source frequencies.

Digital Coding

In digital coding, the information is coded into a series of pulses. In binary coding, there exists only two states, the "1" and "0" states. If the pulse is present, it is in a "1" state. Absence of a pulse is in a "0" state. Each state is allowed to occupy a short time duration Δt on a clocked time scale. Since the pulse amplitude is unvarying, a system of coding the information onto a pulse train must be adopted for transmitting and receiving the message. There are many ways to code the pulse train. If the pulse width is varying according to the information, it is called *pulse-width modulation* (PWM). If the pulse position is varying, it is called *pulse-position modulation* (PPM). If the pulse amplitude, position, and width are all unchanged, as with pulse-code modulation (PCM), a special coding scheme must be worked out. The scheme that is suitable for optical fiber applications is the PCM system.

Any prearranged coding system can be used to code the message. For example, let's assume that a clocked pulse train is continuously sending pulses. A word length may consist of any number of time slots. Within this time interval a combination of "1"s and "0"s is sufficient to construct the message one desires to send. For example, if the word length is limited to three time slots (or bits), numerical numbers from 0 to 7 can be sent by this pulse train, such as 111 for 7, 101 for 5, and so on (as $2^3 = 8$). This is a binary code. A long pulse train containing the binary number 101111 can be interpreted as the message of the number 57 by reading 3 digits at a time. It is important that the sending and receiving ends use the same synchronized time scale; otherwise, error results. With 16- or 32-bit word length, many complicated messages can be coded into the scheme.

In binary digital signaling, a string of discrete-amplitude rectangular pulses might resemble that shown in Fig. 8.6a. It represents the message 10110100 (an 8-bit message) as it appears at the output of a digital computer. This waveform, consisting of a simple on–off sequence, is said to be unipolar because it has only one polarity and is synchronous. All pulses have equal duration, and there is no separation between them. It is also called the no-return-to-zero (NRZ) type of pulse, since each pulse completely fills up the allocated time slot. This waveform indicates that a nonzero dc component exists. Now, dc components are difficult to transmit and also represent a waste of

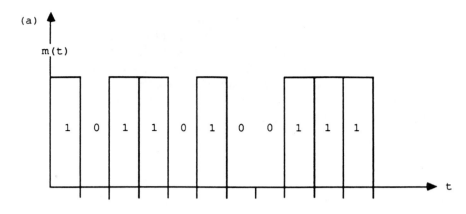

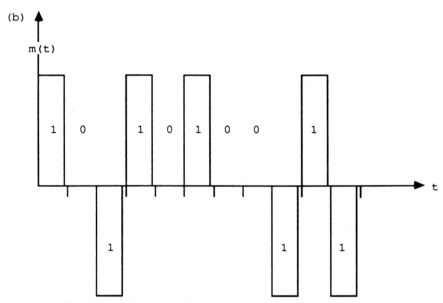

Figure 8.6 Binary digital code waveform: (a) unipolar non-return-to-zero pulses; (b) polar return-to-zero pulses. [After D. A. Pinnow [8] © IEEE, 1970.]

power. Moreover, synchronous signals require time coordination at transmitter and receiver ends. Other schemes intended to eliminate these disadvantages have been tested. One such scheme is the bipolar (two polarity) return-to-zero code shown in Fig. 8.6b. Here, "1" and "0" are represented by alternating positive and negative pulses. If the number of positive and negative pulses are equal, the dc components average to zero. Otherwise, the dc component is still greatly reduced. Notice also that since each pulse does not fill up the allotted time slot, we call these the "return-to-zero" (RZ) signals. The spaces between pulses make the signal self-clocking so that close time coordination of the transmitter and receiver is not so important. But valuable transmission time has been wasted. Other schemes are also possible.

If the system is linear and distortionless, and the output signal undergoes no degradation in transmission, then an arbitrarily large signal rate can be achieved by using very short pulses. But a real fiber system has finite bandwidth and distortion, causing the pulse to spread and even overlap. This is known as *intersymbol interference*. It limits the maximum signal rate. If we let n be the signal rate, then $\tau = 1/n$ is the pulse duration. Nyquist [11] estimated that the signal rate cannot be larger than twice the bandwidth to avoid intersymbol interferences. Thus, $n \le 2B$ agrees with the pulse resolution rule $\tau_{\min} \ge \frac{1}{2}B$.

Digital Coding of Analog Signals

In analog modulation, the modulated parameter varies continuously and acquires any value corresponding to the range of the message. When the modulated wave is adulterated by noise, there is no way for the receiver to discern the exact transmitted value. Since continuous modulation gives a redundant message, if we allow only a few discrete values to modulate the carrier and if the separation between these discrete values is large compared to the noise perturbation time, the effect of random noise can be virtually eliminated at the receiver end. This is the basic idea of pulse-code modulation.

Analog messages can be transformed into digital form by sampling, quantizing, and coding by an analog-to-digital (A/D) converter. Sampling is done via the switching operation. During a period of the input waveform, a switch periodically switches between "on" and "off" portions at a rate of $f_s = 1/T_s$. The output $X_s(t)$ then consists of short segments of the input $X(t)$, with the amplitude of each segment representing the actual signal at that instant. This operation is called *unipolar chopping*. The output $X_s(t)$ is designated as the sampled wave, and f_s is the sampling frequency.

How often do we have to sample in order to sufficiently represent the original signal? It has been shown [1] that if the signal is band-limited, the minimum sampling frequency should be $f_s \ge 2B$, and if $X_s(t)$ is filtered by an ideal L–P filter, the output signal will be proportional to $X(t)$, the input signal.

Next, the amplitude information is coded into the binary codes for transmission. This operation is called *encoding*. First, the amplitude of the original signal is divided into a number of quantum levels Q, for each code word must uniquely represent one of the possible quantized samples. Let ν be the number of digits in the code word, each having one of μ discrete states. Since there are μ^ν different possible code words, we require $\mu^\nu \ge Q$, or

$$\nu = \log_\mu Q \qquad (8\text{-}15)$$

In binary PCM, for which $\mu = 2$, then $Q = 2^{\nu}$. Thus, the higher Q is (i.e., the more quantum levels we assume), the more digits per code word need to be used. Figure 8.7 illustrates these operations in a binary PCM. If $Q = 8$, it corresponds to eight distinct levels as $-\frac{7}{8}$, $-\frac{5}{8}$, $-\frac{3}{8}$, $-\frac{1}{8}$, $+\frac{1}{8}$, $+\frac{3}{8}$, $+\frac{5}{8}$, and $+\frac{7}{8}$.

From Eq. (8-15), since $\nu = \log_2 8 = \log_2 2^3 = 3$, each code word must contain three digits in each unit time. The sampling points are indicated by the dots at an interval of four unit times: at $t = 1_{\text{ms}}$, $Q = -\frac{5}{8}$, the amplitude code number is 1, and the code is 001. At $t = 2$, $Q = -\frac{1}{8}$, the amplitude code number is 3, and the code is 011. At $t = 3$, $Q = \frac{5}{8}$, the amplitude code number is 6, and the code is 110, and so forth. The binary code is thus $\underline{001}\ \underline{011}\ \underline{110}\ \underline{110}\ \underline{000}$, and the $f_s = \frac{1}{4} \cdot 10^{-3} = 250$ Hz. The PCM waveform of the pulses is shown on the lower portion of the figure as RZ pulses as each pulse width is less than the allotted time.

The information of the waveform that is contained in the coded number of 15 bits is transmitted sequentially through the fiber. At the receiving end, the code bits are read 3 bits at a time and the amplitudes are interpreted accordingly.

We choose to show the resultant pulse code in "polar return-to-zero" form because other forms such as the unipolar signal contain a nonzero dc component that carries no information and is a waste of

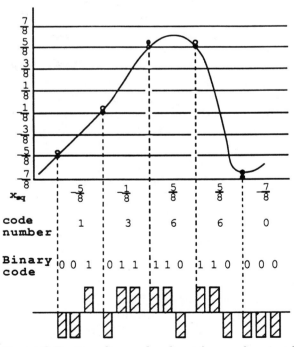

Figure 8.7   Sampling and coding of an analog signal.

power. Besides, unipolar pulses usually cause the baseline of the waveform to drift and might be unsuitable for transmission.

The basic requirement for a PCM system is time synchronization of transmitted signals. This is usually done by providing time information in the transmitted signal. However, such time information can easily be lost. The "polar return-to-zero" system shown in Fig. 8.7 overcomes this difficulty by providing signals for self-clocking. Other coding schemes may use ternary line codes rather than binary ($\mu = 4$) by grouping the binary digits in blocks of two to reduce the clock synchronization problems. But the bandwidth requirement will increase likewise.

Optic Fiber Receiver

The function of a receiver is to recover the information carried on the lightwave installed by the modulation process to a transmitter. It is shown as the final block of the system diagram in Fig. 1.1. Although a detector is the most important element of a receiver, a practical receiver could include additional active and passive components to make it work better. A typical receiver for an optical fiber system is shown in Fig. 8.8. Here a PIN diode or an APD photodiode converts the incoming optical signal into an electrical signal and injects it into a preamplifier. A voltage gain of 1000 is often needed. Going into the second block of Fig. 8.8, which consists of an attenuator, another amplifier, and a feedback network, the signal is further amplified. A low-pass or a bandpass filter is also included, which can be adjusted to improve the linearity of the amplifier and to achieve a maximum signal-to-noise ratio of the system. The final block contains the demodulator or decoder where the original signal is reclaimed.

The performance of receiver in an optical fiber system depends not only on the properties of the photodetector that we discussed in Chapter 7 but also on the characteristics of the amplifier and other components that follow the photodetector. In the following sections, we discuss briefly the function of each block and its design considerations.

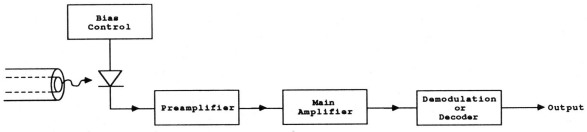

Figure 8.8 Block diagram of typical optical fiber receiver.

Preamplifier

The preamplifier ranks number one in importance in the design of an optical fiber receiver. Besides the required voltage gain it should provide, limiting the noise generation by the amplifier is the most important consideration. For a carefully chosen photodiode, such as a PIN, the receiver performance is often limited by the noise of the preamplifier.

There are three approaches to design a preamplifier: the resistance-loaded input amplifier, the high impedance (HZ) amplifier, and the transimpedance (TZ) amplifier.

The resistance-loaded amplifier is the simplest to implement. The signal current from a photodetector is fed through a standard resistor, usually a 50-Ω one to match the input impedance. The voltage generated across this resistor serves as the input to the amplifier. The noise performance of this amplifier is determined by the thermal noise generated through the 50-Ω resistor, which is relatively poor. But it has the advantage of providing wide bandwidth easily. In fact, it might be the only alternative that provides the gigabit-per-second bandwidth for high-bit-rate applications.

A high-impedance (HZ) preamplifier is perhaps the most sensitive amplifier for optical fiber applications. It is sensitive because a high-impedance input amplifier requires less signal current from the photodetector for high voltage gain. However, it is equally sensitive to other disturbances such as the undesirable electromagnetic interferences. Also the high input impedance together with the unavoidable parasitic capacitances, produce a large RC (resistance $\times$ capacitance) time constant that promotes intersymbol interference that requires compensation by equalizing networks. On the plus side, a high-impedance preamplifier can easily be integrated with a photodetector to make a front-end unit that can be handled nicely.

The transimpedance (TZ) preamplifier is the most popular one in current design. The input to the amplifier is a signal current instead of a voltage as in other amplifiers, and its output is a voltage. This is accomplished by a feedback loop extended from the output to the input of the amplifier through a feedback resistor R_f. The feedback effectively reduces the input resistance of the amplifier by its open-end loop gain, thus producing an effective current input. This arrangement eliminates the need for equalization or compensation for intersymbol interferences as in a HZ amplifier.

The TZ preamplifier is the most desirable design because it provides larger bandwidth, improved dynamic range, and simplicity in implementation [12].

In some designs, an integrated front-end unit consists of a pigtailed photodiode and an FET or MOSFET to further improve the receiver performance [13].

The Main Amplifier

The objective of this block of circuit is not to influence the signal-to-noise ratio of the amplifier, but to improve the overall performance of the receiver. It provides a wider bandwidth (the 3-dB BW) of 2 GHz; a wide dynamic range of, say, 40 dB; and an input reflection of less than -20 dB. This block may contain series and shunt feedback circuits, frequency compensation, attenuation unit, and AGC voltage to control the dynamic range. A lowpass or bandpass filter is used to achieve the maximum signal-to-noise ratio.

The Demodulator or Decoder

The final block of Fig. 8.8 consists of a demodulator or decoder. In an analog system, the signal may have been modulated by frequency-shift keying (FSK), amplitude modulation (AM), or frequency modulation (FM). The receiver must use appropriate demodulation schemes to recover the original signal. In a digital system, a decoder must be used.

The Bias Voltage and Power Supply

Bias voltage supply to the photodetector is very critical in limiting the noise in the preamplifier, particularly if an APD is used. In this case, the bias voltage must be tightly regulated against temperature and voltage fluctuations. A 70% fluctuation in voltage supply for an APD operating on a bias of 140 V could produce a gain variation of 10 to 1, or 2.5 V^{-1}. Temperature change effects the cutoff frequency of the diode and the receiver performance. For PIN photodiodes, the bias seldom presents a problem. For the rest of the amplifiers, decoupling circuits must be provided to limit the noise.

The Noise Problem

It is important to keep the noise of the front-end unit as low as possible. The sources of noise in the photodiodes were discussed in Chapter 7. Here, let us recall that for a well-designed PIN diode, quantum noise is a limiting factor. Preamplifiers add another noise factor that is mostly Johnson noise. Equation (7-16) is applicable. For an APD, Eq. (7-29) is also applicable. We notice that here the signal-to-noise ratio can deteriorate by increasing the multiplication factor M such that an excess noise factor F is introduced; M must be limited to $25-40$ in order to reduce noise within tolerable limit. Effort to reduce noise in APD has been successful by splitting the avalanching and collecting regions as in SAM–APD discussed in Chapter 7. For a photoconductive detector, the noise is again the thermal noise. Equation (7-36) applies.

Receiver for Coherent Detection

In a coherent optical fiber receiver, the signal carrying incoming lightwaves is heterodyned or homodyned just like radiowaves by a local lightwave generator to produce a microwave beat frequency for further processing. The processing includes amplifying, channel filtering, and detection. It has been established that heterodyning can increase the sensitivity of the receiver by one to two orders of magnitude. At the same time, the possibility of using sharp microwave filters to separate neighboring signals allows hundreds of thousands of channels of closely spaced wavelength to be carried on one fiber [14]. We shall return for more discussion in the section on coherent fiberoptic communication systems (below).

System Design

Design parameters for an optical fiber communication system can be grouped into three categories: the fiber, the light source, and the detector. Although they can be designed individually for the best performance in each category, the final choice is somewhat interdependent. For example, the attenuation of an optical fiber does not decide whether a SMF or MMF should be chosen; the choice is quite dependent on the types of light source to be used. If an LED is chosen as a carrier generator, the fiber could be an MMF. If a laser is chosen, either SMF or MMF can be used. These interrelationships among parameters often make the decision more difficult.

A Guide to the Initial Choice

It is usually a good practice to grasp the general picture of a system design before engaging in the actual design process. We prepare a list, as shown in Table 8.2, where various components such as the fiber, light source, and detector are listed for various distance and speed requirements of a given system. There are eight columns, and each is characterized by a required distance range and speed of the system. Under each column, a list of components usually associated with this particular requirement is suggested. The purpose of this list is only for the initial choice. The final decision will be made after detailed investigation and perhaps through some optimization processes.

Column I in Table 8.2 is dedicated to interoffice applications where

Table 8.2 A Guide to the Initial Choice of Optical Fiber Systems

	I	II	III	IV	V	VI	VII	VIII
	L < 1 km	L < 10 km		L < 100 km		L < 1000 km		L < 10 m
	B < 2 Mbit/s	B < 30 Mbit/s	B < 100 Mbit/s	B < 1 Gbit/s	B < 100 Gbit/s	B < 5 Gbit/s	B < 100 Gbit/s	B < 100 Gbit/s
Source	LED (GaAs)	LED (AlGaAs)	LD(mm) (AlGaAs)	LED, LD(mm) (InGaAsP)	LD(sm) (InGaAsP)	LD(sm) (InGaAsP)	LD(sm) (DSM or C)	LED (GaAs)
Wavelength (μm)	0.85	0.8–0.9	0.8–0.9	1.3	1.3	1.55–1.65	1.55–1.65	0.6–0.8
Fiber	MMF, SI (silica)	MMF, SI (silica)	MMF, GI, or SI (silica)	MMF, GI (silica)	SMF, SI (silica)	SMF, SI (silica)	SMF, SI (silica)	MMF, SI, plastic
Detector	PN or PIN (Si)	PIN or APD (Si)	APD (si) or (GaAs)	APD (Ge)	PC or PIN (InGaAs,Ge)	PIN (InGaAs,Ge)	PC, PIN, heterodyne (GaInAs,Ge)	PD (Si)
Amplifier	FET	FET	Bipolar	Bipolar	—	—	—	—
Repeater	—	—	—	Yes	Yes	Yes	Yes	—
Applications	Interoffice data transmission, communication	Intracity telephone data	Intracity systems	Intracity telephone, television data	Intracity systems	Long-haul underwater	Long-haul	Medical

Abbreviations: Ld—laser diode, LED—light-emitting diode, PD—photodiode, PC—photoconductor, mm—multimode, sm—single mode, SMF—single-mode fiber, MMF—multimode fiber, SI—step index, GI—graded index, PIN—p–i–n photodiode, APD—avalanche photodiode.

the distance is short, usually within 100 m, and the speed requirement is moderate. Very reliable systems can be built at a very reasonable cost. Such a system uses LED as the light source, which operates at 0.83-μm wavelength on a step-indexed multimode fiber. For the detector, this system uses a silicon photodiode or even an avalanche photodiode if needed. Both GaAs LEDs and silicon PIN detectors are very reliable and are low in cost. The lifetime of an LED runs up to 10^8 h, and the lifetime of an Si detector is also infinitely long. Although an MMF may have a larger loss, the distance is too short to be of concern. So is the dispersion. The receiving system may also work with an FET amplifier as needed. Only connectors and couplers are needed for interconnections; no splices are necessary.

For long-haul and high-data-rate operations, the choice is again limited. Column VII in Table 8.2 lists the required components. Here, a point-to-point link can be established that utilizes the ultra-low-loss SMF operating in the 1.55-μm range. If the zero-dispersion point can also be shifted to within this wavelength range, a truly large information-carrying-capacity system can be realized. (This is the so-called third-generation fiber.) For the light source, however, only a highly stabilized single-mode single-frequency InGaAsP laser can be used for realizing all the advantages of this SMF operation. PIN photodetectors made with InGaAsP semiconductor materials may be adequate, but a special arrangement to reduce the noise output is required. Photoconductors may have comparable noise property at this wavelength, yet the structure is much simpler than the photo-PIN diodes. Its application for detection at longer wavelengths is on the rise. Another trend is to use heterodyne receivers. With a stabilized single-frequency laser, optical waves can be detected just like radiowaves (single frequency) using the heterodyne principle, which can boost the sensitivity over 10 or even 100 times.

Column VIII in Table 8.2 lists another interesting application of the optical fiber system. For example, medical applications of optical fibers involve the observation of irregular situations inside a human body or monitoring the operation in progress by fibers to probe the area of interest. For these applications, only a few meters of fiber length is needed but the strength of the fiber must be high in order to endure the bending and twisting in the manual operation mode. An all-plastic fiber or silica fiber with a heavy plastic cladding may be the best choice. The length is too short to cause concerns of fiber loss, although all-plastic fibers may have 400-dB/km loss. A single LED that emits red light and a silicon PIN diode are sufficient to accomplish the task.

The use of long wavelength, typically at 10.6 μm from a CO_2 laser, has attracted much interest in modern medicine for application in laser angioplasty. Special fiber glass such as the fluoride glasses are used. This topic will be discussed further in Chapter 12.

Between these extremes, the choice of components for different fiber lengths and speed requirements is quite involved, indeed. This is because there can be many combinations of components, many of which require trade-offs. Often the selection processes involve iteration and computer optimization. The final choice is usually determined by the ultimate cost. We subdivide these applications and list these in columns II–VI (Table 8.2) inclusive.

To facilitate the analysis and eventually the optimization of the system components for a required system, one must familiarize oneself with the components and do some computations to compare the results in order to select the best system parameters.

Fiber Parameters

As we have seen in Chapter 3, the transmission characteristics of optical fibers and the loss and dispersion in fibers depend on the structural parameters. These are the core diameter, the cladding thickness, the refractive-index difference between the core and the cladding, and the refractive-index profile of the core in an MMF. Other parameters such as cabling loss, bending loss, and microbending loss may also affect fiber performance. Since designers usually select the fiber for use in a particular system from among many manufacturers rather than fabricate the fiber themselves, a thorough knowledge of the existing fibers will suffice for choosing the most suitable fiber to use.

While the actual choice depends on the availability of a manufacturer's catalog, Table 8.3 may serve as a guide for selection. In this table, general characteristics of most often used fibers are listed. (In

Table 8.3 Fiber Parameters and Characteristics

	Wave-length Range λ (μm)	Fiber Core Diameter (μm)	Refractive-Index Difference Δ (%)	Dispension		Fiber Attenuation (dB/km)	Unrepeated Distance L (km)	λ_c η_c
				Chromatic (ps/nm·km)	Modal (ns/km)			
SMF,	0.85	5–10	0.01	100		2–5		0.7 low
V < 2.4	1.3	9–10	0.01–0.7	0–1		0.4–0.8	100–30	1.2 low
	1.55	4.5–11	0.01–0.45	0–1		0.2–00.54	220–130	1.0 low
MMF, plastic, SI	0.6–0.8	200–400	—	—		400	—	Med
GI, MMF,	0.85	50	1		0.15–5	3–5		Med
V > 2.4	1.3	50–60	1		5	0.6		Med
MMF	0.85	50	1–2		10–20	1.4		Med
V > 2.4,	1.3	50–60	1–2			0.5		Med
SI	1.55	50	1–2			0.26		Med

actual design, one has to consult the manufacturer's data.) In this table, we have divided the fibers into two groups: SMF and MMF. The parameter used to classify this is given by V, discussed in Chapter 2. For $V < 2.4$, it is an SMF, whereas for $V > 2.4$, it is an MMF. SMF is always of the step-index (SI) type, while the MMF may be either of a step-index type or a graded-index (GI) type. All fibers listed in this table are silica fibers except the all-plastic fiber in one case as indicated. We list λ, the operating wavelength range; $2a$, the core diameter; λ_c, the cutoff wavelength $[=(V\lambda/2.405)]$ for HE_{11} mode); Δ, the index difference $[= (n_1 - n_2)/n_1]$; α, the fiber attenuation (in decibels per kilometer); the dispersion—chromatic dispersion (in picoseconds per nanometer-kilometer) for SMF or modal dispersion (in nanoseconds per kilometer) for MMF; and the relative coupling efficiency from the source to the detector. The numbers are averages from reported experiments and serve only as a guide for the choices. One should consult manufacturers' data in an actual design.

Step-index fibers are relatively inexpensive. They can have relatively large numerical aperture NA values. But SI fibers suffer from high intermodal dispersion compared to GI fibers of the same core diameter. Single-mode fibers are all SI-built with a zero-dispersion point appearing at about 1.3 μm for silica fibers. This point can be shifted to longer wavelengths by other methods as were discussed in Chapter 3.

Multimode fibers can be either SI- or GI-built. The refractive-index difference Δ in MMF must be higher than 0.5% in order for the material dispersion to cancel the waveguide dispersion. A higher Δ value also helps to reduce the bending loss. But a higher Δ fiber may cause high splice loss due to more imperfections introduced in the processing.

The attenuation and dispersion data are again averaged values derived from published data.

One should notice that the all-plastic fibers have extremely high loss (ca. 400 dB/km) and, therefore, their application is limited to short links only.

For power transmission, another class of fiber has been introduced, the fluoxide glass fibers. These are used for transmission over very short distances in the wavelength range 2.0–10.6 μm. Fiber loss and dispersion are of no concern. They are typically large fibers with a core diameter approaching 1 mm or so. They are not listed in this table. They will be discussed in Chapter 12.

Light Source Parameters

As mentioned earlier, the most suitable light sources for optical fiber systems are semiconductor LEDs and LDs, because of their size compatibility with the fibers and their reliability in performance. We do not include the Nd:YAG laser in this discussion because of its extremely long fluorescence lifetime (230 μs), bulkiness, and high power consumption.

The parameters that characterize a light source are the emission wavelength, the emission spectral width, the optical power, the driving current, the threshold current for the laser, and the lifetime. Other properties such as the spectral coherence, the noise factor and its sensitivity to temperature variations, and cost are also important.

Table 8.4 summarizes the state-of-the-art characteristics of the light sources now available. These are grouped into LED, laser diode (LD), and special LD, which includes the distributed Bragg reflection (DBR), distributed feedback (DFB), distributed single-mode (DSM), and C³ lasers and the tunable lasers. The approximate numerical values or range of values given are believed to be reasonable estimates from presently available data. Materials for fabricating these devices are also listed.

Some general remarks for each group of light sources are given in the following paragraphs.

The LEDs

For light-emitting diodes, the light power available at the fiber end is usually small because of its poor coupling efficiency. The spectral width is large since the light emission is incoherent. The GaAs LEDs are particularly attractive in the applications for short-wavelength systems (0.8–0.9 μm). This is because the technology of GaAs processing has matured and reliable devices can be fabricated at a reasonably low cost. A life expectancy on the order of 10^8 h can be achieved. Multimode fibers with large NA can be used advantageously with LEDs in this wavelength range.

The use of LEDs and MMF for 1.3-μm-wavelength applications becomes increasingly attractive with improved LEDs made from InGaAsP/InP semiconductors. In the 1.3-μm wavelength range where the zero-dispersion point occurs for most silica fibers, the effect of modal noise, partition noise, and reflection noise due to coherent interaction in MMF are minimized [11, 15, 16].

The Semiconductor Lasers

We conveniently divide semiconductor lasers into two groups: the conventional lasers and the special lasers. Conventional lasers are ordinary injection lasers that produce multimode oscillations. Their spectral width is relatively large but much smaller compared to that of LEDs. The output light power is large and can replace LEDs in many applications whenever increasing light power is needed. With the reduced spectral width of the emitted light, the bandwidth can be

Table 8.4 The State-of-the-Art Light Sources

	Material	Wavelength Range (m)	Spectral, Width (nm)	Power (dB$_m$)	Bandwidth	Coupling to Fiber	Driving Current (mA)	Estimated Life	Remarks
LED	GaAs	0.85	50–200	−1.3–−0.3	<100 MHz	Poor	100–300	10^8	Lambertian
	GaAlAs	0.75–0.9	20–30	0.7	<100 MHz	Poor	100–200	10^8	Lambertian
	InGaAsP	1.3	120	−0.3–0.3	50–200 MHz	Poor	100–200	10^7–10^8	Lambertian
MM	GaAlAs	0.8–0.9	2–3	0.7–1.7	1000 MHz	75%	30–150	10^5–10^6	Gaussian beam
	InGaAsP	1.3	8–10	0.48–1.7	1000 MHz		20–100	10^6	
	InGaAsP	1.55	10	−2.2	2 GHz		100–125	10^5	
LD SM	GaAlAs	0.85	2–3	0.48–1	1–2 GHz	75%	70	10^5	
	InGaAsP	1.3	1.5–10	−7 to −1.2	140 MHz		30–70	10^4	
	InGaAsP	1.55	10	−8			25–50	10^4	
DBR	InGaAsP	1.5–1.65	Small		120 GHz			10^3	
DFB	InGaAsP	1.5–1.65	Small		240 GHz			10^3	
C^3	InGaAsP	1.3	Small		2 GHz			10^3	
	InGaAsP	1.5–1.65	Small						

increased considerably, as BW varies as $1/\Delta\lambda$. In the 1.3-μm wavelength range, lasers become the dominant source of light power [17].

In the 1.5−1.65-μm range, where silica fiber loss is lowest, system designers are troubled by spectral broadening and mode-jumping phenomena in conventional lasers. Special lasers are designed to alleviate these problems. They are stabilized single-mode, single-frequency lasers such as the distributed single-mode lasers (DSM LDs) [18], the cleaved coupled-cavity laser (C[3]) [19], and many others [20]. The spectral width of these lasers is narrow, and the coupling efficiency to the fiber is good. But they all suffer from the "laser failures" characteristic of long-wavelength lasers. Besides, special lasers all require special construction involving many complex arrangements of many epitaxial layers for various functions such as current restriction, waveguiding, and distributed-feedback mechanism. In InP-based lasers, the reliability is a major concern. The high rate of change of the threshold current with temperature is still not well under control. The life expectancy of complex lasers is well below that of the LEDs.

In recent years, a new type of laser has been developed to satisfy the requirement of the coherent optical communication system at 1.5-μm range. Lightwaves are heterodyned with a local laser oscillator to produce a microwave beat frequency that can more easily be amplified, channel-filtered, and detected. This can increase the sensitivity of an optical fiber system by one to two orders of magnitude. Sharp microwave filters can be designed to separate neighboring signals, thus allowing hundreds or thousands of channels of closely spaced wavelengths to be carried on one fiber. Such design has been used in wavelength multiplex systems. The essential component of this system is a tunable laser that can be monolithically fabricated with other components to function as a local oscillator. Two of these tunable lasers have been reported at the 19th International Semiconductor Laser Conference in Kanazawa, Japan, October 14−17, 1986. One by Marata et al. has a tunable range of 1 μm to 1.5 μm, and a bandwidth of 120 GHz [21], and the other, by Yoshikani [22], uses the DFB structure of an InGaAsP/InP laser. Its tuning is over 2 nm and it has a bandwidth of 240 GHz.

Detector Parameters

The property of a detector is characterized by its sensitivity, speed of response, bandwidth, and noise level as discussed in Chapter 7. The sensitivity can be expressed in terms of either the responsivity or the minimum detectable power. Table 8.5 presents the characteristics of some selected photodetectors in terms of their wavelength, bit rate (response time), normalized detector sensitivity, dark current, and junction capacity. The detectors are grouped into types, such as the

Table 8.5 Photodetector Characteristics

	Material	Wavelength range (μm)	Quantum Efficiency Q (%)	Sensitivity (dB$_m$)	Response Time (ns)	Dark Current (nA)	Junction Capacitance c (pF)
Photo-conductor	InGaAs	1.3	<50	−35.9	1		0.8
	InGaAs	1.51	<50	−30.3	0.5		
PIN diode	Si	0.85	<90		1	10	<1
	Si	0.9	83		3	5000	3
	Ge	1.0−1.6	>85		0.1−0.5	10	<1
	InGaAs GaAlAs	1.3−1.55		−30−40	1.8−.83		.5
	InGaAsP	1.3−1.5	55−60				
APD	Si	0.6−1.1	85	−54 gain 25	2	200	2
	Ge	0.6−1.6	>85	−22 → 40 Gain 4	0.25−2.4	5000	0.8
	GaAs		50	−40		10−50	6
	GaInAsP	1.0−1.7	80	50−60 Gain 7−8		00	1.5
	GaAlAsP	1.0−1.4	90	50−60	0.1	10−40	

photoconductor type, the PIN type, the APD type, and the heterodyne type. Materials used to fabricate these devices are specified, and quantum efficiency is listed for reference.

The PIN photodiodes [23−25] are the most common type of detector used for optical fiber systems. The Si PIN diode is suited for the 0.8−1.0-μm wavelength range, while the Ge and InGaAsP PIN diodes are useful for longer-wavelength systems. APD of both materials can be built to increase the sensitivity, but the bandwidth will be compromised with a higher multiplication factor M. In a practical design, it is customary to impose a limit of $M = 25$. Ge and InGaAsP diodes are very noisy as a result of high leakage and dark current. Often, complex structures are designed to combat the problem. At longer wavelengths, the photoconductor, which is much simpler in structure and has noise characteristics comparable to those of the PIN diode, is becoming increasingly attractive.

Heterodyne detection is also a new experimental technique for optical fiber systems, although its use in radio and microwave frequencies has been in existence for more than 50 years. Only after special lasers with single mode, single polarization, and extreme stabilization become available can this method be used. This had to await the recent invention of an equally pure and stable tunable laser.

System Considerations

We are now ready to assemble the components to form a system that satisfies the design specifications. From Table 8.2, the initial choice has been made. Going through Tables 8.3, 8.4, and 8.5, we notice that the fiber, the light source, and the detector types are decided. These decisions are based on the following considerations. For the source emitter and detector, the ability to handle the signal bandwidth is the first consideration. The choice of fiber type should be guided by the signal distortion due to dispersion rather than by fiber loss, because fiber loss can be compensated either by increasing the emitter power or by using a more sensitive detector, or both. Degradation of a signal must be minimized or prevented by choosing the right kind of fiber. As dispersion or time delay is (almost) proportional to the fiber length, only long-distance links in very high speed operations need be of concern.

The Power Budget

After the initial choice, we next propose a number of checks. The first is the power budget check [26]. This check is intended for comparison of various combinations of components for a system. Figure 8.9 shows a typical power budget plot. On this plot, the ordinate is the optical power in dBm. (The dBm is a unit if power referenced to 1 mW.) A power of 10 mW becomes $10 \log^{10} (10/1)$ or 10 dBm. The abscissa in-

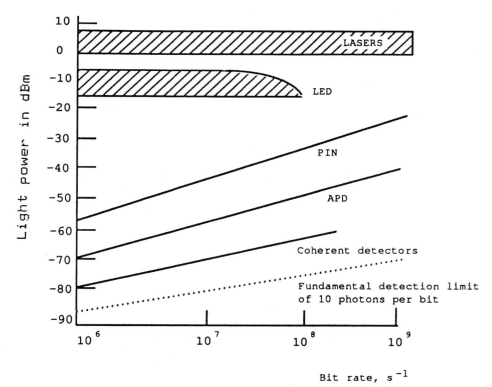

Figure 8.9 The power budget diagram modified to cover future laser power and heterodyne receivers.

dicates bit rate. The working range of both lasers and LEDs is approximately drawn in the respective shaded areas, representing available sources corresponding to the range of power level and bit rate. The available detectors, either PIN or APD and coherent detectors, are shown as individual sensitivity lines as indicated. All these lines are calculated on the basis of achieving a 10^{-9} bit error rate. The dashed line at the bottom is the fundamental achievable detection limit based on the theoretical limit of 10 photons per bit for the detector sensitivity. The use of this chart can be illustrated in the following example. If we wish to operate the system at 10 Mbit/s with an LED/PIN diode combination, a vertical line is drawn on the 10 Mbit/s that intersects the PIN line and the LED block. In the worst case, that is, using the lower limit of the power given by the LED, the power difference between this lower power reading and the PIN line intersection is about 30 dB. This means that this combination can tolerate a maximum loss of 30 dB. If a fiber with a loss factor of 5 dB/km is used, and if we allow for a safety margin of 5 dB, a total loss of 10 dB/km would give a maximum range of 5 km. On the other hand, if a laser–APD combination were used, a similar line would intersect the laser block and the APD line, giving a minimum tolerable power limit of approximately 60 dB. Using the same fiber with the same margin of safety, the maximum range now increases to 11 km. Similarly, if a coherent detector were used, in combination with a single-mode, single-frequency stabilized powerful laser and an ultra-low-loss fiber (0.16 dB/km) at its minimum dispersion wavelength, the power difference would be 80 dB and a range of 468 km could be covered.

The maximum usable fiber length between the two terminals of an optical fiber system depends on two factors: fiber loss and fiber dispersion. In a low-bandwidth system, fiber loss is usually the limiting factor, whereas in a high-bit-rate system, fiber dispersion becomes the limiting factor. An example should clear this point: Suppose an SI-type fiber has a fiber loss of 1 dB/km, and a signal power margin of 40 dB can be tolerated. On the basis of the fiber loss alone, a cable length of 40/1 or 40 km can be used. However, if the cable dispersion is given as 1 GHz · km (1 ns/km) and a signal bandwidth of 100 MHz is required, then the maximum usable cable length becomes $10^9/100 \cdot 10^6$ = 10 km; thus, the cable is dispersion-limited. A step-index fiber suffers high intermodal dispersion, while in a graded-index fiber, this dispersion is greatly reduced.

The System Response Time

A second check is the system response time. The total system response time τ_s can be expressed in terms of the combination of the response times of its components: the transmitter response time τ_t, the fiber response time τ_f, and the receiver response time τ_r as

$$\tau_s = (\tau_f^2 + \tau_t^2 + \tau_r^2)^{1/2} \qquad (8\text{-}16)$$

On the other hand, the required system bandwidth can be expressed in terms of a time constant τ_B.

A system is rated as adequate if $\tau_B > \tau_s$.

The Power Margin

From the power budget diagram shown in Fig. 8.9, we estimate the power margin. Let P_l be the emitter power that can be read from the diagram with either an LED or laser light source; P_n is the detectable power from either the PIN or APD line; and $P_f = L\alpha$ is the total fiber loss, a product of the fiber length and the unit loss factor α. Then an adequate system design should have a power margin equal to or greater than 10 dB, or

$$\text{Power margin} = P_l - P_f - P_n \geq 10 \text{ dB} \qquad (8\text{-}17)$$

We wish to work out an example to illustrate the use of the response time and power margin to determine how a system is rated:

Given: Transmitter   LED
      Power   10 dBm
      Rise time $\tau_t = 6$ ns
      Coupling loss 10 dB

   Fiber   GI MMF
      Length   80 km
      Fiber loss $\alpha = 0.5$ dB/km; total loss = $0.5 \times 80 = 40$ dB
      Dispersion   0.125 ns/splice; 3 splices total 1.5 dB

   Receiver   Si APD
      Power -59 dBm
      Rise time $\tau_n = 3$ ns
      Losses   2 dB

   Allowances   5 dB

If a system bit rate is 20 Mbit/s with an NRZ signal, this corresponds to $\tau_B = 0.7/20$ Mbit/s $= 35$ ns.

The first check on the power budget is:

$$\text{Power gain:} \quad 10 + 59 = 69 \text{ dB}$$

$$\text{Power loss:} \quad 10 + 40 + 1.5 + 7 = 58.5$$

$$\therefore \quad 69 - 58.5 = 10.5 \text{ dB}$$

The second check on response time is

$$\tau_B = 35 \text{ ns}, \qquad \tau_t = \underline{10^2 + 6^2 + 3^2} = 12.04$$

$$\therefore \tau_B > \tau_t$$

Both checks indicate a satisfactory design.

System Performance

In the preceding section, we have defined the system performances of a digital communication system. First, the signal that arrives at the receiver is set at a sufficient amplitude to achieve a specified bit error rate. Then, the bandwidth is selected to ensure that the fiber dispersion and other interferences do not cause an unacceptable level of distortion. In practice, however, other factors such as noise or source jittering may add to the pulse distortion, which, in turn, causes the degradation in the bit error rate of the system. A more meaningful reassurance of the performance of a practical system is highly desirable. This can be done by observing the eye pattern.

Eye pattern is a display of the received digital signal on an oscilloscope that is being triggered at the data rate. The appearance of the pattern resembles a human eye, from which the name is derived. It is the result of superimposing on a scope traces of a wide variety of input pulses in the corresponding time interval. The persistence of the cathode-ray tube (CRT) blends together all allowed signal waveforms to form the eye pattern. For an ideal noise-free binary signal with no distortion, a three-level digital NRZ signal 111 may resemble that shown in Fig. 8.10a. The opening at the center, called the *eye*, has a height H and a width B. If noise, jitters, and interferences were added, the pattern in Fig. 8.10b would result. Notice the change in the eye opening. The new height becomes h and the new width becomes b. As h is less than H and b is less than B, the eye is closing.

The height of the opening indicates what margin exists over noise. If h/H is 50%, an equivalent S/N degradation of $-20 \log(1 - 0.5) = 6$ dB exists. The horizontal dimension of the opening indicates the range of correct sampling time and the amount of peak distortion due to intersymbol interference. The sensitivity of the system to timing errors is revealed by the rate of closing of the eye or the slope as sam-

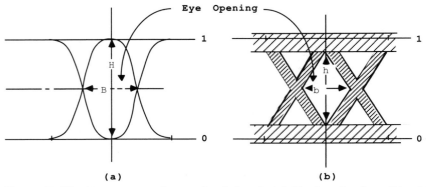

Figure 8.10 Eye pattern of a received signal: (*a*) ideal, noise-free; (*b*) with noise and intersymbol inferference.

pling time is varied. The eye pattern of a linear system is usually symmetrical about a centerline, which is the sampling time. Any asymmetry indicates the existence of nonlinearity in the system.

The Repeater

The repeater is a substation used to extend the transmission range of a communication system. It may contain a receiver, an amplifier, an equalizer, and finally an emitter to resend the signal to another length of fiber. Of course, the power supply and the synchronization signal must also be included. Besides amplifying the signal to compensate for the fiber loss, an optical equilizer is required to cancel the small residual chromatic dispersion of the fiber and to reshape the distorted pulse due to an intersymbol interference effect. An optical equalizer is a short section of an SMF that has a strong dispersion property in a rather compact form (usually < 1 km). At the operating wavelength, its dispersion takes a sign opposite that of the fiber and serves to compensate for the total dispersion. Thus a repeater station is a costly addition.

In optical fiber system design, we wish to investigate the unrepeater distance first before we consider adding repeaters. The unrepeater distance is a function of many variables. Figure 8.11 [27] shows a typical graph. The relationship among various parameters can be easily compared. Here, the unrepeated distance is plotted against the bit rate for various groups of detectors at different operating wavelengths. All calculations are based on a BER of 10^{-9}. Three wavelength ranges are marked and grouped as the 0.85, 1.3, and 1.55 μm, respectively. The input power is assumed to be set at -6 dBm. Solid curves are for a straight calculating without reservation. The dotted curves are with a 10-dB system margin to cover the fiber loss, including the splice loss (2.5 dB/km for $\lambda = 0.85$ μm, 0.8 dB/km for 1.3 μm, and 0.5 dB/km for $\lambda = 1.55$ μm, respectively).

We notice that for all calculations, the repeater span decreases with increasing bit rate, as expected. At high bit rate, the dispersion effect is more damaging. Next, for the $\lambda = 0.85$-μm group, the repeater span is only several tens of kilometers, while for the 1.55-μm group, it increases to several hundred kilometers. With proper choice of a detector, the maximum distance can easily be increased by a factor of 2.

For the longer wavelength groups, detectors using GaInAsP and GaInAs give consistently higher repeater span readings than that made of Ge. This is because in recent designs using compound semiconductor materials, efforts have been exerted to reduce the leakage current in the diode.

Experimental results report 1.6 Gbit/s using 15-km SMF without a

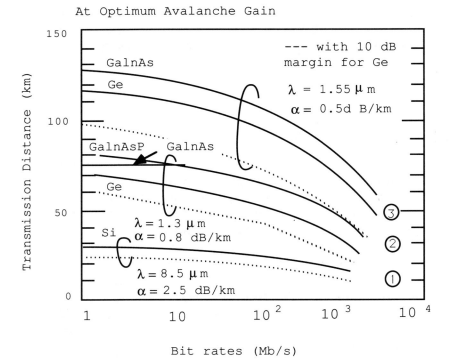

Figure 8.11 Unrepeated fiber length as a function of bit rate for various wavelength and source–detection combinations. [After Y. Suematsu [27] (1983) © IEEE, 1983.]

repeater, 23 km at 1.2 Gbit/s, and 30 km at 800 Mbit/s, all at the zero-dispersion wavelength of 1.3 μm [28]. Other reports include the use of SMF at 1.5 μm with 800 Mbit/s [29, 30], a 290-km span for SMF at 400 Mbit/s [31], and a 84-km span for an SMF with 42 Mbit/s [32]. Also at $\lambda = 1.3$ μm, a 100-km span has been reported using an MMF with 100 Mbit/s [28]. More recent advancements of increasing span are forthcoming with heterodyne reception [33].

A few words might be added regarding the repeater design. For digital communication the incoming signal from an optical fiber is first detected, amplified, equalized, or reshaped. The processed signal is then used to modulate an optical source to resend the information. A low-noise diode, either PIN diode or APD, can be used for detection. The amplifier that follows must be designed to operate on a capacitive input impedance. Low noise is again an important factor in amplifier design. An equalizer is used to compensate the frequency roll-off due to dispersion. Sometimes a feedback amplifier or transimpedance amplifier should be considered. The optical light source for resending the processed signal is usually a high-current, low-impedance device. This makes the driver circuitry a fairly complicated design problem.

Expanding System Capacity by Multiplexing

One way to increase the system capacity without modifying the fiber itself is to employ the technique of multiplexing. Multiplexing technologies include wavelength-division multiplexing (WDM) [34] and time-division multiplexing (TDM) [35].

Wavelength-Division Multiplexing

When a laser diode with a spectral width of 10 nm operating at a wavelength of 1.3 μm is used as a lightwave carrier in a fiber-optic transmission system, plenty of bandwidth remains unused. Even with an LED light source that has a spectral width 10–30 times greater than that of a laser, many additional channels can be accommodated within the operating wavelength range of the fiber. Thus, a dramatic increase in information-carrying capacity becomes possible if simultaneous transmission of many carriers of slightly different wavelengths, each modulated by its own information, can be multiplexed and transmitted over the same fiber. This is known as *wavelength-division multiplexing* (WDM). Figure 8.12 shows the transmission characteristics, the transmission loss versus the wavelengths of an *n*-channel WDM scheme.

There are *n* channels, each with its own light source at $\lambda_1, \lambda_2, \ldots,$ λ_n; $L_1, L_2, \ldots, L_n$ are the minimum transmission losses corresponding to each channel. The transmission loss curve rises very sharply on both sides of the passband. The wavelength separation between channels depends on the sharpness of the loss curves. The sharp rise in loss characteristic enables more channels to be multiplexed in the same available bandwidth. The criterion is based on the requirement

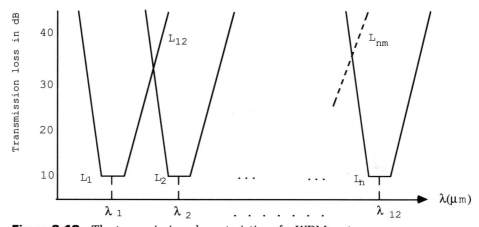

Figure 8.12 The transmission characteristics of a WDM system.

of the isolation loss, denoted by L_{mn}, between adjacent channels. Higher isolation ensures less interference from source instability and cross-talk between neighboring channels. For a laser source, a channel separation of several tens of a nanometer and an isolation of $10-30$ dB are usually adequate.

There are two different ways to implement a WDM: unidirectional and bidirectional, as shown in Fig. 8.13a and 8.13b, respectively.

In the unidirectional scheme, different light sources are multiplexed by a multiplexing coupler to launch the light to an optical fiber for transmission. At the receiving end the wavelengths are separated by a demultiplexing coupler, which feeds each wavelength to a separate detector to reclaim its own information (Fig. 8.13a). In a bidirectional system, the fiber is used to transmit information in both directions simultaneously. The coupler at each end separates the incoming from the outgoing signals.

The most important element in implementing a WDM system is the multiplexing coupler. Both the multiplexing coupler at the transmitter end and the demultiplexing coupler at the receiving end may use the same designing principle, but the design considerations may be quite different. This is because the detectors used in conjunction with demultiplexing couplers are wideband devices. Interchannel isolation is required (high L_{mn} isolation loss). Otherwise, cross-talk between neighboring channels may become excessive. At the transmission end, no such problem exists.

The design of the most widely used multiplexing couplers in a WDM system can be either a dispersive device such as a prism or a

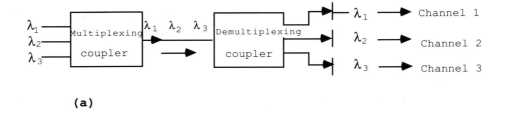

(a)

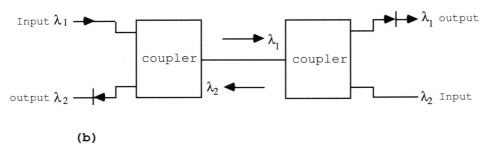

(b)

Figure 8.13 Two ways to implement WDM systems: (a) unidirectional; (b) bidirectional.

grating [36] or a filter-based device [37]. Because the dimensions involved in these devices are small, one can take full advantage of the integrated optics technology. These devices will be covered in Chapter 13.

Time-Division Multiplexing

A scheme more suitable for data transmission multiplexing is the time-division multiplexing (TDM). In data transmission, pulse-code modulation (PCM) is usually used. A pulse is characterized by its short duration followed by a period of silence before it repeats. If the pulse duration lasts one microsecond and repeats at a repetition rate of 10,000 times per second, there are 100 microseconds between pulses. Allowing a maximum time duration of 10-μs period, one could send along the fiber tenfold more information than before. This is called *tenfold multiplexing*. Of course, an identification signal for each channel should also be sent along to the receivers to recognize the appropriate time interval for each of them to reclaim its information. The transmitter of each terminal will have sole use of the optical fiber for the duration of one message, and during this message time, all other terminals will be receiving. The time sequence can be set by either a rotational switching system or a center controller that allocates the time slots for each channel. Other schemes are also available.

Coherent Fiber-Optic Communication

Early in the introductory remarks in Chapter 1, we stated that in today's optical fiber communication systems, on—off intensity modulation and direct photodiode detection are most commonly adopted. This is because a stabilized light source with true single-frequency property was hardly available. In these systems, the receiver sensitivity is limited by the detector and amplifier noises. Through technological development in recent years, lasers with narrow spectral width (~1 Å) and highly stabilized frequency became available. Much research has been focused on developing coherent communication systems. Impressive results have been reported. We wish to briefly describe these new developments.

The Modulation Format

Coherent communication systems employ external modulation and heterodyne or homodyne detection instead of the intensity modulation and direct detection used in ordinary optical fiber systems. Direct driving current modulation to lasers usually causes instability in laser frequency due to loading. External modulation light sources can accomplish amplitude, phase, or frequency modulation. Theoretically, it has been shown by Yamamoto and Kimura [38] that phase modulation offers the best receiver sensitivity and the simplest modulation configuration. This type of modulation scheme is called *phase-shift keying* (PSK). External PSK can be accomplished by using electrooptic crystal such as $LiNbO_3$ to produce changes in the refractive index Δn when the crystal is subject to electrical field of the modulating signal. A phase shift $\Delta\phi = (2\pi \Delta n \, 1/\lambda)$ can be obtained over a relatively short interaction length l.

The external *amplitude-shift keying* (ASK) scheme uses two phase-shift signals and converts them into intensity variations by constructive and destructive interferences of the phase-shift signals. Examples of these modulations will be given in Chapter 13.

Frequency-shift keying (FSK) is the third modulation format for the coherent system.

A laser source usually requires a frequency stability scheme to maintain a stable frequency. But a slight modulation in mode resonance frequencies can be achieved by a small change in the refractive index. Electrooptic or acoustooptic effects can be used to provide the frequency shift required for FSK.

With the advanced technology of integrated circuits, monolithic integration of sources and detectors of different wavelengths on the same chips becomes available. An array of 6 DFB lasers, coupled to a single-channel waveguide, has been applied experimentally for multiplexing [39].

Optical amplifiers that provide 25.5 dB gain with only 1.5 dB of ripple was reported by Saitoh and Mukai recently [40]. This value is considerably better than the optical amplifier gain achieved earlier. This is a highly desirable addition to the coherent optical communication system.

Heterodyne and Homodyne Detection Schemes

In heterodyne or homodyne detection schemes, a photodiode is used to mix the incoming light signal with a much stronger signal from a local oscillator laser at the receiver. In this manner, optical phase information is preserved and can be used for information transmission, thus making PSK possible.

In heterodyne detection, the frequency of the local laser oscillator ω_{lo} is set to be slightly different from the signal frequency ω_s such that the difference frequency $\omega_{if}(= \omega_{lo} - \omega_s)$, the intermediate frequency, is in the microwave range. Signal processing can then be carried out in the microwave frequencies, where excellent designs are readily available. In a homodyne detection scheme, the local laser frequency is set to equal the signal such that $\omega_{if} = 0$. The requirement that the signal and local oscillators have equal frequencies generally leads to very stringent criteria for optical sources. A locking mechanism such as injection locking [41], phase-locked loop [42], and other schemes [43] must be employed to secure frequency equality.

Representing the electric field of the optical signal as

$$E_s(t) = e_s(t)e^{-i\omega t} + e_s^*(t)\,e^{i\omega t} \tag{8-18}$$

Then the optical power incident on the detector is

$$P_s(t) = e_s(t)\,e_s^*(t) = |e_s(t)|^2 \tag{8-19}$$

where e_s^* is the complex conjugate of e_s.

In coherent optical fiber communication, the optical field incident on the detector system is the sum of the weak signal and strong local oscillator, such that

$$E(t) = E_s(t) + E_{lo}(t) \tag{8-20}$$

and

$$p(t) = p_s(t) + p_{lo}(t) + e_s(t)\,e_{lo}^*(t)\exp\{i[\theta + (\omega_{if}t)]\} \tag{8-21}$$
$$+ e_s^*(t)\,e_{lo}(t)\exp\{-i[\theta + (\omega_{if}t)]\}$$

where $\omega_{if} = \omega_{lo} - \omega_s$.

The information-bearing portion of the signal is obtained by taking the real part of Eq. (8-21), which is

$$p(t) = p_{lo}(t) + 2[p_{lo}p_s(t)]^{1/2}\cos(\omega_{if}t + \theta) \tag{8-22}$$

for heterodyne detection and

$$p(t = p_{lo} + 2[p_{lo}p_s(t)]^{1/2}\cos\theta \tag{8-23}$$

for homodyne detection.

The mean signal power can be computed by taking the time average, thus

$$\bar{p}_{het} = 2p_{lo}p_s \tag{8-24}$$

$$\bar{p}_{hom} = 4p_{lo}p_s \tag{8-25}$$

Compared to direct detection, where

$$\bar{p}_{dd} = p_s^2 \tag{8-26}$$

One can immediately conclude that since $p_{lo} > p_s$, the power gained by using heterodyne or homodyne detection can be many times larger than that from the direct detection process. Also, it is interesting to note that homodyne detection can produce twice the power produced in the heterodyne process. But the stringent requirement for oscillator frequency stability in the former procedure may offset this advantage.

For more detailed discussion, refer to the recent work by Basch [44].

Conclusion

In this chapter we have shown how to gather information for optical fiber system design and steps for initiating the actual design. Tables listing the most important properties of various components are presented as examples. More detailed and up-to-date data should be used in any actual design. The processes of optimization, including the use of computer techniques, are omitted for the readers to develop their own programs.

References

1. H. Taub and D. L. Schilling, *Principles of Communication Systems*, Chapter 13. McGraw-Hill, New York, 1971.
2. R. W. Dawson and C. A. Burrus, Pulse behavior of high radiance small-area electro-luminescent diodes. *Appl. Opt.* **10**, 2367–2369 (1971).
3. R. L. Harman, J. C. Dyment, C. J. Huang, and M. Kuhn, Continuous operation of GaAs-Ga$_{1-x}$Al$_x$ as double heterostructure laser with 30°C half life exceeding 100 hours. *Appl. Phys. Lett.* **23**, 181–183 (1973).
4. P. K. Cheo, *Fiber Optics—Devices and Systems*. Prentice-Hall, New York, 1985.
5. F. S. Chen, Modulations for optical communications. *Proc. IEEE* **58**, 1440–1457 (1970).
6. I. P. Kaminow and E. H. Turner, Electro-optic light modulators. *Proc. IEEE* **54**, 1374–1390 (1966).
7. I. P. Kaminow, *An Introduction to Electro-optic Devices*. Academic Press, New York, 1974.
8. D. A. Pinnow, Guidelines for the Selection of Acoustic Materials, *IEEE J. Quantum Electron*. **QE-6**, 223 (1970).
9. L. Levi, "Applied Optics," Vol. 2, Chapter 14. Wiley, New York, 1980.
10. D. Chen, Magnetic materials for optical recording. *Appl. Opt.* **13**, 767 (1974).
11. H. Nyquist, Certain factors affecting telegraph speed. *Bell Sys. Tech. Journ.* **3**, 324–326 (1924).
12. R. G. Smith and S. D. Personick, Receiver design for optical fiber communication systems. In *Semiconductor Devices for Optical Communication* (H. Kressel, ed.), Chapter 4. Springer-Verlag, Berlin, 1982.
13. D. J. Jackson, J. Y. Josefowicz, D. B. Rensch, and D. L. Persechini, Detectors for monolithic optoelectronics. *Proc. SPIE—Int. Soc. Opt. Eng.* **839**, 161–164 (1988).
14. R. A. Linke and P. S. Henry, Coherent optical detection. *IEEE Spectrum* **24**, 52–57 (1987).
15. Y. Okano, K. Nakegawa, and T. Ito, Laser mode partition noise evaluation for optical fiber transmission. *IEEE Trans. Commun.* **COM-28**(2), 238–243 (1980).
16. O. Hiroto and Y. Suematsu, Noise properties of injection lasers due to reflected waves. *IEEE J. Quantum Electron.* **QE-15**(3), 142–144 (1979).
17. J. E. Midwinter, Studies of monomode long wavelength fiber system at the British Telecom Research Laboratories. *IEEE J. Quantum Electron.* **QE-17**(6), 911–919 (1981).
18. W. T. Tsang, R. A. Logan, and J. A. Ditzenberger, Ultra-low threshold, graded-index waveguide, separate confinement CW buried-herterostructure lasers. *Electron. Lett.* **18**, 845–847 (1982).
19. W. T. Tsang, R. A. Logan, and J. A. Ditzenberger, Ultra-low threshold, graded-

index waveguide separated confinement, CW buried heterostructure lasers. *Electron. Lett.* **18**(19), 845–847 (1982).

20. C. E. Harwitz, J. A. Rossi, J. J. Hsieh, and C. M. Wolfe, Integrated GaAs-AlGaAs double-heterostructure lasers. *Appl. Phys. Lett.* **27**(4), 241–243 (1975).

21. F. Heisman, A. C. Alferness, L. L. Buhl, G. Eisemstein, S. K. Korotky, J. J. Veselka, L. W. Stulz, and C. A. Burns. Narrow-linewidth, electro-optically tunable InGaAsP-Ti: LiNbO$_3$ extended cavity laser. *Appl. Phys. Lett.* **51**, 164–166 (1987).

22. R. A. Linke. Beyond gigabit-per-second transmission rates. *Tech. Digest of Conf. on Opt. Fib. Comm.*, Reno, NV. Paper WO3 (1987).

23. T. P. Lee, C. A. Burrus, and A. G. Dental, InGaAsP/InP pin photodiodes for light-wave communications at the 0.85–1.65 μm wavelengths. *IEEE J. Quantum Electron.* **QE-17**, 232–238 (1981).

24. G. E. Stillman, L. W. Look, N. Tabatabair, and G. E. Bulman, InGaAsP photo-diodes. *IEEE Trans. Electron Devices* **ED-30**, 364–381 (1983).

25. T. P. Pearsall and J. F. Pearsall, Photodetectors for optical communications. *Opt. Commun.* **2**(6), 42–48 (1981).

26. J. E. Midwinter, Optical communication systems, current research and future systems. *J. Chin. Inst. Commun.* **6**(1), 4–16 (1985).

27. Y. Suematsu, Long-wavelength optical fiber communication. *Proc. IEEE* **71**(6), 692–721 (1983).

28. J. Yamada, S. Machida, T. Kimura, and H. Takata, Dispersion-free single mode transmission experiments up to 1.6 Gbit/s. *Electron. Lett.* **15**(10), 278–279 (1979).

29. J. Yamada, M. Sarawateri, K. Asatani, H. Tsachiya, A. Kawana, K. Sugiyama, and T. Kimara, High-speed optical pulse transmission at 1.59 μm wavelength using low loss single-mode fibers. *IEEE J. Quantum Electron.* **QE-14**(11), 781–800 (1978).

30. W. Albrecht, G. Elge, B. Enning, G. Wolf, and G. Wenke, Experiences with an optical long-haul 2.24 Gbit/s transmission system at a wavelength of 1.3 μm. *Electron. Lett.* **18**(17), 746–748 (1982).

31. K. Sakai, K. Utaka, S. Akiba, and Y. Matsushima. 1.5-μm range InGaAsP/InP distributed feedback laser. *IEEE J. Quant. El.* **QE-18**, 1272–1278 (1982).

32. M. M. Boenke, R. E. Wagner, and D. J. Will, Transmission experiments through 101 km and 84 km of single-mode fiber at 274-Mbit/s and 420 Mbit/s. *Electron. Lett.* **18**(2), 892–898 (1982).

33. W.-T. Tsang, R. A. Logan, and J. A. Ditzenberger, Ultra-low threshold, graded-index waveguide separate confinement, CW buried-heterostructure lasers. *Electron. Lett.* **18**(19), 845–847 (1982).

34. T. Miki and H. Ishio, Viabilities of the wavelength-division multiplexing transmission system over an optical fiber cable. *IEEE Trans. Commun.* **COM-26**(7), 1082–1087 (1978).

35. R. Watanabe, Y. Fujii, K. Nosu, and J. I. Minowa, Optical multi/demultiplexors for single-mode transmission. *IEEE J. Quantum Electron.* **QE-17**(6), 974–982 (1981).

36. D. C. Flanders, H. Kogelnik, R. V. Schmidt, and C. V. Shank, Grating filters for thin-film optical waveguides. *Appl. Phys. Lett.* **24**, 194–196 (1974).

37. S. P. Bandettini, Optical filters for wavelength division multiplexing. *Proc. SPIE—Int. Soc. Opt. Eng.* **417**, 67 (1983).

38. Y. Yamamoto and T. Kimura, Coherent optical transmission systems. *IEEE J. Quantum Electron.* **QE-17**(6), 919–936 (1981).

39. K. Aiki, M. Nakamura, and J. Omeda, Frequency multiplexing light source with monolithically integrate distributed-feedback diode lasers. *IEEE J. Quantum Electron.* **QE-13**, 220–223 (1977).

40. T. Saitoh and T. Mukai, reported by S. J. Campenella in *Technology 1987. IEEE Spectrum* **24**, p. 43, No. 1 (1987).

41. K. Kikuchi *et al.*, Degradation of bit error rate in coherent optical communications due to spectral spread of the transmitter and local oscillator. *IEEE J. Lightwave Technol.* **LT-2**(6), 1024–1033 (1984).

42. W. R. Leeb, H. K. Philipp, A. L. Scholtz, and E. Bonek, Frequency synchronization and phase locking of CO_2 lasers. *Appl. Phys. Lett.* **41,** 592–594 (1982).

43. D. W. Smith *et al.*, Application of Brillouin amplification in coherent optical transmission. *Opt. Fiber Commun., Tech. Dig.*, pp. 88–89 (1986).

44. E. E. Basch, ed., *Optical-Fiber Transmission*, Chapter 16, pp. 503–532. Macmillan (Sams & Co.), New York, 1987.

9

Fiber Optic Communication Networks

Introduction

Fiber optic technology entered the telecommunications field only around 1970. Yet, its rapid progress is almost unparalleled by any other scientific or technological development in history. Although most current applications of optical fiber technology have been concentrated in the field of telecommunications, many other fields of application are quickly opening up, including data transmission, home entertainment, remote sensing, medicine, and industry. Many countries are participating in this race. Among the forerunners of fiber optic research and installations are the United States, Japan, and the United Kingdom, followed closely by France, Germany, Italy, and many other countries.

Any description of the existing fiber optic telecommunication systems of a particular type and those used in a specific country would run the risk of becoming outdated very rapidly. Instead, we shall review the present status of fiber optic technology in the telecommunications field briefly and speculate on the future of optical fiber developments.

Past and Projected Milestones of Fiber Optic Technology

We shall borrow a graph suggested by Kao and Basch [1] with modification. In this milestone diagram (Fig. 9.1) the repeater spacing and the bit rate are plotted against the past and the projected future of fiber optic technologies for the years 1975 to 1995. Also indicated on this

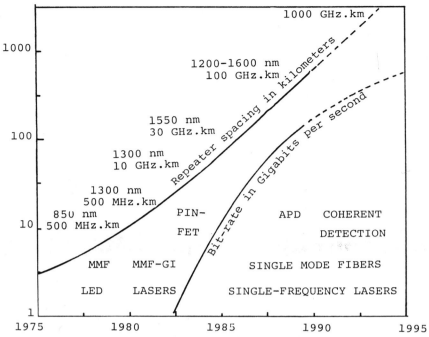

Figure 9.1 Past and projected milestones of fiber performance.

diagram are the various component developments that enter the picture at various stages.

In Chapter 1, we cataloged the optical fiber development stages as the first, second, and third generations in accordance with the wavelengths used that correspond to the spectral loss characteristics in a silica optical fiber. First-generation fiber utilizes the 0.85-μm wavelength range with a typical fiber loss of 2–4 dB/km and a bandwidth–distance product of about 500 MHz·km. We reached the milestone of 10 km around 1980 with an LED light source, a silicon photodetector for direct detection, and a multimode fiber (MMF) for transmission. This combination is widely used in most short-range fiber communication systems because it is simple to install, inexpensive, and provides reliable service. The performance of first-generation fiber exceeds conventional coaxial cable in channel capacity, the speed of operation, and repeater spacing. Only the fiber loss is still relatively high and the dispersion is large for high-speed service. Second-generation fiber utilizes the 1.3-μm wavelength range of the fiber where the loss is lower (ca. 0.7 dB/km). The possibility of the zero-dispersion property makes this range very appealing. Systems built with this fiber become attractive for medium-range communications and are developing very rapidly at present. However, for long-haul systems in the future, the use of third-generation fibers that operate within the 1.55-μm window is highly anticipated. Fiber loss as low as 0.14 dB/km for a pure-silica fiber is now available at this wavelength range. A quick look at the chart in Fig. 9.1 also reveals the fact that component developments contribute greatly to the progress in fiber optic technology milestones. These contributions modify our classification of the fiber generations. We shall arbitrarily subdivide the second- and third-generation fibers into 2a, 2b, and 3a, 3b to emphasize these effects.

The shift of the operating wavelength from 0.83 to 1.3 μm allows a substantial increase in repeater spacing. As can be seen from Fig. 9.1, MMF with a laser source pushed the repeater spacing to around 50 km in the early 1980s. The bandwidth–distance product of this combination is around 500 MHz·km. We name this stage the *generation 2a*. Longer repeater spacing soon imposed very stringent requirements on fiber bandwidth because the fiber dispersion effect increases with the fiber length. Graded-index (GI) fibers were developed to reduce modal dispersion. With these GI fibers the repeater space was pushed upward to around 100 km and the bandwidth–distance product could reach 10 GHz·km.

However, graded-index fibers are more expensive to manufacture and the production yields of high-bandwidth GI fibers are relatively low. At the same time, the technique for producing low-loss single-mode fibers (SMF) with the zero-dispersion property has advanced sufficiently to make SMF preferable to MMF at this wavelength (1.3 μm). The combination of low loss (0.4 dB/km for SMF vs. 0.7 dB/km for MMF) and the zero-dispersion property further extend the repeater spacing. We name this stage *generation 2b*.

Two obstacles prevent the full utilization of the bandwidth potential of a fiber at the 1.55-μm wavelength range. First, although an SMF may have its lowest attenuation coefficient at 1.55 μm, its dispersion coefficient is not zero at that wavelength. Thus, pulse distortion may be serious enough to limit the bandwidth or the bit rate of the operation. Second, ordinary lasers generate many longitudinal modes that form a spectral width $\Delta\lambda$. Even though the total laser power could be kept constant, the power distribution among the modes may fluctuate widely. These fluctuations cause partition noise and contribute to an increase in the bit error rate. Even a single-mode laser may have a spectral width of 3 nm. If a fiber dispersion coefficient of 17 ps/(nm·km) is operated with a laser of spectral width of $\Delta\lambda$ = 3 nm at λ = 1.55 μm, after traveling through 100 km, a pulse spread of about 5 ns can be expected.

The solution to these problems leads to the operation of a fiber system into the third-generation fiber. In order to solve these dispersion problems, a special fiber design known as *dispersion-shifted fiber* has been attempted. One attempt is to make the material and waveguide dispersions cancel each other around 1.55 μm, where fiber loss is the lowest [2]. Another attempt is to design fibers with an extended range of near-zero dispersion as mentioned in Chapter 3. The primary advantage of these approaches is that it is possible to apply multimode laser technology where the fiber loss is low. But the bit-rate–distance product is still limited by the spectral width of the laser. In order to achieve maximum distance limited only by the fiber loss, single longitudinal mode lasers have been designed. These include the distributed-feedback (DFB) lasers, and the cleaved coupled-cavity (C³) lasers discussed in Chapter 6. When these objectives are fulfilled, we will have reached milestone stage 3*a* in Fig. 9.1.

The bit-rate–distance product or the repeater spacing can be further raised with existing single-mode laser and SMF by improving the detection system. This is the coherent detection system in which a local (tunable) laser is used to frequency-mix with the incoming signal so as to produce a microwave frequency with which further processing can be done more efficiently electronically. However, the heterodyne or homodyne process requires an extremely stable laser light source. Optical-phase information is preserved and can be used for information transmission, so that phase-shift keying (PSK) becomes possible.

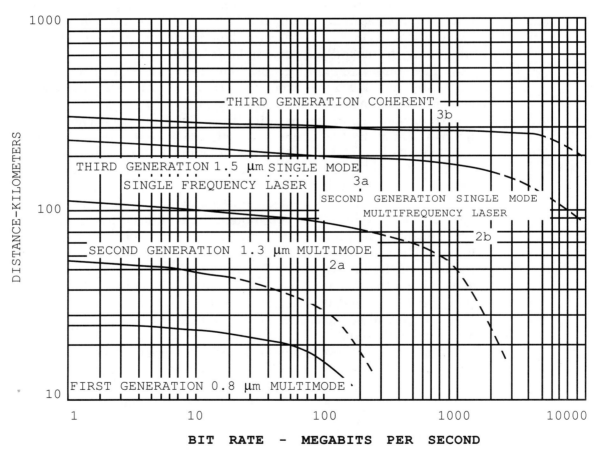

Figure 9.2 The performance of fiber systems (repeater distance vs. bit rate) of the various stages or types of development. [After J. E. Midwinter [3] (1985) © IEEE, 1985.]

An estimated gain in receiver sensitivity of 20 dB could be achieved by this method. We shall designate this system as the *final third-generation fiber system 3b,* which is our immediate future goal of optical fiber communication.

The performance of these fiber system classifications can be shown clearly by an estimate plot of the distance (in kilometers) versus the bit-rate curves defining each stage as shown in Fig. 9.2 [3].

Network Development

In the initial stage of fiber optical system development, the primary objective has been to extend the transmission rate (or bandwidth) and the range of telecommunications link as outlined in the preceding section. Augmented by advancement in optical fiber, light source, and detector technologies, high-speed, large-capacity, long-range, point-to-point transmission systems have been deployed rapidly. Further increases in channel capacity, although modest, are also accomplished through the use of multiplexing techniques. These systems as they stand are rather traditional compared to the new network development to be described in this section.

Local-Area Networks and Metropolitan-Area Networks

Since 1985, a new and significant trend in communication system design has been unfolding. It may be called the "network of networks." It is a system that combines a high-speed, large-capacity fiber optic system with a widely distributed highly flexible local telephone switching network to form a local-area network (LAN) system. The deployment of this system becomes widespread almost immediately. Practically all modern cooperations employ interoffice communication and data transmission systems. Industrial plants use this system to coordinate production processes and gain control of every step of the manufacturing processes. Speeds of up to 10 Mbit/s are widely deployed. An extension of this network to cover metropolitan areas soon developed; this is termed the metropolitan-area network (MAN). Speeds of up to 100 Mbit/s are contemplated. Further increase in speeds of up to Gbit/s is the objective of current designs.

Local-area networks initially developed from local computer communication networks. The technology of linking computer gears located within a geographically restricted area relies heavily on the existing telephone switching technology to interconnect each other at first. As the advantages of these computer networks become fully realized, for example, computer users can share expensive resources and databases, factories can exercise process control centrally, offices can send electronic mail efficiently, and so on; hence, the demand for

LANs has increased rapidly. This has been followed by the demand for higher speed of transmission. The existing copper-wire pairs and coaxial cables can no longer satisfy these demands. New LAN installations all use optical fibers for the transmission media. Now, it is possible to expand the local radial distance of 30 km and a speed of Gbit/s with an optical fiber bundle. In comparison, the copper-wire pairs and coaxial cables can reach only a few kilometers, with a speed of only up to 100 Mbit/s.

Local-area network schemes adopt certain network topologies: the bus, ring, star, or graphic. These are spatial patterns formed by the network's nodes and connecting links. In a bus scheme, each station is connected to a common running bus by a tapper. Signals from one station branch out from the tap in both directions, thus reaching every station in the network. In a ring scheme, a chain of signal repeaters are connected to form a circular ring. Each station launches its message into the ring via the repeater. Both schemes require a means of allocating an access to the common transmission medium. A technique known as *time-division multiple access* (TDMA) is usually used. In a star scheme, all stations are connected to a center hub by optical fibers in a radial fashion. It is widely used in long-distance networks and in connection with centralized local networks. The graphic scheme is an unconstrained system used in decentralized networks.

The local distributed intelligence center of a LAN adopts an open system architecture that allows users to select equipments from many suppliers if they wish to. The network architecture is built in multilayers. Each layer is built to perform a specific function using standardized interfaces. Computer programs are used to interconnect various interfaces for certain types of service. Examples of early deployment of these systems are described in the next section. Readers are advised to read a recent book edited by Kummerle [4], and a special section on telecommunication in [5] for more detailed discussion.

Integrated-Service Digital Networks

The network networking expands rapidly into an even more ambitious scale. The environment that favors this new expansion comes from two sides, service demands and technological advancements. The dynamic growth of data communication in the last decade has promoted further service demands, which has, in turn, stimulated more technological developments, and so on. The following is a list of the demands of services and the technological advances that promote the network networking. Demands in services can be grouped into the following categories:

1. Broad services, including voice, data, video, and other optical information services
2. Flexible service capability and multiservice integration
3. The need to share expensive resources between users, such as printers and other instruments
4. The need to share accurate information on common data buses
5. Customer-controlled services to enable free selection of services by users
6. Iterative services; a two-way communication system to enable the suppliers and users to work together to reach agreement

With the proper technical capability, one could make the deployment of the network networking possible and profitable. Here are just a few:

1. Single-mode fiber will replace copper-wire pairs and even coaxial cables to provide high-speed, large-bandwidth communication links.
2. Inexpensive but reliable light sources and detectors have become available to operate the optic fiber systems.
3. Digital technology, because of its intrinsic freedom from noise impairment, offers potential for the carriage of different types of information, simplifies maintenance, and reduces the cost of installation.
4. Computer-controlled switching programs provide flexibility and reliability to automate many network functions.
5. Integrated optics add to the simplicity and economy of the system structure.
6. International standardization facilitates the design of network interfaces.

These technological advances, which are present in the network today, are leading to a more versatile and responsive network, the integrated-services digital networks (ISDNs).

ISDN Specifics

Integrated-service digital networks are designed to take full advantage of modern digital network technology, utilizing the fully developed telephone network structures to achieve extremely flexible service capabilities, including transport on a circuit-switched, packet-switched, and private-line basis. Fundamental to ISDNs are standard network interfaces, the basic and primary interfaces. The basic interface provides two 64-kbit/s channels for transmission of information plus one 16-kbit/s channel for the transmission of both customer information and signal data for a total of 144 kbit/s. The primary interfaces are designed for higher bit rates (at 1.54 Mbit/s in the United States and 2.05 Mbit/s in Europe), divided into both information-bearing and signaling channels. Completed ISDNs will be able to transmit high-speed data, facsimile, telemetry, signaling, and slow-motion video digitally over the same connections as voice, either separately or simultaneously over the same fiber links [6].

Installation of ISDN systems in the United States began in November 1986. Regional companies of Bell Telephone Company, Mountain Bell, Northwestern Bell, and Pacific Northwest Bell installed a Northern Telecon DMS-100 switch linking their regions. In March 1987, four other trials were carried out at Phoenix, Arizona; Portland, Oregon; Minneapolis, Minnesota; and Illinois. They used a GTE GTD 5 switch or an AT&T 5E55 switch. Pacific Bell also successfully completed a test of the ISDN Project Victoria recently, which allows a single telephone line to carry simultaneously two 32-kbit/s voice channels, a 9.6-kbit/s medium-speed data channel, and four 1200-bit/s low-speed channels. In Danville, California, 35 miles from San Francisco, 200 residential customers used the system to access six information services.

Research is under way to develop ISDNs to operate on 100-Mbit/s capability, to include such broadband services as enhanced-quality video, high-definition television, and very high speed data transmission.

Examples of the Existing LANs and MANs

Optical fiber communication systems can be divided into three groups according to the distance to be covered: the local system, the intermediate-range system, and the long-haul system.

Local Systems

A local system covers a communication network in the immediate vicinity of the stations located in a factory, a bank, or a small community. The use of optical fiber systems can be profitable in two respects: (1) to replace the overcrowded telephone communication lines and control and sensing links and (2) to establish fast multichannel database links between computer terminals.

Established metropolitan telephone communication systems utilized pairs of copper wires embedded in conduits under the city streets. With the increasing demands of wideband transmission such as cable TV, coaxial cables have also been used to link networks with broadcasting stations. However, most cities have experienced an overcrowding situation even with the present load conditions. Further improvement in the quality and quantity of the service can be reached only by replacing the existing wire pairs and coaxial cables by an optical fiber where bandwidth is wide enough to accommodate the combination of services. How soon the existing system will be replaced by an optical system depends on economical factors rather than any technological problem. But any new installations will undoubtedly adopt the optical fiber system.

For database systems, copper-wire systems are entirely unsatisfactory because of the limited bandwidth and high cost. The fiber optic system becomes the only choice.

Since 1976, many companies throughout the world began installing optical fiber local distribution networks. In the continental United States, more fiber optic telephone links were installed during 1984–1986 than in all the years previously, and the trend is continuing.

Practically all local distribution optical fiber systems have one thing in common. Most are of the first-generation system; that is, they practically all use MMF (SI or GI), LED, or multimode LD for the light source and Si–PIN (silicon/p–i–n) diode or Si–APD (silicon–avalanche photodiode) for the detector and operate on 0.82–0.85-μm wavelength. At these short ranges, fiber losses and dispersion do not matter except for very high bit rate services. LED, Si–PIN photodiode, and MMF step index offer inexpensive combination of system design yet provide adequate reliability and long-life service.

Since 1977, many optical fiber communication systems have been installed in the United States, Japan, and Europe. We shall cite just a few to show how valuable experiences have been gained from the design and installation of these systems.

For the United States, we cite the example of the "Chicago Project" because of the unique environmental conditions it encountered. After the first successful experimental optical fiber system in Atlanta, Georgia in 1976, the Chicago Project was conceived. It was designed to carry commercial traffic, including telephone channels, interoffice trunks, a picture (video) phone, and a 4-MHz black-and-white video conference service. The highlights of the design are shown in Table 9.1.

The unique experience during the installation of this network was that during the installation, the work crew had to pull the cables through about 32 manholes in the congested Chicago metropolitan area, some of which were partially filled with water. It was a real test for this new technology. Engineers learned that in designing a system, a sufficient margin must be allowed to account for the extremely hostile environmental conditions and rough treatment. The service of this network has been operating smoothly without any outage since its installation.

For Japan, we cite the "Higashi-ikoma Optical Visual Information System" as an example because it is believed to be the first interactive optical fiber system designed to serve a community. It is a two-way

Table 9.1 The Chicago Project Data

	System
Bit rate	44.7 Mbit/s
Total length	2.5 km
Channels	144
	Fiber
Type	Graded-index ribbon structure
NA	0.23
Dimensions	Core diameter 55 mm, cladding diameter 100 mm
Fiber Loss	4 dB/km
Dispersion	2.3 ns/km
Splicing loss	0.8 dB
	Source
Type	AlGaAs DH Laser
Wavelength	0.82 mm
Power	−3 dBm
	Detector
Type	Si–APD
Sensitivity	−54 dBm

Data taken from the Third International Conference on Integrated Optics and Optical Fiber Communication, San Francisco, April 27–29, 1981, *Tech. Digest.*

interactive cable TV service built for the suburban town of Osaka. The complex system consists of computer-controlled video services to about 160 home subscribers and 8 local studio terminals, using a total of about 400 km of fibers with no repeaters. Subscribers may select any TV stations and key-in any desired programs, such as for education or entertainment. A two-way picture phone (each home is equipped with a TV camera and a TV monitor) provides shopping opportunities with participating commercial organizations. The system provides 36 channels, each 6 km long, for the distribution network. In addition, there are 26 fibers 400 m in length, 18 fibers 500 m in length, 6 fibers 5.5 km in length, 4 fibers 1.5 km in length, and 2 fibers 3.1 km in length for interconnections. Plastic-cladded optical fibers with NA = 0.25 and a fiber loss of 16 dB/km are used. A total of 30 × 168 matrix video switches are used for the interconnection. The light source is an LED (0.82 μm). A sound carrier at 6 MHz is provided. The LED output power is about 1 mW. Si–PIN diodes are used for detection. The detection sensitivity is −27.9 dBm for a video with S/N = 52 dB.

In the United Kingdom, field trials of the first-generation optical fibers started in 1977. Two lines, one involving an 8-Mbit/s optical fiber over a 13-km route and the other, a 140-Mbit/s fiber over a distance of 8 km between the British Post Office Research Center at Martleshaw Heath and the Ipswitch Telephone Exchange in Suffolk, England, were installed [7]. Both fibers are of Cornings graded-index type with a fiber loss of 4.5 dB/km installed into the existing ducts of the old telephone systems. The 8-Mbit/s line uses an LED at 0.82-μm wavelength as the light source. The output power is approximately 65 μW. Silicon–APD was used for the detector. It is followed by Si–FET amplifiers with a gain of 40 and a receiver sensitivity of -59.7 dBm. It carries about 120 telephone channels. The 140-Mbit/s line is powered by a double-heterojunction (DH)–GaAlAs stripe geometry laser source with a wavelength of 0.84 μm and a GaAs FET amplifier. The gain is 80, and the receiver sensitivity is about -43 dBm. This system carries about 1920 telephone channels. Subsequent improvements in fibers have reduced the average attenuation and dispersion. No repeaters are used in the system.

In Germany, the Heinrich Hertz Institute was engaged in a number of basic research and development programs on wideband information handling systems [8]. The prototype system uses optical fiber highways to provide high-bit-rate digital communication between various terminals. Each customer shares a time slot in the program whose receiver decodes the timeframe of the passing digital highway and processes it for its own use. The highest data rate on the subscriber-loop highway is 280 Mbit/s, which allows the use of color TV, telephone, and a variety of data services.

In France, a fiber link to bring picture phone, TV, and FM radio was installed in Biarritz, a seaside resort town in southwestern France, to serve about 5000 subscribers [9]. The optical fiber used is a MCVD graded-index 50–125-μm fiber with a fiber loss of 2.5 dB/km at 0.85-μm wavelength. The subscriber link is composed of a laser–PIN link from the center to the subscribers. From subscriber to center, it is an LED–APD link. The system has been in full service since 1986.

In Elie, Manitoba, Canada, another interesting experiment on a rural area optical fiber subscriber system has been carried out [10]. A particularly difficult environment that involves providing economically viable wideband service in extreme local climate ranging from $-40°C$ in winter to $+30°C$ in summer is a real test to the fiber system. The area is very thinly populated with an average distance between a private home and a station of about 5 km. The estimated total installation cost to the customers averages $2000–$3000 per home. The optical fiber cable to the subscriber contains three fibers, one for each direction of transmission and one as a spare. Fiber loss is about 4 dB/km, and the bandwidth–distance product is 600 MHz·km. An LED is used in both directions as the light source. A laser diode is used for transmission to the subscriber in some long loops (>3 km). The wavelength of the LED is 0.84 μm and that of the LD, 0.93 μm. The different wavelengths allow bidirectional transmission on a single fiber through the use of biconical-taper optical directional couplers. Although wide-range temperature variations are experienced, no major problems have been reported. The only question is whether it is economically viable for these installations.

In 1985, GTE Corporation of Stamford, Connecticut, began to set up an experimental fiber optic LAN station to replace the existing TV cable system in Cerritos, California. It is intended to display the video world of the future. It includes

Transmitting movies of one's choice right into one's TV set at any time of day

Providing a picture-phone possibility by connecting one's telephone to a videocamera and a TV set

Setting videocameras to monitor a sleeping infant from a neighbor's house

Since 1985, LANs and ISDNs have been deployed rapidly throughout the world. Eventually, a local telephone subscriber should theoretically be able to order anything from anywhere in the world at a reasonable cost. This includes high-quality voice transmission, selective entertaining programs, high-resolution image transmission, and even iterative shopping.

Optic Data Bus for Local Data Communication

Besides the local telephone communications and TV services described in the previous section, another important application of optic fibers in local communication is the transmission and handling of data in local systems. In recent years, an enormous amount of data has been compiled in scientific, medical, business, and industrial databases that need immediate attention. The data bus is a communication system that interconnects a number of terminals (computers) over a common channel (the bus) to keep track of the flow and storage of these data and to process them. Conventional data buses use twisted-wire pairs and coaxial cables. The speed of data processing is usually limited by bandwidth. When parallel multiwire data buses are used, the data rates are limited to less than 10 Mword/s. The number of switches and connectors required for parallel operations add to the complexity of the system and reduce the reliability even further. The replacement by fiber optic systems offers many advantages, and it suddenly becomes urgent that these conversions take place. A high-rate, high-capacity optic fiber data bus that is capable of handling in excess of 100 Mbit/s can offer a tenfold reduction in cost, 1000-fold increase in the amount of information return, and 100-fold improvement in data access time. A single fiber might replace a 4-, 8-, 16-, or 32-word parallel data bus with the information being transmitted in serial form at a correspondingly high bit rate. This reduces the number of cables and connectors used and improves the reliability of the system [11].

In this section, we shall cite one application of an optical fiber data bus system for the management and control of an electric power system in Japan. For a large system involving many power-generating facilities and substations, the transmission of information for system protection, supervision, and control becomes extremely important. Existing communications have been transmitted by microwave links in combination with various power-line carriers and communication cables to maintain reliability and quality at high levels. Cities such as Tokyo and Osaka are connected with a high-voltage (275–400-kV) power network. The existing microwave links meet with increasing interference due to electromagnetic radiation and high-building blockage. Fiber optic communication systems offer a perfect replacement with added advantages such as higher speed and safe operation. Except for intercity communication links, other uses of optical fibers involve only short lines, less than a few kilometers in length. Thus, fiber loss is not a serious concern. Fibers with a transmission loss of 4 dB/km are normally sufficient. To increase the data rate and to use the fiber links fully, both time-division multiplexing (TDM) with 6.3-Mbit/s rate and frequency-division multiplexing (FDM) with 600 channels are installed. The fibers are contained in cables laid either in city conduits or on rural aerial distribution poles. All-optic fibers use thermal-fusion joints, are powered by multimode lasers at 0.89 μm, and use Si–APD as detectors. The system reliability recorded $BER = 10^{-11}-10^{-12}$, which is far better than can be obtained with a microwave link. The lifetime of each laser is more than 10,000 h [12].

The success of this system is attributed to the system organization. At the center is a central processing unit. Incoming data from various sources are first collected, sorted, and stored in a memory bank. The processor then distributes the data to various divisions such as the power system protection division, the supervision and control division, the intercomputer division, and the general communication division. Each division then acts on the order and reports back to the central processor to complete the mission.

This general scheme can be adapted to any closed-loop networks in other applications.

Long-Haul Systems

Under this heading, we report all long-haul systems, including land cables and underwater or submarine cables where the distance is large and repeaters must be used to extend the range of operation. Repeaters are considered as the most expensive items in a system and are even more expensive if maintenance and repairs are considered. Existing undersea coaxial cables may require repeaters every 5–8 km.

A tenfold increase in repeater spacing may make an enormous scoring in both installation and maintenance costs. This makes the use of optical fibers for long-haul transmission even more attractive and worthwhile. A typical optical fiber cable may require a repeater every 50 km at a data rate of 140 Mbit/s. At a higher speed service, the saving is even more impressive.

In this section we shall discuss first the general approaches to the design of optical fiber long-haul communication systems. This is followed by some examples of the existing and proposed systems around the globe. It is meant to show the trend, not exclusive cases, of the system design.

General Approaches to the Design of Long-Haul Optical Fiber Communication Systems

At present, there are in general three approaches to the design of a long-haul optical fiber system. The first approach is to use a second-generation fiber operating at a wavelength of about 1.3 μm where the dispersion is zero (silica fibers) and to develop a system using LEDs and Ge–PIN diodes. This system is relatively inexpensive to install and operate and is very reliable. To maximize the repeater spacing up to 50 km, SMF with a loss factor of less than 0.5 dB/km and a dispersion coefficient of less than a few picoseconds per kilometer-nanometer are required. Replacing the LED by an InGaAsP laser (driven by a GaAs FET circuit) and using Ge–APD or InGaAs–APS as detector followed by a low-noise FET amplifier could boost the repeater spacing even higher.

The second approach is to use an ultra-low-loss fiber (0.16 dB/km) operating at 1.55-μm wavelength and a precisely graded fiber to minimize the multipath dispersion effect. LED and PIN diodes can again be used to achieve an inexpensive and reliable operation. Again replacing the lasers and detectors would increase the bit rates of transmission and repeater spacing. But modal noise would also be increased. The design and graded indexing of the fiber could be done only by sacrificing the attendable minimized fiber loss and with added cost.

The third approach is to use a fiber designed for the second approach but to use a narrow-spectral-width laser source, preferably a single-frequency stabilized light source at 1.55-μm wavelength. The performance of this combination could be improved by 10–100 times compared to the second one. (Coherent detection could also improve the range and performance.) The obstacle to this approach is that construction of these laser sources is at present very expensive and the laser is relatively short-lived.

A few examples of the developing fiber optics communication systems around the globe could increase our awareness of the rapidly developing technology. Although most of these systems are still in the testing stage, the experiences gained from these test systems are most valuable.

Land Fiber Cables

Field trials of land fiber cables have been reported in two wavelength regions, 1.3 and 1.55 μm. The principal characteristics of these fiber link experiments are listed in Table 9.2 [13–16].

A summary for the main features of the optical cable transmission system for Japan's Integrated Network System (INS) has been released by Nippon Telegraph and Telephone Corporation. This is shown in Table 9.3.

Submarine Fiber Cables

A more exciting development is the testing and construction of undersea optical fiber communication systems. Under international cooperation, many companies from different countries are participating

Table 9.2 Field Trials of Land Fiber Cables—Principal Characteristics

	1.3 μm		1.55 μm	
	A	B	C	D
Bit rate (Mbit/s)	400	140–650	2000	140
Fiber type	SI	SI	SI	SI
Δn	0.35 ± 0.05%	0.31 ± 0.05%	0.01	~0.45%
Core diameter (μm)	—	9.6 ± 0.4	4.5	8
$\lambda_c(\mu$m)	1.17 ± 0.07	1.2 ± 0.05	1.0	1.2 ± 0.05
Cladding diameter (μm)	12.5 ± 3	—	—	—
Fiber loss (dB/km)	0.8	0.68	0.54	0.34
Splices used	—	—	2 fused splices	10 fused splices
Source	Laser	InGaAsP–BH	InGaAsP–BH	InGaAsP–BH, injection-locking
$\Delta \lambda$ (nm)	20	1.5	—	0.2
Launch power (dB$_m$)	0	−7 to −8.7	−2.2	−8
Power margin (dB)	—	—	1.4	1.1
Detector	GeAPD	InGaAsP–PIN	$P^+n\,n^-$Ge–APD	InGaAs–PIN
Amplifier	—	FET	Si bipolar transfer	FET
Reference	[13]	[14]	[15]	[16]

Table 9.3 Main Features of Japan's INS Optical Fiber Cables

Systems	Trunk					Submarine Nonrepeater			Repeater
Capacity	Small F-6M	Medium F-32M	F-100M	Large F-400M	Very large F-1.6G	Small FS-1.5M	Medium FS-6M	Large FS-400M	Large FS-600M
Information bit rate (Mbit/s)	6,312	32,064	97,728	397,200	1,588,800	1,544	6,312	←—— 397,200 ——→	
Line bit rate (Mbit/s)	12,624	64,128	111,689	445,837	1,783,300	3,088	12,224	←—— 445,837 ——→	
Telephone channels	98	480	1,440	5,760	23,040	24	96	←—— 5,760 ——→	
Wavelength (μm)	1.2/1.3	←———————————— 1.3 ————————————————————→							
Optical fiber	←—— GI ——→		←—— SMF ——→			←— GI —→		←——— SMF ———→	
Light source	←——— InGaAsP Laser diode ———————→					special laser diodes			
Detector	←—— Ge APD ——→		←— InGaAsAPD —→			←——— Ge APD ——————→			
Repeater spacing (km)	20	10,20	10	←—— 40 ——→		25–45	20–40	←——— 40 ———→	

Adapted from a table supplied by Nippon Telegraph Telephone Corporation.

Table 9.4 Main Features of Submarine Optical Fiber Communication Systems

	United States	United Kingdom	France	Japan
Bit rate (Mbit/s)	280	140–280	140–280	260–400
Wavelength (μm)				
a	1.3	1.3	1.3	1.3
b	1.55	1.55	1.55	1.55
System length (km)	8000	7500	10,000	10,000
Sea depth (m)	7500	7500	7,650	8,000
Fiber loss (dB/km) (Cable, splices)	SMF (w-type) <1	SMF (w-type) <1	SMF <1	SMF (w-type) <1
Cable strength (kN)	80	90	80	75–100
Repeater spacing (kN)	25–50	25–50	25–50	25–50
Repeater power (W)	4	3–5	5	3–5
Laser redundancy detector	InGaAsP–BH ≤3 standby InGaAs–PIN FET	—	≤3 standby	≤1 standby
Designed reliability (years)	8	10	15	10

in these developments. Table 9.4 summarizes the features of some projects under construction [17, 18].

The work on the installation of the eighth submarine transatlantic telephone cable (TAT-8), under the sponsorship of the AT&T Corporation of America, the British Telecon, and the French Telecommunication Administration, continues. This 6600-km-long cable connects Tuckerton, New Jersey with a branching repeater off Europe that allows one leg of the cable to enter the United Kingdom at Widemouth and another leg to enter France at Penmarch. The usable capacity of the system is to be 560 Mbit/s, which is the equivalent of up to 40,000 telephone calls. Four-digit services are available among the three countries, operating at 56 and 64 Mbit/s and at 1544- and 2048-Mbit/s data rates.

This system has been in operation since December 1988.

Other connections under construction and their scheduled opening dates are

Between San Francisco and Hawaii—1989
Between Florida and Colombia (South America)—1990
Between Florida and Jamaica—1990
Between Italy, Greece, Turkey, and Israel—1990
Between Spain and Italy—1991
Between Los Angeles, Australia, and New Zealand—1991

Early in 1986, AT&T signed formal agreements with 20 companies in eight nations—Australia, Canada, Great Britain, New Zealand, The Philippines, South Korea, Taiwan, and West Germany—to construct two fiber optic submarine cable systems that together would span the Pacific Ocean by the early 1990s, thus replacing the existing coaxial cables already linking these countries.

Conclusion

The general trend in the development of optical fiber technology can be seen from the plots in Figs. 9.1 and 9.2, which show the milestones and the progressively increasing capabilities in terms of repeater spacing and bit rates [16]. Should the new search for ultra-low-loss fibers at longer wavelengths ($> 1.7\ \mu$m) be successful, and new light sources and detectors be found, the capability of optical fiber communication systems could be extended even further.

The examples given in the text are for reference use only. New data are surfacing every day. Readers should consult recent publications for more current information.

The business of installing long-haul optical fiber communication systems, including submarine cables, is progressing very rapidly. It is all very impressive and exciting. But the big business could lie in the development of local data bus systems that are applicable and useful in every branch of service. Every business could use a good, fast data system to help in organization and decision making.

References

1. C. Kao and E. Basch, Introduction to fiber optics. In *Optical-Fiber Transmission* (E. Basch, ed.), p. 1. Macmillan (Sams & Co.), New York, 1987.
2. T. D. Croft, J. E. Ritter, and V. A. Bhagavatula, Low-loss dispersion-shifted single mode fiber manufactured by the OVD process. *IEEE J. Lightwave Technol.* **LT-3**(5), 931−934 (1985).
3. J. E. Midwinter, Current status of optical communication technology. *IEEE J. Lightwave Technol.* **LT-3**(5), 927−930 (1985).
4. K. Kummerle, ed., *Advances in Local Area Networks*. IEEE Press, New York, 1987.
5. Special section on Telecommunications: The Next Ten Years. (Contains four papers.) 1222−1274, Proc. *IEEE*, September 1986.
6. Special section on Integrated Optics and Optoelectronics. Contains five invited papers, pp. 1472−1535, *Proc. IEEE*, **75**, November 1987.
7. R. W. Berry, D. J. Brace, and I. A. Ravanscroft, Optical fiber system trials at 8 Mbit/s and 140 Mbit/s. *IEEE Trans. Commun.* **COM-26**, 1020−1027 (1978).
8. K. Fassganger, U. Haller, H. J. Matt, and H. Ohnsorge, Experimental system for an integrated communicators network using optical links. *Frequenz* **32**, 165 (1978).
9. G. Lentiez, The fibering of Biarritz. *Laser Focus* **17**(11), 1125−1128 (1981).
10. K. Y. Chang, Fiber-optic broad-band integrated distribution-Elie and beyond. *IEEE J. Select. Areas Commun.* **SAC-1**, 439−444 (1983).
11. D. R. Porter, P. R. Couch, and J. W. Schelin, A high speed fiber optic data bus for local data communications. *IEEE J. Select. Areas Commun.* **SAC-1**, 479−488 (1983).
12. F. Aoki and H. Nabeshima, Optical-fiber communications for electric power companies in Japan. *Proc. IEEE* **68**(10), 1280−1285 (1980).
13. E. Iwahashi, First field trial of long-haul transmission system using single-mode fiber cable at a 1.3 μm wavelength. *IEEE J. Quantum Electron.* **QE-17**(6), 890−896 (1981).
14. C. J. Todd, J. R. Stern, and K. J. Beales, *Dig. Tech. Pap., Opt. Fiber Commun.*, *1982*, Paper TUDD7 (1982).
15. J. Yamada, A. Kawana, H. Nagai, T. Kimura, and T. Miya, Gbits/s optical receiver sensitivity and zero-dispersion single-mode fiber transmission at 1.55 μm. *IEEE J. Quantum Electron.* **QE-18**, 1537-1546 (1982).
16. D. J. Malyon and A. P. McDonna, A 102 km monomode fibre systems experiment at 140 Mbit/s with an injection locked I-52 laser transmitter. *Electron. Lett.* **18**(11), 445−447 (1982).
17. P. K. Runge and P. R. Trischitta, The SL undersea lightguide system. *IEEE J. Select. Areas Commun.* **SAC-1**, 459−466 (1983).
18. R. E. Wagner, S. M. Abbott, R. F. Gleason, R. M. Paski, A. G. Richardson, D. G. Ross, and D. R. Tuminaro, Lightwave undersea cable system. *Proc. Int. Conf. Commun.*, *1982* (1982).

PART FOUR

Optical Fiber Sensors, Passive Applications, and Integrated Devices

In this part, Chapters 10–13, we introduce optical fiber sensors, miscellaneous passive applications of optical fibers, and integrated optical fiber devices. Optical fibers offer the same advantages to sensors as they do to telecommunication systems: high information capacity, imperviousness to electromagnetic interferences, capability of working in hostile environments, low cost, and light weight. In some respect, however, the criteria for a good sensor is just opposite that of a good optical fiber, such as a lightguide. While a good lightguide requires that its characteristic parameters remain unaffected by environmental changes, an optical fiber sensor capitalizes the variability of the parameters by the environment. These demands may seem contradictory, yet the parameters in a conventional fiber leave enough variability for use also as a sensor. Moreover, special sensitive coating can always be added to the fiber to render it more sensitive to a particular type of application.

In Chapter 10 we put together the physical effects that can be utilized to affect the properties of an optical fiber. In Chapter 11 we describe the pure fiber sensors and remote fiber sensors, respectively. In Chapter 12, many passive applications of optical fiber, such as those in medicine, industry, and commercial systems, are introduced. Finally in Chapter 13, integrated optics is briefly discussed. The author feels that integrated fiber optics is the optical frequency analog of a microstrip circuit in microwave. These devices may have considerable potential in many applications, although the discussion here will be only introductory in nature.

10

Physical Phenomena for Optical Fiber Sensors

Introduction

Optical fiber sensors represent a new branch of optical fiber engineering that has been developing rapidly in recent years. An optical fiber sensor is a length of fiber that modulates the light passing through it when exposed to the changing environment we wish to sense. Besides the optical fiber, it has a light source such as a light-emitting diode (LED) or a laser at one end and a photodetector at the other end of the fiber to register the changes. We divide the optical fiber sensors into two types: the pure optical fiber sensor, where the fiber itself is the sensing element, and the remote sensor, in which the fibers are used only to bring light to and from a separate sensing device.

In pure fiber sensors, the measurand interacts directly with the light traveling in the fiber. The resultant light can be either intensity-, phase-, or polarization-modulated within the fiber. There generally are no optical interfaces at the modulator site. The feed and return fibers may also impose modulation of the light passing within them, however, giving misleading information or an error. Selecting a proper fiber and arranging for a suitable interaction region become important considerations in the design. A scheme of detection that favors self-canceling of these difficulties is preferred.

A remote sensor involves the use of a special sensing element that is sensitive to the environment one wishes to probe. Subjected to the environmental change, such as temperature and pressure, the sensing element modulates the light leading to and from the sensor by the

279

fibers. The detection process that follows can be calibrated in accordance with the respective changes. The sensing element is also called a *transducer*, as usually a change of energy form is involved.

Intensity modulation is the most commonly used method and is easy to implement. Phase modulation is by far the most sensitive and accurate technique if used in combination with an interferometric scheme. Polarization modulation is unique, as the state of polarization of the light through the fiber is affected by the environment. Other methods are also possible.

Both single-mode fibers (SMF) and multimode fibers (MMF) can be used as sensors. Multimode fibers are used for less demanding applications; single-mode fibers are used when extreme sensitivity and accuracy are demanded. In the case of a sensor where polarization modulation is used, only SMF fiber is recommended.

In this chapter, we bring out the physical concepts and phenomena that have to be exploited for making an optical fiber into a practical fiber sensor. We defer discussion of the practical implementation of these sensors to the next chapter.

Fiber Birefringence

One important parameter that characterizes an optical fiber is the refractive index n. In the simple theory, n is introduced as a pure material constant that is isotropic and single-valued. This is, however, an oversimplified picture. We notice that all fibers are birefringent. The refractive index is actually a tensor quantity. The implication of this complicated representation is that the propagation of light in this medium will have propagation constants that vary in different directions. To simplify, let us illustrate a two-dimensional example. Let n_x and n_y be the refractive indices in the x- and y-directions, respectively. Then the propagation constants in the x- and y-directions become β_x and β_y, respectively. The difference in β values, or $\beta_y - \beta_x$, is a measure of the birefringence of the medium (fiber). Many fiber sensors are built based on the birefringent nature of the fiber.

Birefringence of a fiber can either be intrinsic or induced. Intrinsic birefringences are those inherited from manufacturing processes of the fiber caused by, for example, the core ellipticity, cladding eccentricity, and so on. Induced birefringences are those induced by external forces during operation. These include thermal stress, mechanical bending, twisting, and other phenomena. For optical fiber sensing, induced birefringence constitutes a useful phenomenon that one wishes to take advantage of, whereas one would like to minimize or compensate for the intrinsic birefringence as much as possible.

Birefringence can be linear or circular. Both effects can be used to build sensors. But the linear birefringence is most useful in optical fiber sensors.

The State of Polarization

The state of polarization (SOP) of the electromagnetic wave is characterized by the orientation of the electric-field vector representing the propagating wave along a transmission path. The wave is said to be linearly polarized if a single E-field vector, perpendicular to the direction of propagation, can represent the wave. Of course, any vector representing a linearly polarized wave can be resolved into a pair of mutually perpendicular components, such as the x- and y-components of the electric field, traveling in phase along the direction of propagation (z-direction). This is shown in Fig. 10.1a. If the two components along the x- and y-axes are equal in amplitude, but are optically 90° out of phase, then the resultant wave is a circularly polarized light. This state is shown in Fig. 10.1b. Circularly polarized states are distinguished by the sense of rotation of the vector. The

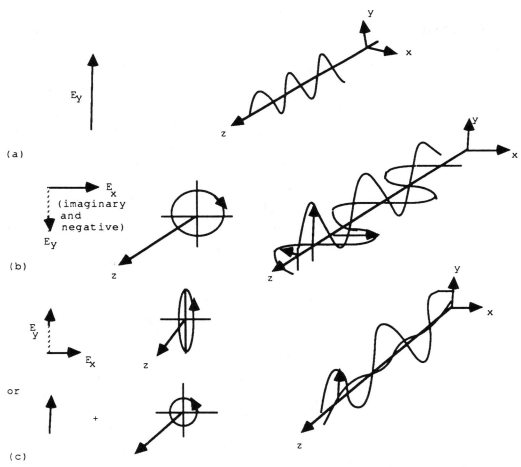

Figure 10.1 The state of polarization of lightwaves: (a) linear; (b) circular; (c) elliptical.

state shown in Fig. 10.1b represents a right circular polarization, as the resultant total polarized vector rotates clockwise when viewed toward the source. If the relative phases of E_x and E_y of this right circularly polarized light is shifted by 180°, left circularly polarized light will result. Linearly polarized light is specified by its amplitude. Circular polarized light is specified by its amplitude and the sense of rotation (left or right). The orientation of a reference x- and y-axes is immaterial.

If x- and y-components of unequal field amplitudes are added, the result is an elliptically polarized light as shown in Fig. 10.1c. This wave is now characterized by its amplitude and ellipticity and by the orientation of the major axis of the ellipse with respect to a reference axis. It is sometimes convenient to consider elliptically polarized light as the sum of a circular component and a linear component.

When polarized light passes through a medium possessing birefringence, the output light will have its state of polarization changed. The detection of a change of the state of polarization constitutes a useful measure of the cause that brought about these changes.

The Electrooptic Effect

In the section titled "External Modulation" (Chapter 8) we have introduced the electrooptic effect as a means of externally modulating the source light for transferring information to the fiber. Since the operational principle of optical fiber sensors utilizes the same mechanism to affect the information, the electrooptic effect becomes one specific choice to implement the modulation.

Studies on the application of an electric field to a crystal such as lithium niobate ($LiNbO_3$) showed that the refractive index of the material changes differently for different polarization and propagation directions of the field through the crystal [1]. The change in refractive index expressed in terms of $\Delta(1/n^2)$ (Eq. 8-15) is reproduced here as

$$\Delta\left(\frac{1}{n^2}\right) = \alpha_{ij}E + \beta_{ij}E^2 \qquad (10\text{-}1)$$

Depending on the crystal structure and orientation, either the longitudinal or the transverse electrooptic effect may be observed. The former is known as *Pockel's effect*. Here, linear birefringence is introduced as a result of applying a field along the wave-propagation axis. This effect is linear with the applied field as indicated in the first term on the right-hand side of Eq. (10-1). This effect can be used to sense a voltage [2].

There is a second effect that is proportional to the square of the field intensity as indicated in the second term of Eq. (10-1). It is called the *Kerr effect*. Although the Kerr effect is not considered as a good modulator because of its nonlinear field dependence, it is extremely fast, and has been used as a switch known as the Kerr cell for optical switching up to 10 GHz in measuring systems [3].

A short discussion on both types of effects will be given in this section.

Pockel's Effect

Pockel's effect has been used as an external modulator for optical fiber systems as discussed in the section on external modulation in Chapter 8. Since a sensor requires a similar modulation process to affect the birefringence of the crystal, Pockel's effect can be applied to it as well. The use of Pockel's effect for a voltage sensor will be described in Chapter 11. For now, let us investigate the property of a crystal for this purpose. Obviously, we wish to choose a crystal that has the most strong Pockel's coefficient. But we must also consider the availability and the cost of the crystal. From Table 8.1, we noticed that the electrooptic coefficients are listed as the scalar quantity α, without the double subscript α_{ij}, indicated in Eq. (10-1). This is because by proper choice of the crystal orientation and the direction of lightwave propagation in the crystal, the sensor coefficient can usually be reduced to a single coefficient as shown in the table. Lithium niobate ($LiNbO_3$) and lithium tantalate ($LiTaO_3$) crystals have the highest Pockel coefficient. However, these materials are difficult to grow into usable sizes and they are very expensive. For this reason, the next best candidate is potassium dihydrogen phosphate (KH_2PO_4 or KDP). Also useful is the crystal KD*P, which is formed by substituting deuterium (in the above crystal) for hydrogen to become KD_2PO_4. It was found that with this substitution the electrooptic property of the KDP crystal can be improved dramatically. The coefficient α has been increased from 10.6 for KDP to 26.4 pm/V for KD*P. Most important of all, the half-wave voltage that is required for maximum transmission is reduced from 10.2 to 5.82 kV for the KD*P crystal. For longer-wavelength applications, cadmium telluride (CdTe) and gallium arsenide (GaAs) are recommended for $1-28$-μm and $1-4$-μm-wavelength ranges, respectively.

Kerr Effect

The second term on the right-hand side of Eq. (10-1) that represents electrooptic effects is proportional to the square of the electric field. This effect is called the *Kerr effect* in honor of its discoverer, J. Kerr, in 1875. The Kerr effect is an electrically induced birefringence that occurs in all materials. The resultant refractive indices may be denoted by $n_{\parallel}$ and $n_{\perp}$, the parallel and perpendicular components of n arising from the applied electric field. Again by a proper choice of the crystal orientation and lightwave propagation direction, the proportionality constant can be reduced to a single coefficient $K\lambda$ such that the change in the refractive index of the material can be expressed as

$$\Delta n = n - n_0 = K\lambda E^2 \qquad (10\text{-}2)$$

where K is the Kerr constant and λ is the free-space wavelength. Many liquids, as well as crystals, exhibit the Kerr effect. The most often used liquid is nitrobenzene, which has a K value of 24 pm/V^2 at 20°C. Water has a K value of 5.2 pm/V^2, and glass has $K = 1.7-3.0$ pm/V^2.

The Kerr effect is very fast for nonpolar liquids. It has been used to build light switches to turn light on and off when an electric field parallel to the optical axis is applied.

Liquid Kerr cells suffer from the disadvantage of requiring a large power for operation. But by mixing ferroelectric crystals operating at a temperature near the Curie points, such as KTN (potassium tantalite niobate), the Kerr effect can be greatly enhanced, even at reduced voltage. The Kerr effect has been used effectively as a fiber gyroscope [4, 5].

The Photoelastic Effect

Application of stress to a crystal perpendicular to the direction of propagation of a lightwave passing through it will induce an increase in the dielectric constant for the lightwave polarized along the stress direction. In anisotropic materials this effect will become directional and results in birefringence of the crystal [3].

If the mechanical stress is exerted by an acoustic wave, the effect is to change the refractive index caused by the photoelastic effect and is known as the *acoustooptic effect*. This effect was mentioned in Chapter 8, subsection titled "The Acoustooptic Effect," for use as an external modulator. The refractive index change Δn is given by

$$\Delta n = \frac{n^3 p}{2}\left[\frac{2 I_{\text{acoustic}}}{\rho V_S^3}\right]^{1/2} \qquad (10\text{-}3)$$

Here, p is the photoelastic constant, ρ is the density, V_S is the velocity of sound through the material, and I_{acoustic} is the intensity of the acoustic pressure. Values of the appropriate constants for a number of materials are given in Table 10.1. The photoelastic effect can therefore be used as the basis for a number of transducers for monitoring pressure, strain, and other parameters.

The sensing technique may be carried out by measuring either the pure phase changes or the induced birefringence caused by the introduction of an isotropic pressure field. The photoelastic effect may also induce circular birefringence when the medium is subjected to torsional stress. Some of these applications will be described in Chapter 11.

Table 10.1 Photoelastic Constants for Some Materials

Material	Density ρ (g/cm³)	Acoustic Velocity v_s (km/s)	Refractive Index n	Photoelastic Constant p
H$_2$O (water)	1	1.5	1.33	0.31
S$_i$O$_2$ (quartz)	2.2	6.0	1.46	0.21
LiNbO$_3$	4.7	7.4	2.25	0.15
PbMO$_4$	6.95	3.74	2.30	0.28
TeO$_2$ (shear wave)	6.0	0.62	2.35	0.09

Table 10.2 Verdet Constant of Several Materials

Material	V radians/m·T at $\lambda = 0.689\ \mu$m
Quartz (SiO$_2$)	4.0
Zinc sulfide (ZnS)	82
Grown glass	6.4
Flint glass	23
Sodium chloride (NaCl)	9.6

The Magnetooptic Effect

The presence of a magnetic field in some materials may also affect its optical properties. One simple example is Faraday rotation, mentioned in Chapter 8, subsection titled "Magnetooptic Effect," as an external modulator for an optical fiber system. The plane of polarization will be rotated by an angle proportional to the applied magnetic field parallel to the direction of propagation. It is interesting to note that the sense of rotation of the plane of polarization is independent of the direction of propagation, so that the rotation can be doubled by reflecting the lightwave back through a Faraday effect device. The proportionality factor is called the *Verdet constant*. Some representative values of the Verdet constant are shown in Table 10.2. The Faraday effect is small and wavelength-dependent. For flint glass, it amounts to $\theta \approx 1.6°$/m·T at $\lambda = 0.589\ \mu$m.

One of the most successful optical fiber sensors is the Faraday rotation current or voltage monitor [2]. Optical fibers are very attractive for current sensing, especially on high-voltage lines, because of their intrinsic isolation properties. In a fiber where no birefringence other than the magnetooptically induced one exists, a linear state of polarization launched at the input of the fiber will be rotated by an angle proportional to the product of the magnetic field intensity and the fiber length. An optical-time-domain reflectometry device can be built to measure the SOP of the fiber. For example, we cite here a Faraday rotation current monitor [2].

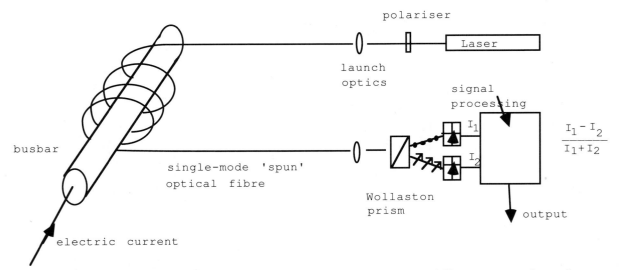

Figure 10.2 A Faraday rotation current monitor. [After B. Culshaw, Optical fibre sensing and signal processing, Peter Peregrinus, p. 105, London, 1984.]

A schematic diagram of this device is shown in Fig. 10.2. Light from a helium neon laser is launched into a single-mode spun optical fiber through a polarizer to a polarization sensor. The sensor consists of a coil of spun SMF wound on a busbar carrying the electric current to be monitored. Polarization of the lightwave through the coil is rotated by the magnetic field produced by this current. The rotated polarization is transmitted back to the signal processor to be analyzed. A spun SMF is used to ensure that the polarization of the transmitted lightwave is rotated only by the magnetic field produced by the current. Other schemes involving the use of two-way propagation have also been reported [6]. A drawback to this method is that any residual linear birefringence on the fiber leads to large sensitivity variations as environmental conditions change. Also, the method is limited to using alternating currents. A second approach is to use such a device as a polarimeter. It is more sensitive and does not have the above-mentioned drawback because the method is self-compensating.

It is impossible to name all physical effects that can be used to make sensors. Just a few have been cited in this section. Many new ideas will certainly develop as the search goes on.

Classification of Optical Fiber Sensors

Optical fiber sensors can be classified according to the mechanism by which the modulation of the lightwave in the fiber is implemented. If the lightwave is intensity-modulated, it is classified as Type *I* (where *I* stands for intensity). Depending on whether the modulation is implemented within or outside the fiber, it is subdivided as Type *I*-1 or

Type *I*-2, respectively. If the lightwave is phase-modulated, it is classified as Type *A* (where *A* stands for phase angle). If the basic Mach–Zehnder interferometric method is used, it is classified as Type *A*-1; if the Sagnac effect is used, it is a Type *A*-2. Basically, a Mach–Zehnder interferometer uses two fiber branches, and a Sagnac effect device uses a single fiber. Other types include the Type *F* for frequency modulated, Type *W* for wavelength-distribution modulation, and Type *P* for polarization modulation types of sensor.

The Intensity-Modulated Type: Type *I*

The intensity-modulated-type fiber sensor [7] is usually the simplest one to build. Environmental changes to be measured are caused to interact with the lightwave within or outside the fiber to affect the light intensity. The detected change is then calibrated in terms of the measurement such as the temperature or the pressure.

The operating principle of Type *I*-1 sensor is sketched in Fig. 10.3*a*. Light entering the sensing region remains within the optical fiber, and the intensity is modulated by, in the case shown, an acoustic wave. The pressure wave affects the transmission properties of the fiber, thus varying the lightwave intensity. Light is then collected by a photodetector, which produces an electrical signal proportional to the intensity [8].

In Fig. 10.3*b*, fibers are used to guide the lightwave to and from an external sensor only. The external sensor may be any transducer that converts the environmental change to lightwave intensity, for example, by reflection or refraction of the lightwave from a surface connected to the fibers. Here light from an LED enters a multimode optical fiber that carries the lightwave to the sensing device. In this device, the intensity of the lightwave is modulated by the stimulus, in the case shown, the pressure on a diaphragm. The returned light is processed as in Type *I*-1 after being picked up by another fiber. This is called *Type I-2 sensor.*

The Phase-Modulation Type: Type *A*

Phase-angle-modulated optical fiber sensors are highly sensitive to environmental changes. Very high resolution measurements are therefore feasible.

The phase-modulation scheme involves the measurement of the light path length of the fiber in terms of the phase angle. The following properties of the fiber can be utilized to effect the modulation:

1. The total physical length of a fiber may be modulated by the application of a longitudinal strain, that is, thermal expansion or the

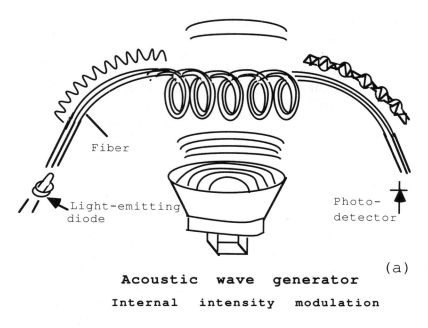

Acoustic wave generator

Internal intensity modulation

(a)

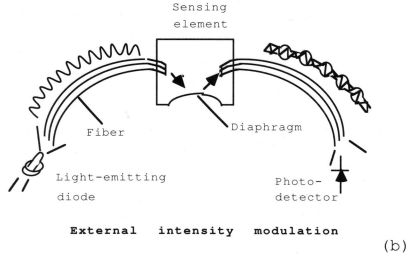

External intensity modulation

(b)

Figure 10.3 Type *I*. Intensity-modulated optical fiber sensors: (*a*) internal modulation type; (*b*) external modulation type. [After T. G. Giallorenzi, *et al.* [7] *IEEE J. Quantum Electron.* **QE-18,** 626–666 (1982) © IEEE, 1982.]

application of a hydrostatic pressure may cause expansion via Poisson's ratio.

2. The refractive index may vary with the temperature, pressure, and longitudinal strain via the photoelastic effect.

3. The guide dimensions may vary with the radial strain on a pressure field, the longitudinal strain through Poisson's ratio, or the thermal expansion.

Variations in environmental parameters other than pressure, strain, and temperature must be converted to cause phase modulation.

A basic Mach–Zehnder interferometric scheme is sketched in Fig. 10.4. Two pieces of optical fibers of approximately equal length, one serving as a reference arm and the other arm exposed to the environmental change, are used in a bridge-like arrangement as shown. Coherent light from a laser diode enters a single-mode optical fiber. A coupler divides the optical power equally between the two arms. Environmental changes would affect the length and the refractive index of the sensing arm, which causes the light to be phase-modulated. The phase shift is converted into an intensity modulation when the two beams are recombined by another coupler. The outputs of the combiner are 180° out of phase and are individually detected and then combined by a differential preamplifier. The processed output of the preamplifier provides an electric signal proportional to the detected stimulus. This type of sensor is classified as Type A-1.

In another Type A-2 optical-phase sensor, only one fiber is used. A typical example is the use of the Sagnac effect to sense rotations. This is shown schematically in Fig. 10.5. Light is split into two beams that travel in opposite directions in an SMF wound around a cylinder (or ring). When the fiber ring rotates, the light beam going in the same direction as the rotation has to travel more than one complete revolution to reach its starting point, which has moved as the light traveled around the ring. Similarly, the beam traveling counter to the

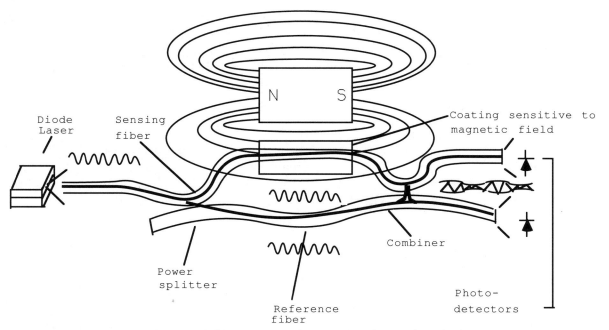

Figure 10.4 Type A-1. Phase-modulation optical fiber sensors. The Mach–Zehnder type. After T. G. Giallorenzi *et al.* IEEE Spectrum, 23, 49 © IEEE 1986.

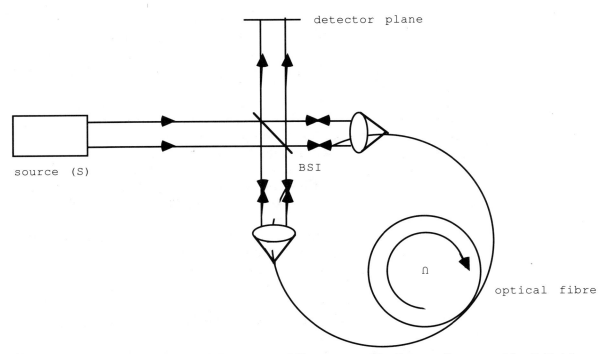

Figure 10.5 Type *A*-2. Phase-modulation optical fiber sensors. The Sagnac effect type. After B. Culshaw, Optical fibre sensing and signal processing. Peter Peregrinus, London, 1984.

ring's rotation reaches its starting point after going less than one complete rotation. As the speed of light is constant, the difference in path length appears as a phase shift, which can be detected by interfering the two beams with each other.

If the angular velocity of a rotating loop is Ω, then the light that is simultaneously injected into the fiber in opposite directions will exit with a phase difference of

$$\Delta\phi = \frac{4\pi LR\Omega}{\lambda_0 c} \sim \Omega \qquad (10\text{-}4)$$

where L is the total fiber length and R is the loop radius. Variation in temperature, pressure, and other parameters may cause a change in the phase angle and thus can be detected and calibrated in terms of the changes. This is the basic working principle of a gyroscope [9, 10].

The Frequency-Modulated Type: Type *F*

In this type of sensor, the Doppler effect of moving objects is exploited. If radiation of a frequency f is incident on a body moving at a velocity v viewed by an observer, then the radiation reflected from the moving body appears to have a frequency f_1, where

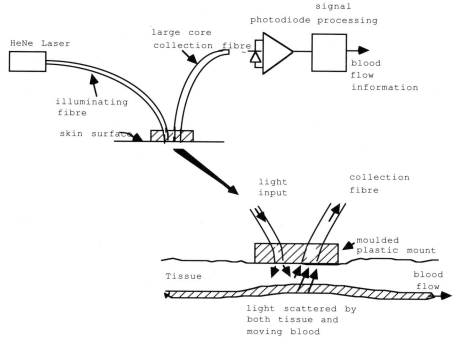

Figure 10.6 Type *F*. Frequency-modulation optical fiber sensors. After B. Culshaw, ibid., p. 123.

$$f_1 = \frac{f}{1 - v/c} \approx f\left[1 + \frac{v}{c}\right] \qquad (10\text{-}5)$$

Doppler shifts provide a very sensitive detection of moving targets and have been used as radars, either electrical or acoustical.

The same principle can be used to build optical fiber motion detectors or speed measurement [11]. Laser Doppler flowmeter [12], laser Doppler velocimeters [13] are just a few examples. The principle of one type of bloodflow sensor is sketched in Fig. 10.6. Two fibers, one used as a light input and the other a large-numerical-aperture fiber for collecting light, are mounted side-by-side on a plastic mount with a flat surface to rest on the surface of the object (e.g., the skin of a vein). Laser light from the input head sends light that penetrates the skin to reach the bloodflow that scatters the light. The returned scattered light is collected and processed to reveal the flow information.

The Wavelength-Distribution Type: Type *W*

Wavelength-distribution sensors are also called *color probes*. This is because many physical phenomena can influence the variation of reflected or transmitted light intensity with wavelength or color of the light. By monitoring the color spectrum of the output lightwave, one can determine the source property quantitatively. One example is the use of a chemical indicator for monitoring pH value [14]. An optical fiber pyromatic probe is another application of this principle [15]. The basic feature of this system is ratiometric measurements of the output power at at least two wavelengths. The ratio should provide information about the measurand. The critical components of the sensor are the source and the spectrometer. The fiber simply serves to feed light to the monitoring region and return the modulated light for analysis.

The Polarization-Modulation Type: Type *P*

A variety of physical phenomena influence the state of polarization of light, as described earlier. We may wish to classify these in terms of the type of polarization modulation one imposes and the physical parameters that may cause this modulation. The birefringence properties of the optical fiber may be affected by a number of physical phenomena of any form, including electrical, magnetic, and mechanical means. On the other hand, external polarization modulation may be introduced by using special crystals such as the KDP and other electroelastic materials [16]. Faraday's rotation is another example. In Chapter 11, we shall describe the detailed implementation of these sensors.

Single-Mode versus Multimode Fibers

Both SMF and MMF can be used as fiber sensors. Many of the less demanding applications will be served by MMF, the technology of which is already fairly well advanced. For some applications where extreme sensitivity is of paramount importance, however, it is likely that the more delicate SMF technology will prevail. Single-mode fibers are more apt to be used in pure fiber sensors where the fiber itself is the sensor that measures the ambient conditions. In polarization modulation schemes, SMF is an important component. By contrast, MMFs are mostly used to transmit and receive modulated light for a remote-sensing system.

Conclusion

In this chapter we have described some physical concepts and effects that can be exploited for optical fiber sensor applications. The list is by no means complete. Any physical effect can be developed to affect a sensor in some way, depending on the ingenuity of the engineers. We have also arbitrarily classified the sensors into several types according to their physical operating principles. Specific examples of sensors are given in Chapter 11.

References

1. J. F. Nye, *Physical Properties of Crystals*. Holt, Rinehart & Winston, New York, 1976.
2. A. J. Rogers, Optical measurement of current and voltage on power systems. *IEEE J. Electr. Power Appl.* **4**(2), 120 (1979).
3. A. Yariv, *Introduction to Optical Electronics*. Holt, Rinehart & Winston, New York, 1976.
4. A. J. Rogers, Distributed sensor: A review. *Proc. SPIE,* **798** Fiber Optic Sensors II. p. 26 (1987). (Society of Photo-Optical Instrumentation Engineers.)
5. R. A. Berch, H. C. LeFevre, and H. J. Shaw, Compensation of the Kerr cell effect in fiber optic gyroscope. *Opt. Lett.* **7,** 282 (1982).
6. J. N. Ross, Measurement of magnetic field by polarization optical time-domain reflectometry. *Electron. Lett.* **17,** 596–597 (1981).
7. T. G. Giallorenzi, J. A. Bucaro, A. Dandridge, G. H. Sigel, J. H. Cole, S. C. Rashleigh, and R. C. Priest, Optical fiber sensor technology. *IEEE J. Quantum Electron.* **18**(4), 626–666 (1982).
8. N. Lagalos, P. Macedo, T. Litovitz, R. Mohr, and R. Meister, Fiber optic displacement sensors in *Phys. Fiber Optics*, p. 539 (1981).
9. R. A. Berch, H. C. LeFevre, and H. J. Shaw, All fiber gyroscope with inertial navigation sensitivity. *Opt. Lett.* **7**(9), 454 (1982).
10. B. Culshaw and I. P. Giles, Fiber optic gyroscope. *J. Phys. E* **16,** 5 (1983).
11. R. B. Dyott, The fiber optic Doppler anemometer. *IEE J. Microwaves, Opt. Acoust.* **2**(1), 13 (1978).
12. G. E. Nilsson, T. Tenlaud, and P. Akeöberg, Evaluation of a laser Doppler flowmeter for measurement of tissue blood flow. *IEEE Trans. Biomed. Eng.* **BME-27**(10), 597–604 (1980).
13. H. Nishura, J. Koyama, N. Koki, F. Kajiya, M. Hironago, and M. Kauo, Optical fiber laser Doppler velocimeter for high resolution measurement of pulsatile blood flow. *IEEE Trans. Biomed. Eng.* **BME-21**(10), 1785 (1982).
14. J. I. Peterson, S. R. Goldstein, and R. V. Fitzgerald, Fiber optic pH probe for physiological use. *Anal. Chem.* **52,** 864 (1980).
15. J. P. Dakin, A novel fiber optical temperature probe. *Opt. Quantum Electron.* **9,** 54 (1977).
16. A. J. Rogers, Polarization optical effects and their use in measurement sensors. In "Proceedings of the Optical Sensor and Optical Techniques in Instrumentation." Institute of Measurement and Control, London, pp. 208–214 (1981).

11

Optical Fiber Sensors

Introduction

In this chapter, pure and remote fiber sensors will be discussed. The format of the discussion will be as follows. The principle of operation of the sensor is briefly outlined. This is followed by specifying the light source and the type of fiber and detector to be used. Then the range of operation and the sensitivity are estimated. The discussion is concluded with a short comment on the shortcomings of this type of sensor.

We classify the sensors according to their application, such as temperature sensor, pressure sensor, and so on. Each sensor is assigned a type designation according to the definitions given in Chapter 10. For each kind of sensor, only one or two examples will be given for illustration. Ample references will be given for further investigation.

Temperature Sensors

Refractive-Index Sensor: Type *I*-1

Temperature may cause the refractive index of a fiber to change. Because the core and cladding differ in composition, their refractive indices change at different rates in response to a temperature change. As temperature rises, the difference in the indices between the core and the cladding becomes smaller, causing more light to leak into the cladding as shown in Fig. 11.1. Figure 11.1a illustrates the fiber at normal temperature. The fiber refractive index is designed to satisfy the total internal reflection condition, and the total light intensity from the input minus losses through the fiber can be detected at the output. The Figure 11.1b graph shows the fiber at elevated temperature. Here the index difference has been reduced so that light can leak into the cladding, causing the output intensity to decrease. A simple thermometer can be built by using a simple light source (LED or LD), a coil of fiber (a step-index SMF or MMF), and a simple photodetector to record the light intensity. When the coil is exposed to a thermal environment, the output intensity meter readings change with temperature and can be calibrated as a thermometer. This thermometer can be used in hostile environments. For example, if the fiber coil is coated to withstand corrosive liquid, the thermometer can be used to monitor the temperature rise in some chemical processes. Coils can also be embedded in electrical generator stator windings to monitor the temperature rise during the generator operation. The range of temperatures between 80 and 700°C has been successfully measured using such a device by M. Gottlieb and G. B. Brandt at Westinghouse. The sensitivity is estimated to be ±1°C. With special fibers having loss peaks at various temperatures (made from different materials), the high-temperature limit can be boosted to over 2000°C.

The thermometer is simple to build, and its accuracy is adequate for many applications.

Interferometric Temperature Sensor: Type A-1

The most sensitive temperature sensors make use of the interferometric method to compare the phase change of a fiber sensor with that of a reference fiber. The Mach–Zehnder method discussed in Chapter

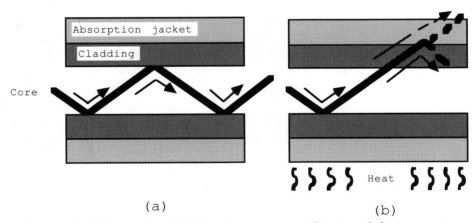

(a)   (b)

Figure 11.1 A simple optical fiber thermometer utilization of the temperature-dependent property of the refractive-index difference between the core and the cladding of the fiber to affect the transmission properties of the light as a temperature indicator: (a) low temperature; (b) high temperature.

10, under "The Phase-Modulation Type: Type A," is well suited for this application.

The range of measurement is estimated to be between -50 and $+200°C$ with an accuracy of $0.1-1.0°$.

Here the sensing arm is a length of SMF that is enclosed and exposed to the temperature change. A temperature change ΔT causes a phase shift $\Delta\phi$ according to [1]

$$\frac{\Delta\phi}{\Delta T} = \frac{2\pi L}{\lambda}\left(\frac{n\ dL}{L\ dt} + \frac{dn}{dT}\right) \tag{11-1}$$

For fused-silica fibers the index change term (dn/dT) dominates. This phase variation $\Delta\phi$ will be detected using the coherent detection method. Laser is used as the source. For a pure-silica guide, the temperature coefficient term is regarded as a constant, $(n\ dL/L\ dt) = 5.5 \times 10^{-7}/°C$, and the temperature coefficient of refractive index dn/dT is $0.68 \times 10^{-5}/°C$. At a wavelength of $0.633\ \mu m$, $\Delta\phi/L\ \Delta T \sim 106$ radians/m $\cdot$ °C. For multicomponent glasses using typical glass fibers, both the temperature term and the refractive-index term can be made to vary widely with the temperature, thus tremendously increasing the sensitivity of the sensor. Sensitivities of $10^{-8}°C$ are achievable [2].

Fluoroscopic Thermometer-Color Probe: Type *I*-2

In this fiber temperature probe, the variation in the phosphorescence spectrum of a rare-earth phosphor [(GdEn)O$_2$S] tip on an optical fiber is used to estimate the temperature. The schematic diagram of this apparatus is shown in Fig. 11.2. Ultraviolet light excites the phosphor, causing it to fluoresce. This phosphor emits most brightly at two wavelengths, one in the red–orange range ($0.62\ \mu m$) and the other in the yellow–green range ($0.54\ \mu m$). In the former range, intensity increases with increasing temperature. While in the latter, it decreases with increasing temperature. Measurement of the ratio of the intensities at these two wavelengths by photodiodes can be calibrated as a thermometer. In a practical system, a resolution of $0.1°C$ and an accuracy of $1°C$ are claimed [3]. An ultraviolet light source and photodetector are used. Multimode fibers work fine with this setup.

This probe can be made into small sizes that can be fitted into electrical equipment, microwave heaters, and even inside the human body.

Blackbody Radiation Probe: Type *I*-2

A most simple fiber thermometer is shown in Fig. 11.3. This device uses no external light source. The blackbody radiation from the probe tip is measured by a photodiode. The total intensity detected by the photodiode depends only on the tip temperature. Resolution of $1°C$

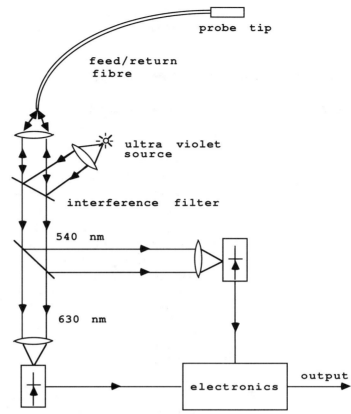

Figure 11.2 Fluoroscopic thermometer: a color probe. After B. Culshaw, *Optical Fiber Sensing and Signal Processing*, Peter Peregrinus, London, 1984.

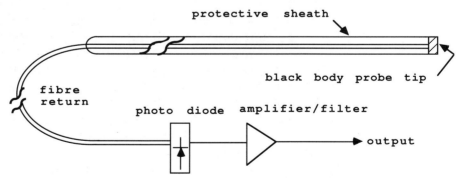

Figure 11.3 A simple blackbody radiator as a temperature probe. [After J. P. Dakin [4] © IEEE, 1977.]

in the range 250–650°C is typical [4]. Multimode fibers can be used for the probe.

Pressure Sensors: Sound-Pressure Sensors

Microbend-Pressure Sensors: Type I-1

Sound waves exert pressure on fibers. One method of detecting and measuring sound pressure is to measure the transmitted light intensity in a section of fiber between two corrugated plates as shown in Fig. 11.4. The plates are adjusted initially to exert little or no pressure on the fiber when there is no sound (or pressure). Acoustic waves push the plates together, creating tiny bends in the fiber. These microbends distort the fiber, allowing some of the light to be coupled to the radiation modes and to be lost. The light intensity at the fiber output decreases as the acoustic pressure is increased. An MMF is usually used. The spatially ridged structure has a spatial period Λ that is chosen to match the difference in propagation constants between suitably chosen modes in the fiber. If the two modes are designated by respective propagation constants β_1 and β_2, then the required value of Λ must satisfy the relation

$$\Delta\beta = |\beta_1 - \beta_2| = \frac{2\pi}{\Lambda} \qquad (11\text{-}2)$$

For an MMF with parabolic index g = 2 (where g is the profile index), typical spatial periodicity for $\Delta\beta[=(2\Delta)^{1/2}/a]$ is in the millimeter range [5].

The sensitivity of the device may be characterized in terms of the rate of change of transmitted power with displacement. For an optimized device, the sensitivity is on the order of 5% per micrometer. A displacement sensitivity in the range of 0.01 mm can be obtained if 1 part in 10^7 resolution criterion is used. The sensitivity can be improved by increasing the optical power level at the receiver, reducing the bandwidth, and increasing the interaction length of the sensor.

Both step-index and graded-index MMF can be used. A laser source is preferred. Mode strippers are used both before and after the fiber sensor to ensure simplified mode coupling.

One important advantage of the microbend sensor is that the optical power remains within the fiber, making low detection threshold possible [6].

For other refinements, see the paper by Lagakos and Bucaro [7].

Sound Interferometric Sensors: Type A-1

An optical fiber interferometric sensor that employs the Mach–Zehnder arrangement is shown in Fig. 11.5. A laser beam is split by a

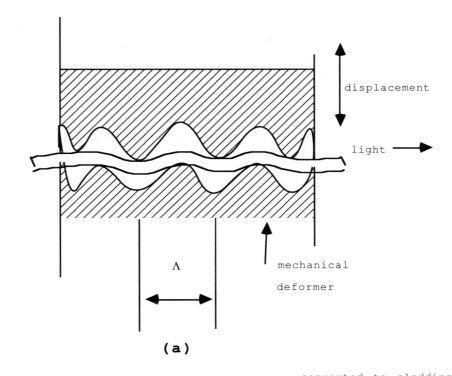

(a)

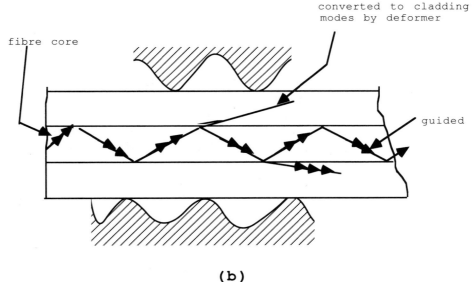

(b)

Figure 11.4 A microbend pressure gauge: (*a*) actual arrangement; (*b*) principle of the microbending sensor. [After T. G. Giallorenzi *et al.* [8] (1982) © IEEE, 1982.]

beam splitter (or an optical coupler). One part of the beam is transmitted into a reference fiber that is kept in a stable environment. The other part is sent through the sensing fiber that is exposed to the acoustic field. As the acoustic pressure distorts the sensing fiber, it changes the effective length and alters the relative phase of light in

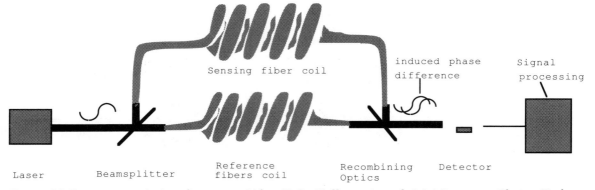

Figure 11.5 An acoustic interferometer. [After T. G. Giallorenzi *et al.* [8] *Microwave Theory Tech.* 472-511 (1982) © IEEE, 1982.]

the fibers. When the two beams are recombined and are allowed to interact on the surface of the photodetector, the phase-modulated signal is then demodulated. This sort of interferometry requires a coherent light source and an SMF.

For a fiber coil 1 m long, a fractional change in length of one part in 2×10^6 produces a 100% modulation of the light beam. Since the fractional change in signal strength observable in a 1-s interval is 2×10^{-8}, a fractional change of 10^{-14} in material properties can be measured for a 1-m length of fiber. If the fiber is longer, even smaller changes can be measured. Thus, an interferometric optical fiber sensor can be very sensitive.

Maintaining a fiber optic interferometer at its maximum sensitivity point is a problem, however. First, the difference in the length of the coils must be maintained at one-fourth the wavelength of light in order to satisfy the phase-quadrature condition, and be accurate to within one-eighth of a wavelength. A temperature change of $10^{-3}°C$ in the sensing coil with respect to the reference coil will change the length enough to result in faulty information unless compensation is made. Second, SMFs allow two orthogonally polarized modes to propagate. The state of polarization (SOP) at the output tends to fluctuate randomly, produces modal noise and thus limits the sensor sensitivity unless polarization control is exercised. Third, SMF sensors designed to respond to a given variable (pressure, magnetic effect, etc.) will usually respond to a temperature change, too, and this may offset an accurate measure of the desired changes. Special coating is usually required to limit the effect of temperature variations. After adding all these controllers, compensators, feedback circuits, and signal processings together, the device can become very complicated.

When the sensor is used as an acoustic sensor, its sensitivity will vary according to the relative size of the loop in comparison to the acoustic wavelength. This is because under different situations, the interaction mechanisms responsible to produce these changes may be different. For example, for a fiber loop with a diameter D formed with fiber of diameter d and an acoustic wave of wavelength Λ, at least two regions can be differentiated. For $\Lambda \gg D$, or in the long-wavelength region, the acoustic wave exerts a varying hydrostatic pressure on the fiber. The sensor sensitivity, defined as $\Delta\phi/(\phi\,\Delta P)$, will involve terms containing the elastooptic coefficients related to the photoelastic effect [8].

On the other hand, if $\Lambda \ll D$, but $\Lambda \approx d$, the elastic strain distribution across the fiber cross section becomes anisotropic and the optical-mode distribution must be taken into account. The strains are no longer uniform, and one must account for the SOP of the optical beam.

The sensitivity of an optical fiber sensor can be increased or decreased dramatically by various coating techniques. For example, a fiber with an inner coating of a soft material (e.g., silicone) and a thick jacket of a material having a high Young modulus can increase the sensitivity by an order of magnitude (increased from 0.45 to 5.8×10^{-4} radians/Pa $\cdot$ m). On the other hand, desensitizing a sensor is also possible by using a stiff metallic coating [9].

Optical fiber sensors are very versatile for use in geometric structures. They can be made to fit any purpose. A single-fiber coil with $D \ll \Lambda$ will serve as a multidirectional sensor. A single-fiber element of length $L \gg \Lambda$ can be a highly directional sensor.

Induced Birefringence Sensors: Type *P*-1

Pressure on optical fibers induces strain, which, in turn, changes the refractive index of the fiber. To exaggerate this effect, the fiber is wrapped around a cylinder that expands and contracts in response to acoustical pressure waves. The resulting strain along the length of the fiber causes a phase shift between two lightwaves of orthogonal polarization traveling through the fiber. This phase shift can be detected by observing the interference of the two waves at the end of the fiber. An example of this sensor is shown in Fig. 11.6. A laser passing through a proper polarizer is focused on a coil of fiber wound on a small cylinder. The strain-induced birefringence due to pressure provides independent propagation of two linearly polarized eigenmodes. The resulting polarization rotation of the light propagating in the fiber is detected as an intensity change when analyzed [10].

The sensitivity of this type of sensor can be rendered comparable to that of the interferometric method described above (section on Type *A*-1, sound interferometric sensors). Polarization-based sensors have many advantages. The light is confined within a single fiber, and the polarization components are simpler to analyze than that in interferometric devices as it involves no reference path.

The system can be modified to sense other effects. For instance, if

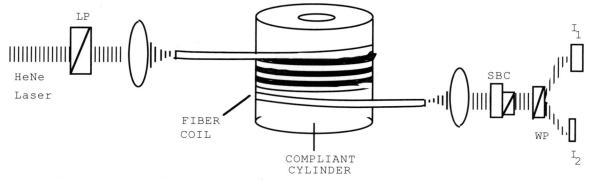

Figure 11.6 An induced birefringence pressure sensor. After S. C. Rashleigh [10].

the fiber is wrapped around a cylinder made of magnetostrictive material, then it can sense any magnetic field or current changes. If the cylinder expands and contracts with temperature, the system becomes a temperature sensor, and so forth.

Only SMF can be used for the sensing coil. Polarizers, one on the input end of the fiber to feed light of one polarization and one at the output end of the coil to separate the polarization components reaching the detectors, are used.

Pressure Sensors: Type *I*-2

A simple pressure sensor suitable for remote sensing is sketched in Fig. 11.7. Here, a pair of fibers with polished ends are enclosed in a box that contains a diaphragm that responds to pressure changes. One fiber serves as the input, and the other functions as the output light pipe. The fiber ends are arranged such that input light can be reflected by a diaphragm and collected by the output fiber. Pressure on the diaphragm displaces it, thus changing the fractional part of light reflected to and collected by the output fiber. A photodetector senses the intensity variation, which, in turn, registers the pressure variation. Linear sensitivity of 0–35 mPa with $\pm 0.5\%$ accuracy can be achieved [11].

Any type of light source can be used for this sensor. MMF is adequate for the light pipes.

Liquid-Level Sensors

Liquid-Level Sensors: Type *I*-1

For monitoring a liquid level an all-fiber sensor can be built. A loop of fiber with a part of its cladding strapped around the loop is sus-

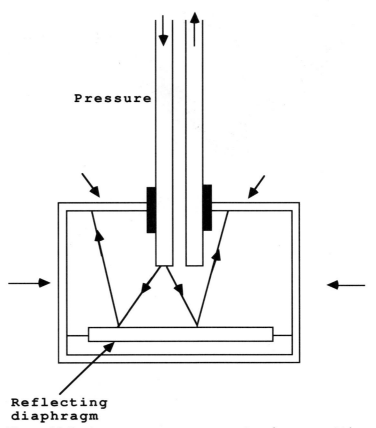

Figure 11.7 A remote pressure sensor: microphone type. [After T. G. Giallorenzi *et al.* [11] © IEEE, 1982.]

pended above the liquid level. Light is directed to pass through it, and its intensity is measured at the output. As the bare core loses more light when it is immersed in liquid than when exposed to air, a sudden change of outcoming light intensity indicates the liquid level. This is an on–off switching type of sensor and can be built very cheaply with an LED light source, an MMF, and a photodetector.

Any liquid residue on the sensor surface (the bare optical fiber) may cause inaccurate level readings, but this can be prevented by a nonwettable coating.

These types of liquid-level sensors have been used to monitor the filling of gasoline trucks, instead of mechanical flotation devices, which were used previously. They can be fixed in place so no mechanical motion is involved. They are safe to use because no electrical contacts are involved.

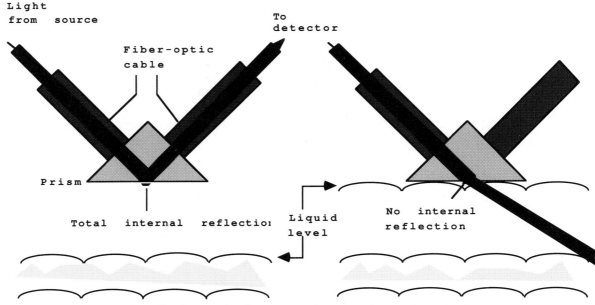

Figure 11.8   Internal total reflection-level sensor. [After J. Hecht, *High Technol.*, July–August, vol. 2 p. 52 (1981).]

Liquid-Level Sensors: Type *I*-2

Gasoline tankers need a level indicator to make sure that the tank is not overfilled. The conventional mechanical flotation system takes a severe beating bouncing around in an empty truck. Use of the optical sensor, which is fixed to the tank wall with no moving parts, is preferable to the conventional indicator. A scheme based on the phenomenon of total internal reflection is shown in Fig. 11.8. One fiber carries light to a prism, and another fiber carries light from the prism to a detector as shown. As long as the prism is not touching the gasoline level (in the air), the incoming light is totally internally reflected from the bottom of the prism and the detector senses full intensity. When the higher index liquid rises up to immerse the prism, it destroys the conditions for total reflection, and light leaks out as shown on the right side of the diagram. The detector, sensing a sudden dip in light intensity, signals the gasoline pump to shut off.

The system works as an on–off switch. Continuous monitoring of the liquid level requires another sensor, for example, an optical radar. The nonelectric nature of the system makes it a perfect candidate for many harsh environmental level-sensor applications.

Flow Sensor: Type *F*-2

The laser Doppler probe proved to be a successful sensor to measure the flow and velocity of liquids [12]. A schematic diagram of its op-

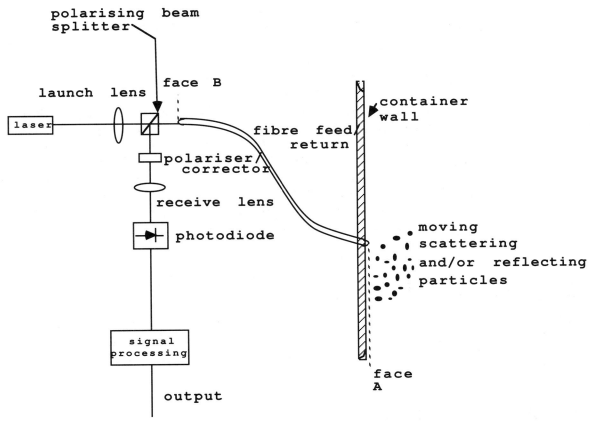

Figure 11.9 A laser Doppler flow sensor. After B. Culshaw, *Optical Fiber Sensing and Signal Processing*, Peter Peregrinus, London, 1984.

eration is shown in Fig. 11.9. A laser source is fed into an MMF via a polarizing beam splitter and launch optics. The other end of the fiber is immersed in a fluid in which the velocity of either the fluid or bodies within the fluid is the parameter to be measured. Light is scattered within this liquid, and some of the scattered light is collected by the fiber and returned to the source end. Since the scattered light is randomly polarized at the laser end of the fiber, the polarized beam splitter returns half of this scattered light to the photodetector. To determine the velocity of the moving flow, one needs a stationary reference. Although the reflection from the far end of the fiber (in the fluid) is weak, it proves to be an adequate stationary reference. The returned signal from the targets depends on the intensity of backscattered radiation, the attenuation within the medium, the receiver area, and the numerical aperture NA of the fiber. The sensitivity of this device is determined by the minimum detectable power returned from the scatters. The ratio of the returned power to the launched initial power is

$$\frac{P_r}{P_0} \sim NAR2\alpha a \qquad (11\text{-}3)$$

where NA is the numerical aperture of this fiber and $R(=\alpha_S/\alpha)$ is an attenuation ratio, where α_S is the scattering attenuation coefficient and α is the attenuation coefficient of the medium. Also, a is the fiber radius. If the medium attenuation is entirely due to scattering, then $R = 1$. For a given minimum detectable return loss of 104 dB, the P_r/P_0 ratio must exceed 4×10^{11}.

This sensor suffers a disadvantage that the distance of penetration into the medium is of the order of a few core radii. Since light that is coupled back into the fiber drops in intensity rapidly, the probe can reliably measure particle velocity in fluid up to a distance of at most only a few millimeters from the fiber end. Within this range, the probe will detect velocities as low as 1 μm/s and up to several meters per second, corresponding to frequency offsets ranging from a few hertz to tens of megahertz [13]. A laser source, a piece of MMF, and a photodetector are used in this setup.

Magnetic Sensors

Two basic approaches exist that utilize the interaction of magnetic field with optical fibers to build sensors: the Faraday rotation approach and the magnetostriction jacket approach. Both use SMF.

Faraday Rotation Sensors: Type P-1

In Faraday rotation sensors an external magnetic field B is applied longitudinally to a fiber to produce a polarization rotation perpendicular to the direction of the beam transmission. The angle of rotation is proportional to $K_V HL$, where K_V is the Verdet constant, H is the magnetic field intensity, $B = \mu H$, and L is the fiber length. For most silica fibers, the Verdet constant is small: $K_V \sim 1.5 \times 10^{-2}$ min/A. Thus, Faraday rotation sensors are useful only for sensing large currents or large magnetic fields. Efforts to dope the fiber with paramagnetic ions to raise the Verdet constant failed to achieve the desired result [14].

The principle of a current sensor using Faraday rotation is illustrated in Fig. 11.10. An SMF is wound around an electrical conductor whose current flow one wishes to measure. Laser light through a linear polarizer is focused on the fiber end.

The magnetic field produced by the current flow causes the polarization of light in the fiber to rotate by an amount proportional to the current (or H) flow. To measure this rotation, the light emerging from the fiber is split into two orthogonally polarized components by a Wollaston prism, each component is detected by a photodetector and the intensities compared. The relative intensity can be calibrated to read currents. The scheme is very much the same as that discussed earlier under "Induced Birefringence Sensors: Type P-1."

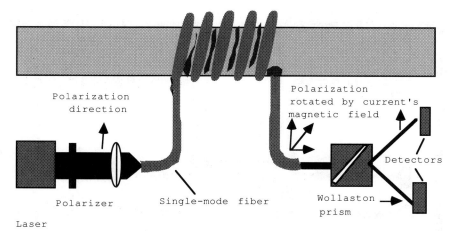

Figure 11.10   A Faraday rotation scheme for sensing electrical current. After A. Yariv and H. Winsor [14].

Currents up to 5000 A have been measured with a response time of about 0.5 μs and an accuracy of about ±1%. For a 10-turn coil of fiber a resolution of 30 A has been reported.

It is essential that the sections of fiber used to lead light in and out of the sensing coil are not birefringent. Otherwise the additional rotation may be cumulative and may affect the accuracy of the sensor. Spun fibers for the lead-in and lead-out are suggested so that the residual birefringent effect will be small [15].

Voltage and Current Sensors: Type *P*-2

A remote voltage sensor that uses Pockel's effect is schematically shown in Fig. 11.11. This apparatus consists of two boxes, a high-voltage line box and a remote measuring box, with fiber connections between them. In the high-voltage line box, a nonlinear crystal such as $LiNbO_3$ is used as the sensing element. When a laser beam (from a laser diode in the remote box) is led into the high-voltage box, it is first polarized and then directed to the crystal, which is under the influence of the high-voltage field. By Pockel's effect, the crystal causes a rotation of the polarized light. The outgoing beam is split into two orthogonally polarized components and sent back to the respective detector in the remote box. The two channels of the returning beam feed these polarized returns to their respective detectors and measure their intensities. The results can be calibrated to indicate voltage.

The advantage of this system is the cost saving over conventional equipment for measuring high voltages and currents (at one-sixth of the cost). The fiber sensor is expected to respond to transients in as little as one nanosecond, and its accuracy might reach 0.1%. This

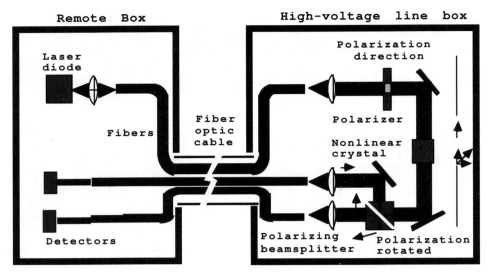

Figure 11.11 A high-voltage sensor using nonlinear crystals. Pockel's effect field sensor. [After J. Hecht, *High Technol.*, July–August, p. 53 (1981).]

sensor can measure up to 500 kV. In comparison, conventional monitors have a response time in the millisecond range and an accuracy of only 0.3%.

By changing the sensing element from a LiNbO$_3$ crystal to a Faraday rotator and using an SMF wrapped around the high-tension line, a current sensor can be built. This scheme resembles a pure-fiber sensor as discussed above under "Faraday Rotation Sensors: Type *P*-1."

Magnetostriction Sensors: Type *P*-1

The magnetostriction approach is more straightforward and is potentially capable of much greater sensitivity in magnetic-field measurements. The fundamental principle is based on the longitudinal strain produced in an optical fiber bonded to or jacketed by a magnetostrictive material. *Magnetostriction* is described as a change in dimension of a ferromagnetic material when it is placed along the axis of an applied magnetic field [16, 17]. The magnetostrictive coefficient is defined as $\Delta l/l$, where Δl is the change in length and l is the total length. Fe, Co, and Ni and their compounds in various proportions are all magnetostrictive materials. Among the various iron–nickel compounds, a 45% nickel exhibits the largest positive magnetostriction. Iron–cobalt alloys with 60–70% Co also offer large values of magnetostriction. These can be used as one arm in a conventional Mach–Zehnder all-fiber interferometer described earlier.

The magnetic sensors can have various configurations: a bulk magnetostrictive cylinder or mandrel with fiber bonded to its circumference as in Fig. 11.6; a nickel coating or jacket applied uniformly to

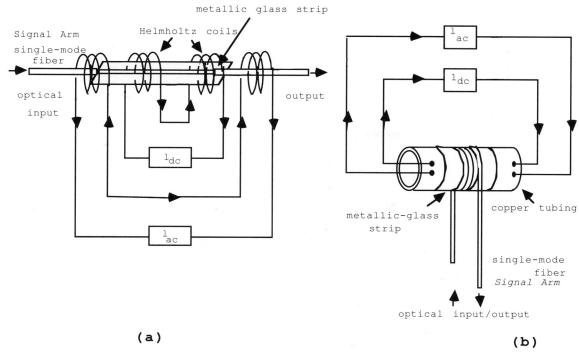

Figure 11.12 Magnetostrictive sensors: (*a*) metallic glass strip coated on SMF; (*b*) SMF ground metallic glass strip on copper tubing.

the surface of the fiber as in Fig. 11.12*a*; or an SMF wound around a cylindrical metallic glass strip as illustrated in Fig. 11.12*b*. The ultimate sensitivity of the fiber magnetic field sensors may be limited by a variety of instrumental, environmental, and material constraints. Temperature and pressure may also cause birefringence in fibers, which adds to the magnetic effect. These effects must be compensated for or subtracted [18].

The magnetostrictive effect is nonlinear. To approach a more linear operation, the magnetometers are usually biased to operate in the linear region.

At low frequencies, the thermal drift in the interferometer introduces error. Thermostabilized circuits with feedback are usually used to minimize this error.

Displacement Sensors

A Moving-Fiber Displacement Sensor: Type *I*-2

A moving-fiber displacement sensor is a multimode fiber sensor. Its structure is sketched in Fig. 11.13 [19, 20]. The input fiber is fixed. The output fiber is free to move or is connected to a diaphragm.

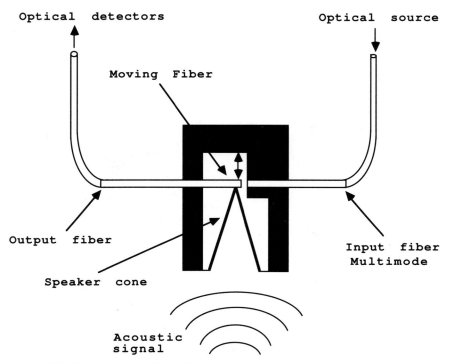

Figure 11.13 A moving-fiber displacement sensor. [After W. B. Spellman, Jr., *et al.* [19], *Int. Soc. Opt. Eng.* (1987).]

The polished end surfaces of these fibers are facing each other squarely. The motion of the diaphragm displaces the fiber transversely with respect to a beam of light from an input fiber, causing the light intensity passing through the output fiber to vary. For a typical fiber with a core diameter of 100 μm, a fairly linear response over a movement distance of 0.005 cm is observed. If one fiber is transversely displaced by 10^{-2} cm, a 100% modulation can be obtained. Under the same (1-Hz bandwidth) condition used to estimate SMF sensor sensitivity, the minimum detectable movement of the diaphragm is 2 × 10^{-10} cm. The dynamic range of this microphone is then 10 dB.

In order for a microphone to work in water (a hydrophone), there is the need to detect sound pressure. However, the high fluid viscosity reduces the range of motion. Therefore, the sensitivity of the device is reduced [8]. A much more sensitive hydrophone is described by the next system.

A Moving Grating Hydrophone: Type *I-2*

A Schlieren or grating sensor is schematically shown in Fig. 11.14 [21, 22]. The sensing element consists of a pair of parallel diaphragms—

DIAPHRAGM

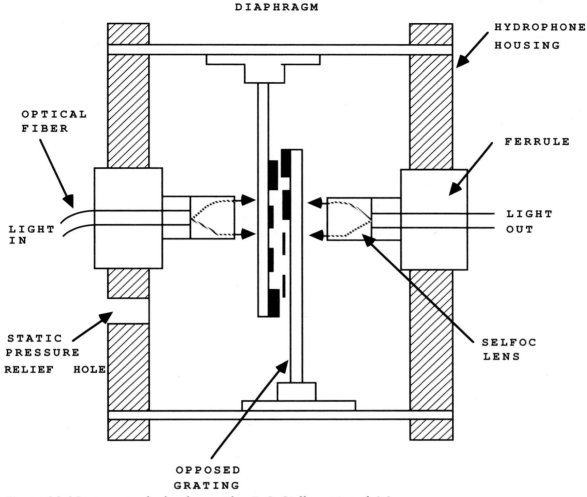

Figure 11.14   A grating hydrophone. After T. G. Giallorenzi *et al.* [8].

one fixed and one movable, mounted on opposite faces of the box. On both diaphragms, there are two transparent plates on which fine absorptive grating stripes are placed, and these gratings face each other. The grated plates are placed perpendicular to the diaphragms and are therefore perpendicular to the light beam sent in and out from the opposite surfaces as shown. When one diaphragm moves relative to the other, the light is modulated as a result of the grating motion. If Γ is the period of the motion of the grating, then the modulation index of the light is given by

$$Q = \frac{2}{\rho c \omega \Gamma} \qquad (11\text{-}4)$$

where ρc is the acoustic impedance of the medium and ω is the angular frequency of the sound frequency [21]. In water, with grating stripes 5 μm wide, the hydrophone is sensitive to sound pressure of less than 60 dB for all frequencies between 100 Hz and 3 kHz.

Multimode fibers are used. Any light source and photodetector are suitable.

Pollution Sensors

Smoke Detection—Optical Radar: Type *F-2*

An optical fiber smoke detector can be built using the principle of radar. A beam of light radiates from the end of a fiber of a microlens system. A similar fiber with a microlens tip collects the scattered lightwave to be analyzed. Smoke causes light to scatter. The variation in intensity of the collected light indicates the presence of smoke particles.

The same setup can be used to detect foreign particles in air or liquid. Different wavelengths or multiple wavelengths may be used to identify different pollutants. A range of 15–1000 ppm with an accuracy of $\pm 5\%$ has been reported.

pH Sensor: Type *W-2*

The optical fiber pH probe is used in remote chemical analysis of acidity of liquid solutions. The basic arrangement of the device is shown in Fig. 11.15. The indicator, in this case phenol red, is used to provide color for small polyacrimide spheres (5–10 μm in diameter) contained within a permeable membrane. The indicator has a transparency that is very sensitive to the pH value in the red, but is virtually independent of pH in the green wavelength. Thus, measuring the ratio of green light to red light transmitted is a direct indication of the pH of the solution in which the sensor head is immersed. This setup uses a tungsten lamp as the source that feeds the light into a container. Another fiber directs the scattered light to a detector through a rotating disk that has two attached color transparent filters, one red and one green. The light falling on the detector is thus switched from red to green during each revolution of the rotating disk. The ratio of the signal during the red period to that during the green period is then related uniquely to the pH value.

This system requires precalibration, since the relative sensitivity of the detector at the red and green wavelengths varies with temperature and aging. A stable dc source is needed to power the lamp. A resolution and repeatability of 0.01 in a pH range of 7.0–7.4 has been reported [23].

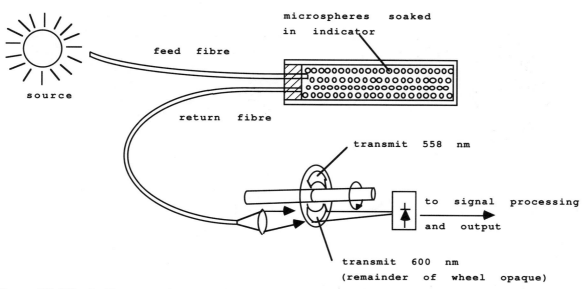

Figure 11.15 A pH sensor. After B. Culshaw, *Optical Fiber Sensing and Signal Processing,* Peter Peregrinus, London, 1984.

Medical Applications

An important application of optical fiber sensors is the medical probe. Such a probe can be small, compatible with living tissue, and require no electric wiring. Already in use are the photometric fiber sensors, which rely on changes in the reflectivity and spectrum of the measurement. Examples of practical devices include a blood monitor that monitors the flow of blood, the laser Doppler velocimeters with fibers to transmit and collect light in arteries, and photometric fiber probes that measure the fluorescence of living tissues in studies of oxidation–reduction reactions. Optical sensors are also being used to monitor intracranial, cardiovascular, urethral, and rectal pressures. Usually the fiber is encased in a catheter and operates by the displacement of a membrane or other part of the tip [24].

An interesting example is the blood–oxygen meter, which calculates the amount of oxygen in the blood by measuring how much light is absorbed at various wavelengths. A schematic diagram of this apparatus is shown in Fig. 11.16. A bundle of fibers (MMF) is used to bring light (can be incandescent or any intense source of light) to the fleshy, translucent portion of a patient's ear through a filter wheel of eight different colors. A second bundle is used to collect light transmitted through the earlobe. Intensity variations at different wavelengths are recorded by the photodetectors. The meter is calibrated through a reference fiber, and the resulting data are used to calculate the oxygen content in the blood [24].

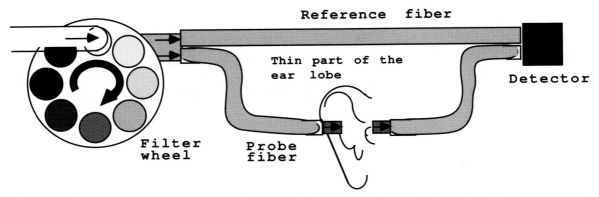

Figure 11.16 A blood–oxygen meter. [After J. Hecht, *High Technol.*, July–August, vol. 1, p. 56 (1981).

Fiber Optical Gyroscope: Type *P*-1

The basic form of an optical fiber gyroscope is an optical fiber Sagnac interferometer discussed in Chapter 10, under "The Phase-Modulation Type: Type *A*." A beam of light from a single optical source is divided into two beams by a beam splitter. They are directed through a multiturn SMF coil from both ends. The fiber coil is rotating about a parallel axis at an angular frequency Ω. Thus the two beams are counterpropagating within the fiber coil. The Sagnac effect causes phase-shift differences 2ϕ, which are nonreciprocal, to appear between the two counterpropagating light beams. One can show that $2\phi = 8\pi NA\Omega/\lambda c$, where NA is the numerical aperture of the fiber, λ is the free-space wavelength, and c is the velocity of light in free space. When these two beams are recombined at the output fiber directional coupler (or beam splitter) and mixed in the detector, the phase difference of 2ϕ leads to an intensity $I = I_0(1 + \cos 2\phi)/2$, which is thus rotation-dependent. The gyroscope sensitivity can be expressed as

$$\frac{I}{I_0} \cdot \frac{dI}{d(2\phi)} = \left(\frac{1}{2}\right) \sin (2\phi)$$

This expression shows that the sensitivity will be small for small rotation rates. For the rotational speeds of interest, the phase differences are in the microradian range. Thus, in order for the phase differences to be measurable, many turns of fiber length are required.

In practice, a simple gyroscope of this construction may encounter many difficulties. It will be unstable and inaccurate. It will be insensitive at a low rate of rotation, just when we need it most. We shall discuss causes of these difficulties and suggest solutions for a better sensor.

Fiber Gyroscope Noise Sources

Noise is the principal source of insensitivity introduced in an optical fiber gyroscope. To be detectable, a Sagnac-effect-based phase-shift difference must produce a corresponding intensity change larger than the quantum noise [25]. Present experimental gyroscopes have sensitivities substantially lower than the theoretically predicted values. This is because many sources of noise exist in practical configurations.

A Sagnac phase shift is a nonreciprocal effect. Accurate measurements of rotation rate require the phase-shift difference between the counterpropagating lightwaves in an optical fiber gyroscope to be virtually eliminated from all sources other than the rotation. This means that the two counterpropagating beams must traverse the same optical path so that they are identically affected by perturbations of their common path, that is, under reciprocal propagation conditions. Even with the best SMF (low loss and minimum birefringence fibers), however, nonreciprocal effects exist. These include the Faraday magneto-optical effect, fiber birefringence, and time-dependent temperature and pressure gradients. The phase shift due to the nonreciprocal Faraday effect is usually small unless strong magnetic fields are applied axially to the fiber. Besides, phase shifts due to magnetic field effect can be minimized by using appropriate shielding. Fiber birefringence is perhaps the major cause of nonreciprocal noise. Single-mode fibers are birefringent and can propagate two modes of orthogonal polarizations with slightly different velocities. These phase differences are very sensitive to environment changes. Ulrich and Johnson [26] suggested that by selecting one particular polarization state for both ends of the optical path of the gyroscope, such that the counterpropagating beams that have not transversed the common path be rejected, the stability of the gyroscope can be improved. This means that the interferometer is operating through a common-point, single-mode, single-polarization filter to guarantee precisely identical polarizations of the two counterpropagating beams at all points. Even then, the amplitude of the output signal can still vary substantially. Time-dependent temperature and pressure-gradient effects arise because the two counterpropagating beams traverse the same region of fiber at different times. During this time interval, the environment may have changed. Thus, for high-precision gyroscopes, stringent control of the environment may be necessary.

Overcoming the Low Slow-Rate Sensitivity

As stated in the introduction of this section, the sensitivity of the Sagnac effect fiber gyroscope is small for slow rotation rates. For a high-sensitivity gyroscope, this is the region in which maximum sensitivity is most desired. There are several ways to convert this least-sensitive region to a maximally sensitive one. One way involves providing a bias. A highly stable $\pi/2$ phase shift is introduced between the counterpropagating beams [27]. The most common approach to achieve bias is to modulate sinusoidally the phase difference between the counterpropagating beams at a low frequency rate. In this way the phase shift can be applied directly to the light in the fiber, thus removing interfacing and reflection problems [28]. The sensitivity of the gyroscope can then be increased to $J_1(2\phi)$ (the Bessel function of argument 2ϕ), which approaches 0.58 for small rotations. An electrooptic crystal, a fiber wound around a piezoelectric cylinder, or a piece of $LiNbO_3$ placed in the optical circuit near one end of the sensing coil, makes an effective nonreciprocal phase shifter [29].

The Dynamic Range

The gyroscope discussed thus far operates with a single frequency, and the output signal varies sinusoidally with an induced Sagnac phase shift. It is found that the dynamic range of this operation is very narrow; that is, the output is linear for only a very small range of Sagnac phase shift. For practical purposes, however, a gyroscope should have a dynamic range extending from 10^{-2} degrees per hour to 10^3 degrees per second.

One proposed approach to increase the dynamic range of the gyroscope is to use feedback loops. A promising scheme known as the *closed-loop* gyroscope using a nonreciprocal phase shifter in the fiber-sensing coil is shown in Fig. 11.17 [30]. A beam splitter divides a light beam and directs each part around a multiturn fiber optic coil through a modal-polarization filter. The polarization filter blocks light that has not followed the reciprocal paths. One beam passes through a phase modulator before propagating through the fiber coil, while the counterpropagating beam passes first through the fiber coil and then through the phase modulator. The reciprocal phase modulator causes a time-varying modulation of the differential phase shift between the two counterpropagating waves. A phase-sensitive demodulator such as a lock-in amplifier is used to measure the ac detector current at the phase-modulation frequency. The closed-loop feedback is added to provide linear response over a wide range of rotation rates. A nonreciprocal phase shifter added to the fiber coil introduces an electronically controlled phase shift between the counterpropagating optical beams to counteract the phase-shift difference due to rotation. The magnitude of the applied nonreciprocal phase shift required to nullify the electronic control signal from the phase-sensitive demodulator is an indication of the rotation rate. The net nonreciprocal phase

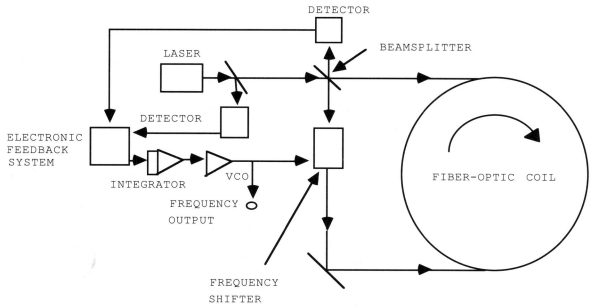

Figure 11.17 A proposed fiber optic gyroscope setup. [After R. F. Cahill and E. Udd, [30] *Appl. Opt.* **19**, 3054–3056 (1980).]

shift due to rotation and the nonreciprocal phase shifter are both zero, so that the circuit achieves linearity without sacrificing sensitivity. The feedback signal to the nonreciprocal phase shifter reflects the Sagnac phase shift and thus the rotation rate.

For other single-mode optical fiber sensor applications, see [31].

Conclusion

In this chapter several pure-fiber sensors in different fields of application have been presented. The field of fiber sensors began to show promise only a decade ago. In this short time, substantial progress has been realized using their unparalleled sensitivity, geometric versatility, and integratability. The remarkable progress has come about in part because of the availability of high-quality fibers and semiconductor sources. While the advantages of optical fiber sensors have been recognized, problems that need further investigation remain. In the intensity-modulation-type sensors, the problems of calibrating drift and biasing sensors need more assurance. The highest performance and the greatest flexibility of fiber sensors are realized in the field of interferometry. Yet, a practical fiber gyroscope still remains to be demonstrated. Interferometric sensors appear to benefit most from fiber-coating technology. By changing the coating on the fiber, the sensor element can be changed from acoustic to magnetic or other types of sensors. For example, soft polymer coatings on fibers are used to enhance the sensitivity of an acoustic sensor and to damp out temperature effects. Metallic coatings can be used to desensitize a sensor to an acoustic field when it is intended to measure current and voltage. This feature should find widespread application in many specialized areas and is worthy of further research.

Remote sensors are usually of the intensity-modulated type. Their construction is usually simple and compatible with MMF. They are adaptable for most applications where extreme sensitivity is not required. The fields of application of this type of sensor are virtually unlimited. If one can design a transducer relating any physical, chemical, or biological change to the intensity of light transmission, one has a new remote sensor. Development of optical fiber sensors started only about 10 years ago. Within this decade, many of the fiber sensors immediately rivaled or even render obsolete previous types of sensors. Magnetic, acoustic, pressure, temperature, displacement, fluid-level, and strain sensors are among the types of fiber sensors being investigated. Medical applications are new additions, and their development is rapid and broad. The integratability of optical fiber sensors with other electronic and optical devices makes them even more versatile for many applications. We have only touched the surface of this technology. The future of optical fiber sensors is in your hands, our good readers.

To show the importance of this subject, the International Society for Optical Engineering (SPIE) sponsors meetings every year to present papers on recent advances in this field. These papers are published in the proceedings of SPIE. On the subject of "Fiber Optic and Laser Sensors," the last three meetings were held in 1985, 1987, and 1988. Two of these proceedings, No. 566 and No. 718, have already been published.

Also, on the subject of "Fiber Optic Gyros", the Tenth Anniversary Conference Record was published as the *Proceedings of SPIE*, Vol. 719 in 1987. *Components for Fiber Optic Applications* was published in the proceedings of the SPIE, Vol. 722, in 1987. The Conference on Components for Fiber Optic Applications III met in 1988. The proceedings of this conference can be expected to be published in 1990.

Readers are advised to read these publications to keep informed of new developments.

References

1. J. E. Midwinter, *Optical Fibers for Transmission*. Wiley, New York, 1979.
2. G. B. Hocker, Fiber-optic sensing of pressure and temperature. *Appl. Opt.* **18**(9), 1445–1448 (1979).

3. The fluoroscopic thermometer: Literature available from Luxtron, 1060 Terra Bella Avenue, Mountain View, California 94043, U.S.A.

4. J. P. Dakin, A novel fiber optic temperature probe. *IEEE J. Quantum Electron.* **QE-9,** 540 (1977).

5. A. L. Harmer, Displacement, strain and pressure transducers using microbending effect in optical fibers. In *Principles of Optical Fibre Sensors and Instrumentation, Measurement and Control,* **15,** 143–151 (19).

6. N. Lagalos, P. Macedo, T. Litovtz, R. Mohr, and R. Meister, Fiber optic displacement sensor. *Adv. Ceram.* **2,** 539–544 (1981).

7. N. Lagakos and J. A. Bucaro, Optimizing fiber optic microbend sensor. *Proc. SPIE—Int. Soc. Opt. Eng.* **718,** 121 (1987).

8. T. G. Giallorenzi, J. A. Bucaro, A. Dandridge, G. H. Sigel, Jr., J. H. Cole, S. C. Rashleigh, and R. G. Priest, Optical fiber sensor technology. *IEEE Trans. Microwave Theory Tech.* **MTT-30**(4), 472–511 (1982).

9. N. Lagalos and J. A. Bucaro, Pressure desensitization of optical fibers. *Appl. Opt.* **20**(15), 2716–2720 (1981).

10. S. C. Rashleigh, Acoustic sensing with a single coiled monomode fiber. *Opt. Lett.* **5,** 392–394 (1980).

11. T. G. Giallorenzi, J. A. Bucaro, A. Dandridge, G. H. Sigel, J. H. Cole, S. C. Rashleigh, and R. C. Priest, Optical fiber sensor technology. *IEEE J. Quantum Electron.* **QE-18**(4), 626–665 (1982).

12. F. Durst, A. Melling, and J. H. Whitelaw, *Principle and Practice of Laser Doppler Anemometry.* Academic Press, New York, 1976.

13. R. B. Dyott, The fiber optic Doppler anemometer. *IEEE J. Microwaves, Opt. Acoust.* **2,** 13 (1978).

14. A. Yariv and H. Winsor, Proposal for detecting of magnetostrictive perturbation of optical fibers. *Opt. Lett.* **5,** 87–89 (1980).

15. S. R. Norman, D. N. Payne, M. J. Adams, and A. M. Smith, Fabrication of single-mode fibers exhibiting extremely low polarization birefringence. *Electron. Lett.* **15,** 309–311 (1979).

16. F. M. Bozorth, *Ferromagnetism.* Van Nostrand, New York, 1951.

17. A. Dandridge, A. B. Tucker, G. H. Sigel, Jr., G. T. West, and T. G. Giallorenzi, Optical fiber magnetic field sensor. *Electron. Lett.* **16,** 408–409 (1980).

18. F. Bucholtz, K. P. Koo, A. D. Kersey, and A. Dandridge, Fiber optic magnetic sensor development. *Proc. SPIE—Int. Soc. Opt. Eng.* **718,** 56 (1987).

19. W. B. Spellman, Jr., and R. L. Gravel, Moving fiber-optic hydrophone. *Opt. Lett.* **5,** 30–31 (1980).

20. W. B. Spellman, Jr., and D. H. McMahon, Schlieren multimode fiber-optic hydrophone. *Appl. Phys. Lett.* **37,** 145–147 (1980).

21. B. W. Tietjan, The optical grating hydrophones. *J. Acoust. Soc. Am.* **69,** 937–997 (1981).

22. T. G. Giallorenzi, D. A. Bucaro, A. Dandridge, and G. H. Cole, Optical fiber sensors challenge the competition. *Spectrum* **26**(9), 44–49 (1986).

23. J. I. Peterson, S. R. Golstein, and R. V. Fitzgerald, Fiber optic pH probe for physiological use. *Anal. Chem.* **52,** 864 (1980).

24. A. M. Scheggi, Optical fibers in medicine. *Proc. Int. Opt. Fiber Sensor Conf. 2nd,* pp. 93–104 (1984).

25. A. Yariv, *Optical Electronics.* Holt, Rinehart & Winston, New York, 1976.

26. R. Ulrich and M. Johnson, Fiber-ring interferometer: Polarization analysis. *Opt. Lett.* **4,** 152–154 (1979).

27. W. C. Goss, R. Goldstein, M. D. Nelson, H. T. Fearnchaugh, and O. G. Ramer, Fiber-optic rotation technology. *Appl. Opt.* **19,** 852–858 (1980).

28. R. Ulrich and S. C. Rashleigh, Beam-to-fiber coupling with low standing wave ratio. *Appl. Opt.* **19,** 2453–2456 (1980).

29. C. H. Bulmer and R. P. Moeller, Fiber gyroscope with nonreciprocally operated fiber coupled LiNbO$_3$ phase shifter. *Opt. Lett.* **6,** 572–574 (1981).
30. R. F. Cahill and E. Udd, Solid-state phase nulling optical gyro. *Appl. Opt.* **19,** 3054–3056 (1980).
31. M. S. Maklad, P. E. Sanders, E. Dowd, and A. Kuczma, Single-mode fibers for sensing applications. *Proc. SPIE—Int. Soc. Opt. Eng.* **718,** 97 (1987).

12

Miscellaneous Passive Applications of Optical Fibers

Introduction

In this chapter, we discuss the applications of optical fiber other than those covered in the previous chapters, that is, beyond those for telecommunications and sensors. To take full advantage of the properties of an optical fiber, such as its lightguiding capability, imperviousness to electromagnetic interference, light weight, and low cost, many new applications have been added. For these purposes, even a special type of optical fiber has been designed to satisfy a certain particular application. For example, fiber bundles and fiber faceplates are designed for illumination and imaging applications. Gradient-index rod lenses are designed for industrial and commercial usages. The field is rapidly expanding. Not a single book can cover these applications adequately. In this chapter, only some isolated examples will be cited just to show how broad this field can be. We begin our discussion with these special optical fibers.

The Fiber Bundle

A fiber bundle is an assembly of optical fibers held together at one end or at both ends while the rest of the fiber body remains flexible. It is designed for illuminating or imaging applications. Depending on how these fibers are assembled, fiber bundles can be classified into noncoherent and coherent bundles [1].

Noncoherent Fiber Bundles

The production of a noncoherent fiber bundle starts at the pick-up drum of a regular optical fiber drawing machine described in Chapter 2. The ideal bundle length should be equal to the diameter of the pick-up drum. For a 10-fiber bundle, the winding machine is stopped after 10 rounds and the fibers are cut from the drum with either one or both ends held together by epoxy resin. The ends are grounded and polished. The finished bundle may then be mounted into a metallic or plastic ferrule for easy handling. This bundle is noncoherent because each fiber in the bundle is treated as being isolated from its neighbors. The light output from each fiber is a function of the light at its input. Since the position of any fiber at the input face is not related in an orderly fashion to that at the output face, the intensity variation at the output face bears no relation with the input. Noncoherent fiber bundles are used for lightguides or illumination purposes. A bundle of fibers increases the area of illumination. Fiber bundles can be stacked up (arranged vertically) to further increase the illuminating area. For a hexagonally packed bundle, the effective area is less than its physical area by a factor $F = \pi/(2\sqrt{3})(a/b)^2$, where $2a$ and $2b$ are core and cladding areas, respectively [2].

The flexibility of a fiber bundle is not limited by the individual fibers constituting the bundle, but is also subjected to different abrasive stresses from neighboring fibers. To prevent the bundle from breakage, it is normally encased in a flexible trunking to limit the bending radius below the critical limit.

A glass fiber bundle can withstand an extremely large temperature variation, typically from -140 to $250°C$, although there is a decrease in transmission as the temperature increases. Plastic fiber bundles are more flexible than glass bundles. Single plastic bundles can be used singly with a diameter of up to 2 mm. But plastic bundles are more easily damaged than glass bundles and show "fatigue effect" after repeated flexing. The operating temperature range of a plastic fiber bundle is also limited.

Coherent Fiber Bundles

Coherent fiber bundles are designed for image-transferring applications. The position of each fiber in the bundle must be identically arranged at both the input and output faces. There are two distinctly different types: the flexible multifiber bundle and the solid faceplate.

Flexible Coherent Fiber Bundles

The process of making a coherent fiber bundle is similar to that of making a noncoherent bundle except that while winding the fiber on the drum, a set of guides and a stylus is used to ensure the position of the fiber on the drum to assume close-wound helical layers. A number of turns of the fiber are then removed from the drum by slitting the helix along a line parallel to the drum axis and with their ends bonded together. We have now a one-dimensional coherent bundle. The bundle is still flexible between the two ends. If a number of these layers are stacked together in an orderly fashion and each layer is accurately registered with respect to the fibers in the adjacent layers at both ends, a two-dimensional image transfer bundle is formed.

The transmission properties of a coherent fiber bundle of interest are the transmission and the resolution. The overall transmission of a coherent bundle is less than that of a noncoherent bundle made with the same glass. This is because a coherent bundle cannot be packed as closely as a noncoherent one because of the restriction of orderly placement of the fibers. Also, a thicker cladding is required on the fiber used for coherent bundles in order to reduce cross-talk between individual fibers.

The resolution of a multifiber coherent bundle depends on the size of individual optical fibers. Fiber diameter can be as small as 5 μm to give a sharp image. But special techniques must be used to handle it. High-quality one-dimensional coherent bundles up to 30 cm wide have been built using 40-μm-diameter fibers. A resolution of up to 50 lines per millimeter with bundle size up to 50 mm in cross section has been reported [3].

Tapered bundles, or conical fiber bundles, offer more interesting features. In these bundles, each fiber constituting the bundle is tapered. The cross-sectional areas of the ends of the bundle, therefore, are different. When the smaller end of the bundle is used for input, the image can be enlarged at the output. To reduce the image, the ends are to be reversed.

Solid Coherent Fiber Bundle: Faceplate

For image-transferring applications, one needs a larger viewing area with precisely positioned fibers across it to produce a high-quality image. Flexibility is usually of no concern. To achieve this, fiber ends must be fused together to form a solid plate whose width could be many times larger than its length. This is called a *faceplate*. The starting point of building a faceplate is a multifiber coherent bundle layer of certain width. A number of these layers are stacked up in hexagonal format to form a unit. Many units can be stacked again to build a faceplate. All end faces are fused to form a solid plate. A hexagonal

format is preferred because for individual fibers of c
tion, its package factor is better than any other forma.

The optical resolution of the image at the exit face c
depends on many factors: the geometric arrangement o.
face, the fundamental coherence of the fiber arrangement, c
of the light spot. To resolve lines, the resolution is further i.
by the azimuthal orientation of the lines compared with the i
metric coherence.

In some applications, the faceplate must be vacuum-sealed.
calls for a complicated interlocking and interstitial filling for vact
tightness.

A number of defects can occur in a fused faceplate, such as (1) bler
ishes, (2) shear distortion, (3) gross distortion, and (4) displaced in
age. Blemishes are caused by defects in the manufacturing processes
They appear as black spots where the light transmission has been cu
to less than 50%. Usually it can be traced to the trapped air bubble:
or dirt in the bundles. In shear distortion, straight lines may appea
discontinuous. This may be caused by twists in the fiber bundles during the manufacturing process. When a straight line appears curved,
it is called a *gross distortion*. Most likely it is caused by an uneven
flow of glass during the fusing process. If the image appears to be
displaced transversely relative to the input, it can be traced to the cut
made on the bundle face. A cutting face not being perpendicular to
the fiber bundle may have caused it.

Optical Fibers for a Wide Wavelength Range

The expanding applications of optical fibers beyond telecommunications in recent years promote interest in the extension of fiber technology into the far-IR wavelength region. Searching for materials that
will work in this wavelength range becomes an intensive research objective. Let us recall the section on fiber loss mechanisms in Chapter
3, where the inherent fiber loss mechanisms were reviewed. We introduced a theoretical fiber loss curve in Fig. 3.13 showing the inherent fiber loss (loss–wavelength plot) that resembled a V-shape. In
the short-wavelength end, Rayleigh scattering loss that decreases as
the inverse fourth power of the wavelength dominates the fiber loss
(shown by the negative slope side of V). On the other end of the spectrum, the resonant absorption of molecular vibrations and their overtones contribute to the exponential increase of fiber losses with increasing wavelength (shown by the positive slope side of V). The hint
for searching materials for fibers to operate in the far-IR region could,
therefore, be looking for heavier-atomic-weight elements.

On the basis of this theory, researches on material searching are
directed into the following areas as described under "Materials for

Long-Wavelength Glass Fibers" in Chapter 2: fluorides of heavy metals, single and polycrystalline halides, and the chalcogenides.

Fluoride of Heavy Metals for mid-IR Range

Glasses of this type appear most suitable for optical fiber applications. They use zirconium tetrafluorides as the primary constituent (ca. 60%) with fluorides of barium, lanthanum, lead, or sodium constituting the remainder. These glasses exhibit a broad range of high transparency and show a minimum absorption. Fiber losses less than 10 dB/km have been reported.

Single and Polycrystals of Various Halides

Single-crystal fibers, such as those made from silver bromide, are transparent in the mid-IR range. But their fabrication into long fibers has presented many problems. Polycrystalline halides such as thallium bromide (KRS-5) exhibit excessive scattering loss, which limits the minimum attainable attenuation. But both materials are candidates for optical fibers in the $6-11$-μm region for optical power transmission or IR imaging applications.

The Chalcogenide Glasses

Chalcogenide glasses are so named because they contain one or more of the chalcogen elements such as S, Se, or Te as in As_2Se_3. They have good mid-IR transmission. For 10.6 μm of CO_2 light transmission, the selenide or mixed selenide–tellenide compositions are most promising.

The Hollow-Core Lightguide

Another interesting development in lightguide is the hollow-core fiber guide. It has been known that the best silica glass fibers have fundamental absorptions in the wavelength range around $8-12$ μm. This renders the silica fiber worthless for transmission in this range. However, we can make a constructive use of this property by building a hollow-core lightguide using the high absorbing silica as cladding. For lightguiding, of course, the refractive index of the cladding material must be less than unity, if the core is hollow and contains air. This can be accomplished by using silicate and germanium glasses. It turns out that many advantages can actually be gained. For example, large index difference and large core diameter can be built into the hollow lightguide to accommodate easy power handling. The cladding may be very lossy, but little power will be lost. Some applications of this type will be discussed later in this chapter.

The GRIN-Rod Lenses

The gradient-index fiber rod (GRIN-rod) lens is the subject of discussion in this section. GRIN has a special property that has attracted wide interest. A single rod cut from a graded-index optical fiber can behave equivalently to a conventional lens.

The radial gradient lens was discovered in 1905 as Wood lens [4]. Fiber GRIN-rod lens has found broad application for building optical coupling devices and microoptical switches. It has also been used as an imaging device. A review of the theory of the gradient-index rod lens appeared in the *Journal of Applied Optics* in 1980 in a paper by Iga [5]. Without going into detail of the theory, we shall give a physical picture of how the lens-like behavior can be visualized.

A generalized representation of the index distribution of a GRIN lens may be written, for an axial ray, as

$$n^2(r) = n^2(0) \operatorname{sech}^2(ar)$$
$$= n^2(0)[1 - (ar)^2 + b(ar)^4 + c(ar)^6 + \dots] \quad (12\text{-}1)$$
$$\approx n^2(0) [1 - (ar)^2]$$

where $n(0)$ is the index at the fiber axis, a is a quadratic gradient constant, and b and c are higher-order aberration coefficients that are usually small and can be neglected. This approximate equation also applies to skew rays.

For a Lambertian source on the optical axis, the image-forming ray path can be calculated within a cone defined by θ as shown in Fig. 12.1a [6]. An equivalent lens group having a similar ray path is shown in Fig. 12.1b. The physical equivalence may not be perfect, but it conveys the idea of the lens-like behavior of a GRIN rod.

The unique transmission property of the GRIN lens is that its magnification is unity. But GRIN offers many advantages over optical lenses. Its desired focal length can easily be obtained by choosing its length; moreover, it has a small diameter (<0.5 mm), can be easily fabricated, has high resistance and high strength, is very flexible, and can be grouped into a linear array or a matrix formation to satisfy any requirements. It can be designed for high resolution by using rods of smaller diameter. These points are all that an image transferring system is asking for. Its application to a copying machine will be discussed in a later section.

In case the image needs magnification, a GRIN lens can be used in

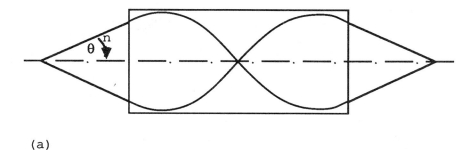

(a)

(b)

Figure 12.1 Similarities between GRIN-rod and conventional lens systems: (a) an approximate ray path calculated by Eq. (12-1) [6] [after J. D. Rees and W. Lame [6] *Appl. Opt.* **19,** 1065 (1980), with permission]; (b) an equivalent lens system.

combination with an image intensifier where a maximum gain of 100,000 has been reported [7].

The compact optical reader (COR) uses GRIN fiber rods with semiconductor sensors, and very large scale integrated circuit VLSI technology has been developed [8].

Other applications of GRIN-rod, including the quarter-pitch coupler, which is able to accommodate a larger beam width and reduce it back to a smaller size or vice versa, is sketched in Fig. 12.2a. A half-pitch GRIN-rod can transfer a focused point at the entrance plane to a focused point at the exit plane, thus making imaging possible. This is shown in Fig. 12.2b.

GRIN-rod couplers, connectors, attenuators, directional couplers, multiplexers, and interference filters have been mentioned in Chapter 8 and will be again in Chapter 13. A good coverage of this subject can be found in a paper by Tomlinson [9].

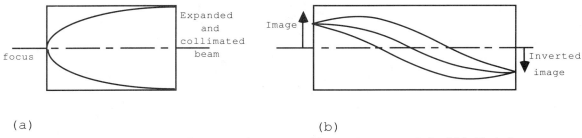

(a)   (b)

Figure 12.2 Two suggested applications of a GRIN-rod lens: (a) quarter-pitch; (b) half-pitch.

Nonsemiconductor Lasers

In Chapter 6, we mentioned that the most suitable light sources for optical fibers intended for telecommunication applications are the semiconductor devices such as the LEDs and the injection lasers. As the power required for these applications is usually small, of the order microwatts to milliwatts, semiconductor devices can serve the purpose very well. Besides, size compatibility between the fiber and the light source plays an important role in this choice. For the applications we shall describe in this chapter, however, the power level that a semiconductor device can deliver may not be large enough. Furthermore, we may need to work into wavelength ranges beyond the capability of silica fibers. Other light sources are needed. In this section, we shall briefly discuss how other lasers work.

Theory of Operation of Other Types of Lasers

Besides the semiconductor LEDs and LDs, we have the gas lasers, the rare-earth lasers, and the chemical dye or liquid lasers. The basic requirements to achieve lasing in any laser are identical: the population inversion, stimulated emission, and resonant cavity. For population inversion we should have the medium in use to have several energy states, preferably more than two. Electrons in the base state are pumped into higher energy states to create population inversion. A situation must be created to favor stimulated emission. And finally, a resonant cavity must be provided to amplify the oscillation. Details for providing pump energy, stimulated emission, and resonant cavity may differ from laser to laser.

The basic energy-state diagrams of a medium that can be used for laser action are shown in Fig. 12.3a and 12.3b. In Fig. 12-3a, a three-level energy diagram is presented. Here, the base level is denoted as E_1; E_3 is the highest level, and E_2 is an intermediate or metastable level. The electronic distribution among these levels follows the normal Boltzmann distribution law, specifically, that E_1 is heavily populated, while E_3 is populated sparsely. External energy must be pro-

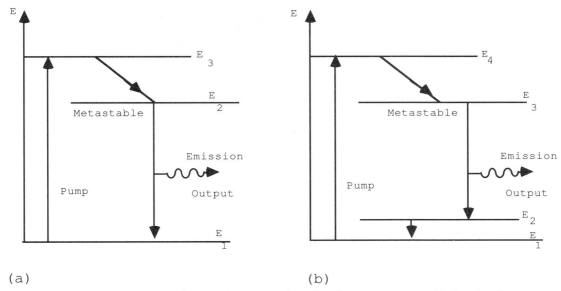

Figure 12.3 Energy states in laser systems: (a) three-level energy system; (b) four-level energy system.

vided to excite the electrons in E_1 to E_3, which immediately relax to E_2, because the excited levels are highly unstable. Level E_2 is a metastable state. The mean lifetime of electrons in this level is relatively long. If electrons are excited from E_1 to E_3 at a rate faster than that decaying from E_2 to E_1, the population of the metastable state becomes momentarily larger than that of E_1, thus achieving population inversion. When the population in E_2 becomes large enough, stimulated emission occurs. All electrons in this state are released simultaneously. Laser energy is released in the form of radiation output at a frequency ν proportional to the energy difference $\nu = (E_2 - E_1)/h$, where h is Planck's constant. Ruby laser is an example of a three-level system.

The disadvantage of the three-level system is that large pump energy is required to excite enough electrons to E_3 in order to achieve population inversion at E_2 as E_1 is so heavily populated. A four-level system partially corrects this, as it requires a significantly reduced threshold pumping power. This can be seen from Fig. 12.3b, where the laser output is between E_3 and E_2; E_3 is now the metastable state and E_2 is the laser lower state, which is much less populated than E_1.

Rare-Earth Lasers

Rare-earth lasers make use of a four-level energy system. For example, neodymium (Nd) or erbium (Er) is used as the active atom in a host material such as YAG (yttrium aluminum garnet) to produce Nd:YAG

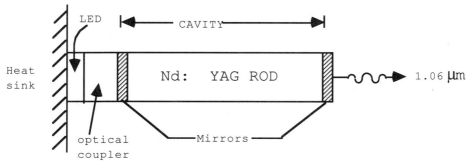

Figure 12.4 A Nd:YAG laser structure.

laser. The pump energy is usually supplied by a built-in LED that couples to the YAG rod by an optical coupler. The ends of the rod are provided with mirrors to form the necessary cavity. Emissions from a Nd:YAG laser is 1.06 mm. Erbium:YAG laser is pumped by a flash lamp. Its emission wavelength is at 2.9 mm. Both have very narrow spectral widths, of about 0.5 nm, which is about four to eight times smaller than those of most semiconductor lasers.

Figure 12.4 shows a rough sketch of a Nd:YAG laser structure. The cavity is formed by two mirrors, one at each end of the rod, which has a diameter of 5–6 mm. An LED that serves as a pump is coupled to one end of the rod through an optical coupler. An emission wavelength of 1.06 nm comes out from the other end.

Erbium:YAG has also been developed for medical use. Its emission wavelength is 2.9 mm. The basic structure is similar to that of a Nd:YAG laser except that the rod is doped with erbium and a flashlight is used for pumping.

Gas Lasers

It can be seen that the use of a flash lamp to pump a laser is very inefficient. Only a small portion of the spectral energy from a xenon flash is used for pumping. Gas lasers use a much more efficient scheme. They use an electric discharge to transfer energy to different levels, as in argon and krypton lasers. In other schemes, mixture of gases are used. When the gas mixture is discharged, collision among different atoms help to transfer energy between them. Helium–neon and N–CO_2 lasers belong to this type. The energy-state diagram of an He–Ne laser is shown in Fig. 12.5. The energy states of the helium atoms are shown on the left and that of the neon atoms, on the right. It is known that the metastable states of He atoms are in close coincidence with the $2s$ and $3s$ levels of the Ne atoms, and that the lifetimes for electrons in these He atoms are relatively long. This allows an

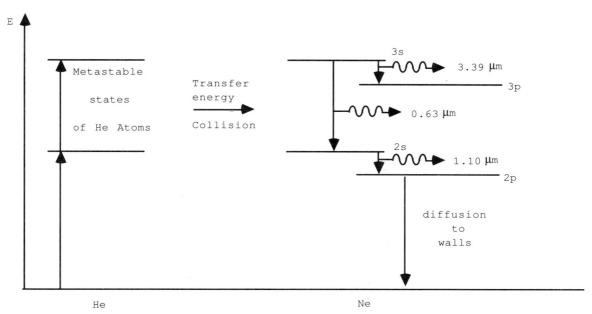

Figure 12.5 Energy states of an He–Ne laser (10:1 ratio).

exchange of energy from the He atoms to Ne atoms to take place when these atoms collide during a discharge, thus increasing the population of the Ne levels. Three important transitions occur between levels: that between a $2s$ and a $2p$ level, resulting in an emission with a wavelength of 1.1 μm; that between a $3s$ and a $2p$ level, with a wavelength of 0.6328 μm; and that between a $3s$ and a $3p$ level, with a wavelength of 3.39 μm. The 0.6328-μm emission is the most popular red light from an He–Ne laser.

For CO_2 laser, the energy-state diagram is similar to that of an He–Ne laser except that nitrogen is replacing the He atoms and CO_2 for the Ne atoms. The structure of a gas laser is sketched in Fig. 12.6. The gas mixture is enclosed in a glass tube whose ends are flattened into plates and are oriented at the Brewster angle to minimize internal reflections. Two external mirrors are aligned, one at each end to reflect the light beam back and forth through the discharge tube to serve as a resonant cavity. The gaseous discharge is initiated by either a dc voltage or a radio-frequency source applied between two terminals placed one at each end as shown. For a typical He–Ne laser, the discharge tube may have a dimension of a centimeter in diameter and several centimeters in length. The gas mixture proportion of 10:1 works very well. Emission of wavelength at 0.6328 μm is used, and CW output power of several tens of milliwatts can be obtained. For CO_2 laser, power in thousands of watts can be generated.

Argon lasers have also been developed particularly for medical applications. Since argon laser emits the green part of the visible spec-

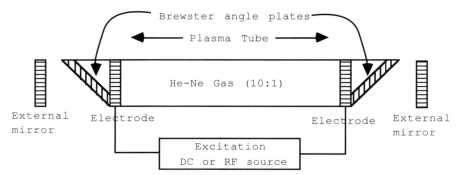

Figure 12.6   A sketch of an He–Ne laser.

trum, its light is readily absorbed by red blood, causing local heating. This heating effect can be used to seal off blood vessels or to attach the retina in an eyeball. Lasing in an argon laser is usually initiated by a radio-frequency discharge. Other gases such as krypton, when mixed with argon in various proportions, can generate wavelengths to cover the whole spectrum.

Liquid Dye Lasers

Liquid lasers consist of dye dissolved in a solvent. By choosing different organic dyes, one can produce a wide range of spectral outputs. The basic structure of a dye laser is similar to a gas laser, except that the solution containing the dye must be circulated through the discharge tube from a reservoir that is kept in constant temperature. Flash tubes are usually used for pumping. Rhodamine 6G is a commonly used dye. Its emission is in the orange region, from 570 to 670 nm. High-power emission is obtainable.

Fiber Optics in Medicine

The properties of optical fiber bundles such as the flexibility and transparency can play an important role in medical science. Diagnostic, endoscopic, and surgical applications use fiber bundles to send light and view areas inside a human body that were previously inaccessible. During recent years, the International Society for Optical Engineers (SPIE) has sponsored annual conferences to present papers regarding fiber optics in medicine and has published the convention records in the Proceedings of the SPIE. A vast amount of materials are now available for reference. We intend to give a bird's-eye view of this field just to demonstrate its importance.

Diagnostic and Imaging Applications

The medical endoscope is designed for internal examination of a cavity inside a human body. It must be able to reach all areas within the cavity and provide adequate illumination and resolution. Prior to the development of optical fiber technology, a rigid endoscope, consisting of a train of lenses, was used. Illumination was provided by a lamp attached to the proximal end and led through a quartz bar to the distal end. The endoscope was large and rigid, which added to the discomfort of the patient. Still its accessibility was limited [10]. Fiber optics changed all this. Modern endoscopes are flexible and small. They can range in diameter from several millimeters down to tens of micrometers. They can provide illumination with LEDs or lasers. Even powerful laser energy can be piped into the area for treatment. We shall describe some typical examples in the following paragraphs.

The gastroscopic endoscope is perhaps the earliest application of fiber optics in medicine. A large flexible trunking about 1 cm in diameter and 100 cm long containing a coherent fiber bundle and a lightguide is used. The lightguide is used to illuminate the area of interest from an external source, and the coherent fiber bundle is used to view the area with an external eyepiece. A remote steering mechanism is incorporated with the setup to locate the area of interest. In a recent design, a miniature camera may be mounted on the tip of the bundle to take either still or motion photographs.

Endoscopes have also been developed with fine optical fiber bundles for viewing more delicate areas such as the cystoscope, perineoscope, and cardioscope.

A colonoscope has also been developed that can be steered to negotiate the bends and turns inside the colon. For this application, bundles of larger diameter can be used.

Optical Fibers in Cardiology

One of the major forms of heart disease is atherosclerosis of the coronary arteries. Atheromatous placques progressively reduce internal lumen of the arteries toward an inevitable total occlusion. Present techniques used to correct this condition include balloon angioplasty and bypass surgery. Instead, an increasing number of surgeons prefer to use optical fiber laser angioplasty (11–15). An optical fiber is inserted into the occluded vascular lesion and a laser beam is used to canalize the artery. For this purpose, a special catheter is built. It consists of three channels: one channel is a fiberscope for visualization; the second, an optical fiber power bundle for laser power delivery; and the third, an open tube to suck out the debris or to flush with

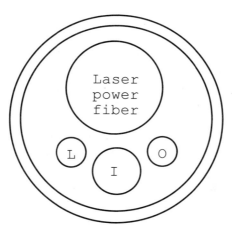

Figure 12.7 A cross-sectional view of a fiber optic bundle for laser angioplasty: L—fiber for illumination; I—fiber for imaging; O—open tube for flushing. Outside diameter 5 mm.

solution. A schematic diagram showing the cross-sectional view of the fiber bundle used in laser angioplasty is given in Fig. 12.7.

There are at least three types of lasers in use at present for laser angioplasty: the argon ion laser, emitting 0.489–0.514 mm in the visible range; the Nd:YAG laser, emitting 1.06 mm in the near-IR range; and the CO_2 laser, emitting 10.6 mm in the far-IR range. The laser beam is used for vaporizing, ablating, or coagulating the tissue under treatment. The difference between the three lasers mentioned is the typical penetration depth in tissue. For the argon laser, the typical $1/e$ distance is 1–2 mm; for the Nd:YAG laser, 3–4 mm; and for the CO_2 laser, only 0.5 mm. Another difference is the absorption selectivity for the blood-containing tissue. Argon laser can be absorbed readily in colored tissue, but for white tissue, its absorption is small. Neodymium:YAG laser does not absorb well. Carbon dioxide laser is highly absorbed by any biological tissue. Total absorption occurs at about 0.1 mm, thus making it easy to use for vaporizing tissue, ablation, or incision. The light-scattering coefficient is small. This could be the ideal laser to use in medicine. The only difficulty is that silica fiber does not work well at this wavelength. Fortunately, fluoride glass has been invented already, which works very well at 10.6-mm wavelength. Neodymium:YAG laser works well with silica glass fiber, but its efficiency is low. Argon laser needs quartz fibers.

For CO_2 laser using fluoride glass fiber, a toxic effect to human tissue may be of concern. It is found that fiber glass made with silver halide crystal, particularly alloys of of AgCl and AgBr, are more attractive for this application. Unclad fibers have been extruded from silver halide crystal at temperatures of 25–200°C through dies of diameter 0.2–10 mm. Several meters of fiber were extruded at a time. These fibers are nontoxic, nonsoluble in water, and flexible. A loss of the order of 1 dB/m at 10.6-μm wavelength has been recorded. Its loss does not increase on bending of the fiber, even if the bending radius is larger than 5 cm. Carbon dioxide laser power of more than 10 W could be transmitted through this fiber.

The attempts to recanalize arteries in humans using laser angioplasty techniques have shown that the channel created by this method is too small, thus facilitating earlier or later reocclusion. A complementary procedure to widen the narrow channel was required. One method to improve the operation is suggested by Geschwind et al. [16]. An artificial sapphire crystal shaped into a semispherical ball is mounted on the tip of the fiber bundle. The ball has a high mechanical strength, low thermal conductivity, and a high melting temperature. Its refractive index is 1.8. Transmission of visible light is excellent and is 85% for IR radiation. In their experiment, a CW Nd:YAG laser at a power level of 40 J or an argon ion laser at 20 J was used. The channel opening with sapphire contact probe was three times wider than that with bare fiber (ca. 2.5 mm vs. 0.7 mm).

Laser Microsurgery through Optical Fibers

A pulsed laser system has been developed to transmit laser energy through flexible fiber optics that is capable of precise ablation of targeted tissue with minimal damage to the surrounding tissue. Pulsed CO_2 laser has been used for microablation of tissue in a wet field when the fiber optic probe is held in contact with the tissue and delivers high power density energy in a pulse of a few milliseconds or shorter duration. Thallium halide- or silver halide-extruded fibers were used. But these systems are not considered as satisfactory for intraocular applications, which require smaller fiber size, greater durability, and larger flexibility. A prototype Er:YAG laser, emitting 2.9-μm radiation has been reported for ablation of tissue with very little damage to the neighboring tissue [17]. They used a 70–170-ms pulse, a power level of 2.4 J/mm², and zirconium fluoride glass fiber of 180–250 μm in diameter to carry the power. This system appears to provide a means of performing precise laser microsurgery to many previously inaccessible sites within the human body. It can be used to ablate both calcified and soft tissues.

In ophthalmology, the laser beam has been used to reattach detached retinas of the eyeball in humans. Fused-quartz fibers of up to 0.3–0.4 mm have been used.

To treat atherosclerosis, the laser power is transferred via a fiber bundle to the point to irradiate and remove the atheroma.

These uses of fiber bundle make it possible to move the bulky and possibly hazardous laser equipment from close proximity to the physician and patient to a more convenient position.

However, the laser power that welds the separated retina or disintegrates the atheroma can also destroy the end face of the optical fiber bundle. To avoid face damage, different schemes have been improvised. One of these uses a tapered bundle. The fiber bundle consists of many fibers about 1 mm in diameter each. They are tapered down to 125 mm in diameter at the output end. Twenty of these fibers form an input bundle of 5 mm in diameter, and the output bundle has a diameter of 0.6 mm. The input face is irradiated by light from an unfocused ruby laser which did no damage to the input face. The output beam is concentrated into a much smaller area by virtue of its passage through the tapered fibers, and a reasonable energy density is achieved at the output face [1].

Fiber Optics in Other Medical Applications

Many applications of fiber optics in medicine use sensors. We shall describe a few examples to supplement those described in Chapter 11. Measurements of temperature, blood velocity, pH value, and other physical parameters can now use optical fibers to obtain more precise and reliable data in medical diagnostics.

An Optical Fiber Thermometer in Cancer Treatment

In cancer treatment, radio frequency or microwave radiation is usually used. It is important to know exactly the temperature of a localized area under treatment without any electromagnetic interference. A small probe made of fiber optics becomes the best choice for this purpose. It can be made very small and is opaque to electromagnetic radiation. A typical example of this thermometer is described in a paper by Brenci *et al.* [14]. A miniature probe using a 200-mm core fiber with a mirrored end face and a 1-cm-long microcapsule of 0.8 mm in external diameter and 0.5 mm in internal diameter is used as a liquid container. It utilizes the attenuation of the light transmitted along an MMF induced by a temperature-sensitive liquid clad (oil). A temperature rise decreases the refractive index and hence increases the numerical aperture of the liquid clad section of the fiber. The light back-reflected by a mirror end face undergoes an attenuation that decreases with temperature. The light source is an 860-nm LED. It is coupled into the fiber through a four-way optical coupler, which provides a reference signal and a direct signal to the sensor. The reflected light is sent backward along the same fiber to the detector. The results can be calibrated in temperature degrees.

A similar arrangement can be used to measure the bloodflow and cardiac output by thermodilution technique.

Laser Doppler Measurement of Coronary Bloodflow Velocity

An accurate way to evaluate the clinical significance of an angiogram would be to measure the coronary reserve by gauging the bloodflow velocity. A laser Doppler coronary bloodflow velocity measuring method has been suggested by Stern [15]. They designed a two-fiber catheter laser Doppler velocity measuring system using heterodyne detection and estimate the velocity by means of analog-computed frequency moments. The light source is a 15-mW He–Ne laser. The laser beam is split into two beams, in one, about 90% is used for illumination and the rest, as a reference beam. The illumination beam is coupled into a fiber that enters the catheter. The other fiber is a 125/50 MMF graded-index fiber that is polished at an angle (45°) at the exit end and is slid into the catheter. The light scattered from blood is collected by a second fiber mounted a precise distance from the first. The reference beam passes through a Bragg cell operating at 80 MHz to shift its frequency by that amount. The intensity of the reference beam is adjusted by a half-wave plate through rotation of its plane of polarization. When this beam is combined with the signal beam, the light is detected and processed to yield the flow velocity. This system permits measurement of relative changes in flow velocity in coronary artery branches, from which the coronary flow reserve can be determined. The small size of the fibers permits their incorporation into existing balloon angioplasty catheters, thus permitting the measurement of coronary flow reserve before, during, and after angioplasty.

An Optical Fiber pH Sensor

In clinical examination of a patient, drawing blood for biochemical tests is usually performed. With the use of fiber optics, many biochemical parameters, such as pH, pO_2, pCO_2, glucose, and ion concentration, can be measured directly on the patient without drawing blood. Bacci et al. described an optical pH sensor that can be used for this purpose [18]. They described one type of optical fiber pH sensor, which measures pH values ranging from 7.0 to 7.5 units using a liquid optrode formed by a dye solution contained within a membrane attached to the end of a couple of optical fibers. By choosing a proper membrane and an appropriate dye, a sensitive pH meter can be built. They chose bromothymol blue for the dye whose absorption wavelength for the base form is 616 nm, while in acid form the absorption shifts toward longer wavelength with the increase of pH. A cellulose acetate membrane is chosen for its high permeability to hydrogen ions and a great capacity to retaining the dye. The light source is a halogen lamp and is followed by a narrowband interference filter with a peak at 603 nm. A beam splitter sends a portion of the beam to the optrode, and another portion to the detector, respectively. The signal from the optrode, which is modulated by dye absorption, is combined with the reference signal to produce the sensor response. It is amplified, filtered, rectified, and sent to a miniprocessor to be plotted out as the response curve. A sensitivity of 0.02 pH units and a response time of about 15 min have been reported. The probe is very promising for blood pH measurement [18].

Fiber Optics in Industry

There are many ways in which fiber optics can be used in industry. Tasks ranging from very simple ones such as illuminating or counting to the most sophisticated process control can all use fiber optics advantageously. We shall omit discussion of the simple ones and describe some of the more complicated cases.

Fiber Optics in Chemical Plants

In chemical plants, the chief use of fiber optics is in process control. Because of its ability to sense physical and chemical parameter changes in most environments during a chemical process, an optical fiber is made to sense these changes and to activate the control mechanism. The most common parameters are the temperature, the pressure, the flow rate, the pH value, the gas density, the refractive index, and the spectral absorption. For each parameter, a sensitive optical fiber sensor is built to detect the change. Many sensors can be used in parallel or in series to monitor the processing. Since they are small in size, impervious to adverse environmental conditions, and most cost-effective, they can be placed in more than one location and thus make continuous monitoring of the processing possible. Fiber optic sensors offer other capabilities that have not existed before. Chief among these is the ability to perform real-time remote spectral analysis at various points in a chemical process.

Miller and Hirschfeld [19] described the design of 12 optrode sensors used in chemical process control for various purposes, including the detection of oxygen, carbon dioxide, copper, nitrogen dioxide, ammonia, sodium, and potassium ions, redox, and uranium ions. Each optrode consists of an optical fiber whose tip is equipped with a special cap sensitive to the gas that is to be detected. For example, to detect uranium ions, the optrode is made of a fluorescent curvette on fiber, and so on.

Fiber Optics in the Aerospace Industry

Fiber optics has two distinct advantages over its electrical counterpart that are particularly attractive to aerospace industry: the light weight and the imperviousness to electromagnetic (EM) interference.

A Boeing 747 aircraft may need tons of copper wires for wiring of the communication equipment, control mechanisms, instrument panel illumination, and even the general lighting. To prevent EM interference, copper wirings for communication and control systems need metallic shielding. Replacing copper wirings with fiber that weighs only one-tenth of that of copper and omitting metallic shields may reduce much weight of an aircraft and thus increase its valuable payload capacity.

It is predicted that future aircraft designers may use much more fiber sensors to monitor engine performance and to gain control of the engine operation even during flight. Spillman [20] has demonstrated successfully that by using a reflective fiber optic sensor on a Boeing 757 aircraft engine, the throttle position of the engine can be detected in the cockpit via a digital encoder wheel. This information is used to control the engine.

A trend can be traced that a significant portion of the development in aerospace industry will be focused toward the incorporation of fiber optic technology into future products. Devices that can be used to sense fluid level, flow rate, pressure, engine speed and torque, strain, and other parameters are under development. Optical fibers can also be multiplexed to perform multiple functions and thus to increase their effectiveness.

Structural failure is a serious problem in aircraft engineering, particularly if the failure were happening during flight. It is suggested that optical fiber sensors be embedded in composition coupons of the aircraft structure. These sensors will be used as a polarimetric strain sensor to monitor strain within a coupon under loading, making safety control of the operation easier.

Fiber Optics in the Automotive Industry

Automotive industry is a big business. An annual production of several million units is an attractive target for the fiber optic industry to strive for. The use of optical fiber in automobiles is on the rise. Already, in the car, optical fiber has been used for instrument panel illumination and some communication equipment wiring.

One obvious application is the visual monitoring of automobile lights by the driver. In this system, a noncoherent fiber bundle is used. The input end of each fiber of the bundle is positioned to accept light from each car lamp, that is, the headlights, brake lights, backup lights, and the directional turning lights. The output end of the bundle is arranged on the dashpot panel to show the respective positions of the lamps. A filter system is used that allows all lights to show up as green dots in normal operation, but a burned-out light shows up as a red dot in order to attract the driver's attention. The system is passive and completely reliable.

But the most significant development might be the use of sensors in automobiles to exercise local control over the power unit of the car, to analyze the exhaust gas, and other controls. Already in the laboratory, the ignition system of an automobile can be monitored by using optical fibers. A single fiber may be inserted in each spark plug of the engine with its output connected to an analyzer to inspect the firing performance of the spark plugs. Or a laser–fiber combination may be used to detect the contents of the firing mixture in the cylinder and to exercise control over the mixture.

Fiber Optics in Military Applications

The same light weight and imperviousness to electromagnetic interference (EMI)—advantages of fiber optics over copper wirings—apply to military applications [21]. Airplanes, tanks, and ships are all candidates for adopting fiber optics. Normal communications use copper wiring subject to EMI. Worse yet, they may leak out enough radiation to reveal one's own location to the enemy. The use of optical fiber maintains true communication silence to the enemy. The security advantage of fiber optic communication systems is very important in military applications.

A guided torpedo or missile with an attached fiber optic cable as a payout line has been proposed for military fighting units. The course of a torpedo or missile released from a submarine or tank can be continuously monitored, or its traveling path could be corrected after the firing took place to ensure a direct hit.

A tactical commanding system consisting of a series of deployable shelters interconnected by an optical fibers star network has been suggested for the military field command. Each shelter is equipped with transmitter, receiver, environment data gathering equipment, and other necessary electronics to provide information to the command post. Commands from the command post can reach all shelters via the star network. Since optical fibers are difficult to tap, the security is increased.

A typical avionics system has been tried out on aircraft. This system contains a Doppler radar, computer, projection map, Doppler panel, and tactical electronics, all linked by optical fiber cables. It supports 13 communication channels, uses only 225 ft of optical fiber cables, and costs only $600 for the cable. To perform similar functions using copper-wire connections would require 115 channels, 1000 ft of copper wire, and a cost of $1800. This system would weigh 30 lb more.

Fiber Optics in Commercial Applications

Again fiber optics can find many applications in the commercial market, from a simple information display unit to a copier machine.

Fiber Optics for Display

The basic principle behind all light-display systems [1] employing fiber optics is to use a complex lightguide whose input is to receive light from one or more light sources and the output is to display the information in any desired manner. The ability of fiber bundles to alter the shape and the distribution of light add to the flexibility of the display. Both noncoherent and coherent fiber bundles can be designed for light display. Digital display is also possible. Many light-display systems are available in the marketplace. We shall describe only two interesting examples, a coded information display and an enlarged image display.

A Coded Information Display

In this system visual information is displayed on a panel and makes sense only to someone who can decode it. The message is fed into the input end of a noncoherent fiber bundle that contains many fibers arranged in a two-dimensional array. The fibers at the output end of the bundle are arranged in a secret pattern such as reversed order of arrangement, in transposed rows and columns, or in any other prearranged manner. The receiver must have the exact reverse of this pattern in order to decode the message.

An Enlarged Image Display

It is often desirable to enlarge the image displayed on a cathode-ray (oscilloscope) tube (CRT) for better observation, although one might use an optical projection system to accomplish this task, which could be bulky and expensive. A coherent bundle could handle this task much better. Suppose we wish to enlarge a 20 × 20-cm scope area to a 2 × 2-m screen. This involves a magnification of 10. We first fabricate a coherent fiber bundle whose input face should cover the 20 × 20-cm area. The choice of the fiber size is determined by the magnification. Assume that at the input end, the fibers of 0.1-mm diameter are bundled and barely touching each other. At the output end, they should be spread out to cover an area of 2 × 2 m, or the pitching-out is about 10 fiber diameters. A 1-mm pitching-out between fiber ends at the output could be enough for the eye not to notice the void spot at a distance of several meters. Many layers are required to build up this screen. They are stacked together, one on top of the other. The spaces between fibers are filled with spacers. With the input face fitted tightly on the scope face by a flat-face–curved-face fiber bundle coupler, a simple enlarger is formed. A sketch of this enlarger may resemble that shown in Fig. 12.8. This scheme could be used in projection television.

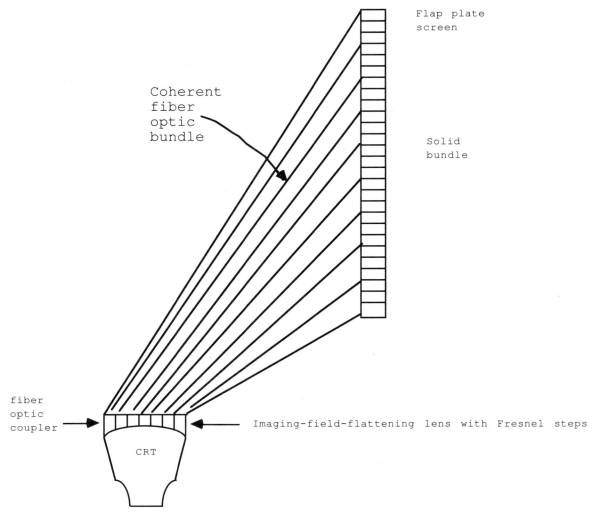

Figure 12.8   A schematic diagram of a coherent fiber bundle image enlarger.

Fiber Optics in the Copier Industry

The photocopier machine [22] is a welcome addition to an office and business establishment with which original documents can be reproduced faithfully and quickly at a reasonably low cost. The demand for a copier is, therefore, very high. The heart of a photocopier is a photoconductive surface on a rotating drum where image transfer takes place. A corona discharge deposits positive charges onto the surface. A lens system focuses the image on the photoconductive surface. Electric charges drain off from the white spots, leaving the image lines still positively charged. The ink or toner in the form of negatively charged particles is then spread over the surface that is attracted to the image. As the drum continues to rotate, it picks up the paper on which the image is to be transferred. The paper and the image-carrying sheet pass together through a set of rollers where the

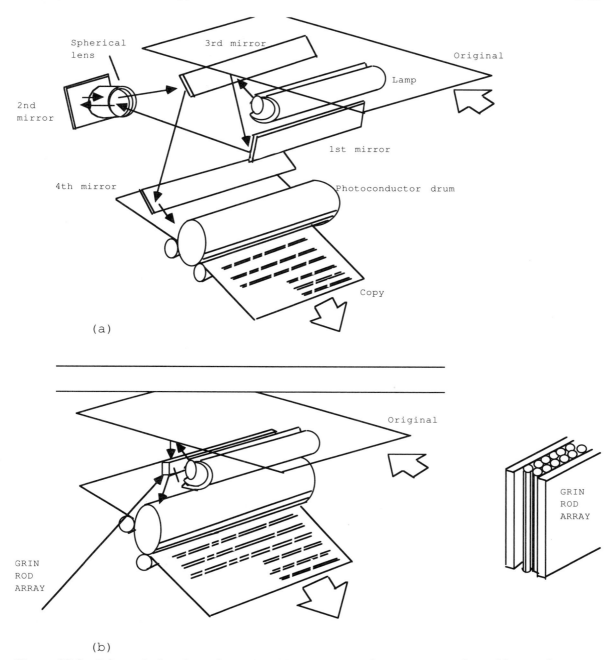

Figure 12.9 Schematic drawings of a copier: (*a*) a conventional copier using spherical lens and mirrors; (*b*) A GRIN-rod-array lens copier. [After C. Lu, *High Technol.*, May, p. 52 (1983).]

inked image is being pressed onto the paper and is then heat-treated to form a copy. To make a sharp copy, the conjugate distance of the lens system must be reasonably long. In a normal copier, this is accomplished by using a spherical lens with a set of four mirrors arranged to provide the necessary distance. This is shown in Fig. 12.9*a*. This lens combination renders the system very bulky and expensive.

Canon, Minolta, and others use Selfoc (self-focused optical fiber) or GRIN-rod lens to replace the spherical lens–mirror combination, resulting in a much more compact unit with even better quality. This setup is shown in Fig. 12.9b. As discussed earlier in this chapter under "Optical Fibers for a Wide Wavelength Range," a GRIN-rod can have the same image behavior as a spherical lens. A linear array of GRIN-rod placed under the copier plate can image the document directly on the drum surface, thus saving all the spaces otherwise needed for the lens system. Besides, GRIN-rod lens has further advantages: (1) it provides uniform image quality at the edge, (2) it can be used for high-speed copiers, and (3) it has high strength and resistance.

The GRIN-rod lens, which has a gradient refractive-index core (usually square-law graded-index profile), is expensive to fabricate. This makes a copier more expensive. A step-index fiber rod can also be used for a direct-contact-type copier. But because of its poor depth of focus, the quality of the copy will not be as good. Also, special coated paper is needed for copying. However, a really low priced copier can be built this way.

Fiber Optics in Instrumentation Industry

The ways in which fiber optics can be utilized in scientific instruments are almost unlimited [22]. Supplemented by its sensing capabilities, fiber optics can be fitted into any scientific instrument design. The applications described in the previous chapters for telecommunication and sensing devices may all be incorporated to build various instruments. Published literature has covered many of these designs. Readers may also use their own imagination and ingenuity to invent more. In this section, we shall mention only a few special examples.

An Industrial Endoscope

As in the medical field, engineers may need to view the interior of an engine cavity that cannot be approached in a straight line, or into a reactor during its operation. An endoscope is an ideal solution. The working principle of an industrial endoscope is similar to that of a medical endoscope. One needs a lightguide to illuminate the inside of the cavity and a bundle for viewing the inside. However, since size may not be a limiting factor, many elaborate features can be added to the design in order to increase the accuracy and to refine the measurement. For example, with the use of coherent fiber optic bundles, a panoramic view of the cavity may be presented. Using multiple bundles to permit a splitted source may provide a stereoscopic image of the object. Also, in cooperation with a miniature or movie camera, the actual operating conditions within the cavity can be observed for further study.

An Interferometric Instrument

One of the most accurate instruments for measuring many physical parameters uses the interferometric principle. In a fiber optic interferometric instrument, a coherent light source (laser) feeds two SMF branches equally via a beam splitter. One branch, which serves as a standard or reference, is kept in a controlled environment, such as constant temperature or pressure, etc. The other branch is exposed to the physical or chemical atmosphere whose parameters are to be sensed. Lights transverse over both paths but experience different phase retardations due to different atmospheres. When they are finally brought together at a combiner, interference fringe patterns result. The in-phase components add to increase the intensity while the out-of-phase components cancel each other to decrease it. The resulting interference pattern can be calibrated to measure the parameter changes. Some of these instruments were discussed in Chapter 5.

Fiber Optic Holography

Optical holography has many uses in industry to produce three-dimensional, highly realistic images unmatched to other photographic methods. It can be used to check for flaws in aircraft structure, boiler welding, and other objects. It can also be used to measure velocities of moving particles.

In a conventional holographic system, several bulky optical components such as the mirrors, beam splitter, spatial filter, and collimating lenses are used in combination with a laser light source. These components are mounted on a large isolated table. A laser beam enters a beam splitter, which divides the beam into two paths, the transmitted and the reflected paths. The reflected beam serves as an object beam and the transmitted one, as a reference beam. The object beam is directed toward the object whose reflection is to appear on a holographic plate as a wavefront. The reference beam bounces back from a mirror and passes through a spatial filter to combine its wavefront with that from the object at the plate. By wavefront reconstruction technique, the holographic image will appear on the plate. Figure 12.10a shows the arrangement of the components of this holographic setup.

In an all-fiber holography, practically all those bulky optical components have been replaced by SMF and fiber optic couplers. In a simple system, laser light is launched into an SMF by a microscopic objective lens. By means of a fiber optic coupler, it is divided into two

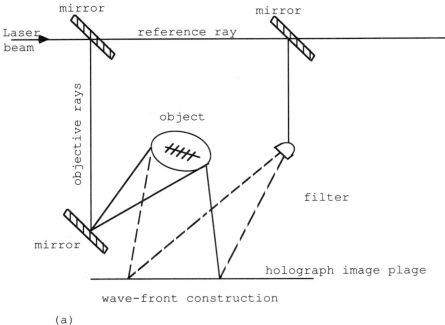

wave-front construction

(a)

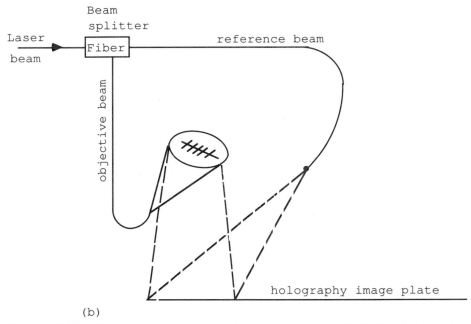

(b)

Figure 12.10 Schematics of (a) conventional holography; (b) all-fiber holography.

beams: the object beam and the reference beam. Notice that there is no need for the mirror to control the beam paths. Also, the SMF needs no spatial filter as high-order modes are not present. The object beam illuminates the object, whose reflected wavefront combines with that from the reference beam to holographically record the image on the holographic plate. This setup is shown in Fig. 12.10b.

One can see several obvious advantages of the fiber optic system over the conventional holography: (1) little space is needed to accommodate a fiber optic holographic system—in fact, several components can be integrated to reduce the space even further; (2) this system is more versatile and can achieve remote holographic interferometry easily; (3) this system can operate in any environmental conditions; and (4) any fringe fluctuation experienced by conventional holography can be easily corrected in all-fiber holography by a simple closed-loop feedback network.

In fiber optic holography, the task of varying the object and reference beam intensities can be achieved by a Mach–Zehnder interferometric and voltage-controlled phase modulator [23].

Another application of fiber optic holography is to view the vibrations of a complicated object and to make measurements of the amplitude at each intersecting point.

In microscopy, the wavefront reconstruction technique has been useful in obtaining a greater magnification without using optical lenses. Moreover, holography can be performed by using one wavelength while reconstructing it with another wavelength. Gabor has suggested a technique for producing a hologram with an electron microscope and reconstructing the image using visible light to take advantage of the fully developed optical technology [24]. By this technique, the unavoidable and uncorrectable aberration on the electron microscope can be recorded and reconstructed at optical wavelength, where the well-developed methods of aberration correction can be used.

Other Applications

Fiber optics may find applications in many other fields. For example, fibers may be used to transport energy. There are two interesting varieties: the solid lightguide and the hollow-pipe energy guide.

Optical Fiber for Solar Energy Transport

Solar energy is a renewable energy source that will become increasingly important in the future. Although solar energy is relatively easy to collect and even to concentrate, as has been done in the past, its use is limited mostly to direct applications. Transporting heat energy to remote locations is not easily done. Recent publications suggest that transport of solar energy for a distance of several meters can be done efficiently by means of optical fibers. Cariou et al. experimented with this idea. They used a parabolic mirror to concentrate and collect the solar energy and fed it to the input face of an optical fiber placed at the focal point of the parabola. Solar energy is transported to a remote furnace or reactor several meters away via the fiber bundle. To make efficient use of the solar energy, special glass fibers must be designed to utilize the full solar spectrum, from 0.45 to 1.30 μm. Using optical fibers of larger numerical aperture, solar energy of up to 3 kW has been transported up to a distance of 10 m with an efficiency of 65% [25].

Hollow-Core Fibers for Energy Transport

Hollow-core fiber glass pipes have been proposed to transport CO_2 laser energy in medical and industrial applications [26]. The core of the fiber is air, which has a relative refractive index of unity. For light-guidance, the cladding glass must have an index of less than unity. Silica glass doped with 27.8% of PbO was found to be useful at 10.6-μm wavelength. The fiber has an inner diameter of 0.6 mm and an outer diameter of 1 mm. A coat of Kynar approximately 20 μm thick was deposited on the inner face for protection. The resultant fiber is quite flexible. For a short length of fiber, power maxima of 18 W and 12 W at 10.6 μm and 9.2 μm, respectively, were reported. Compared to other means of transferring energy, hollow fiber has the advantage of greater flexibility and smaller dimension. The fiber attenuation is about 2 dB/m.

A large-diameter plastic hollow pipe with a sawtooth pattern of prisms molded in was installed in one building in Toronto, Canada, for the purpose of bringing sunlight or other light sources to the deep interior of the building. The prisms reflect light, bounding back and forth down the tube, as if the tube glows all along its length. This system can be adopted to illuminate high-risk areas where the danger of explosion is high, or in areas where light bulbs cannot be easily changed because they are difficult to reach.

Conclusion

In this chapter, we have described some examples of fiber optics in various applications other than those used for telecommunications. These are just a few of an infinitely long list. The fiber optic industry is still young and growing fast. Future applications are waiting to be invented. It is a good hunting ground for everyone.

References

1. W. B. Allen, *Fibre Optics, Theory and Practice*. Plenum, New York, 1973.
2. R. Tiekeken, *Fiber Optics and Its Applications*. Fucal Press, London and New York, 1972.
3. N. S. Kapany, Optical image assessment. *Nature (London)* **188**, 1083–1086 (1960).
4. R. W. Wood, *Physical Optics*. Macmillan, New York, 1905.

5. K. Iga, Theory for gradient-index imaging. *Appl. Opt.* **19,** 1039 (1980).
6. J. D. Rees and W. Lame, Some radiometric properties of gradient-index fiber lenses. *Appl. Opt.* **19,** 1065 (1980).
7. GRIN fiber for imaging. Special issue of *Applied Optics,* April 1, 1980.
8. GRIN-II. Special issue of Applied Optics, March 14, 1982.
9. W. J. Tomlinson, GRIN-ROD lenses in optical fiber communication systems. *Appl. Opt.* **19,** 1127–1138 (1980).
10. E. B. Benedict, Esophagoscopy, gastroscopy and peritonescopy. *Gastroentology* **42,** 17–24 (1962).
11. F. D. D'Amelio, S. T. Delist, and A. Rega, Fiber optic angioscopes. *Proc. SPIE— Int. Soc. Opt. Eng.* **494,** 44–51 (1984).
12. D. Gal, M. Elder, R. Valfden, A. Bayler, H. Newfeld, E. Gaton, M. Volman, S. Akselrod, A. Levite, and A. Katzir, Silver hilide fibers for surgical applications of CO_2 laser. *Proc. SPIE—Int. Soc. Opt. Eng.* **494,** 71–75 (1984).
13. E. B. Carlson, Coronary angioplasty: An alternative to bypass surgery. *Proc. IEEE* **76,** 1193–1203 (1988).
14. M. Brenci, G. Comtorto, R. Ralsiai, and A. M. Schoggi, Optical fiber thermometer for medical use. *Proc. SPIE—Int. Soc. Opt. Eng.* **494,** 13–17 (1984).
15. M. D. Stern, Laser Doppler measurement of coronary blood flow velocity. *Proc. SPIE—Int. Soc. Opt. Eng.* **715,** 132–136 (1986).
16. H. Geschwind, D. Mongkolsma, J. Stern, J. D. Blair, and M. S. Kern, Laser angioplasty with contact sapphire probe. *Proc. SPIE—Int. Soc. Opt. Eng.* **713,** 49–52 (1986).
17. R. F. Bonner, P. D. Smith, and M. Leon, Qualification of tissue effects due to Er:YAG laser at 2.9 μm with beam delivery in a wet field of zirconium fluoride fibers. *Proc. SPIE—Int. Soc. Opt. Eng.* **713,** 2–5 (1986).
18. M. Bacci, F. Brenci, G. Conforti, R. Falciai, A. G. Mignani, and A. M. Scheggi, Model for an optical fiber pH sensor. *Proc. SPIE—Int. Soc. Opt. Eng.* **713,** 88–92 (1986).
19. H. H. Miller and T. B. Hirschfeld, Fiber chemical sensors for industrial and process control. *Proc. SPIE—Int. Soc. Opt. Eng.* **718,** 39–45 (1987).
20. W. B. Spillman, Jr., Industrial uses of fiber optic sensors. *Proc. SPIE—Int. Soc. Opt. Eng.* **718,** 21–27 (1987).
21. D. G. Baker, *Fiber Optic Design and Application.* Reston Publ. Co., Boston, Massachusetts, 1985.
22. J. C. Daly, *Fiber Optics.* CRC Press, Cleveland, Ohio, 1984.
23. J. D. Muhs, M. Corke, R. L. Prater, and K. L. Sweenmay, Fiber optic holography with fringe stabilization and tunable object and reference beam intensities. *Proc. SPIE—Int. Soc. Opt. Eng.* **713,** 105–112 (1986).
24. D. Gabor, Light and information. 109–153 (1961).
25. J. M. Cariou, L. Martin, and J. Dugas, Concentrated solar energy transport by optical fibers. *Adv. Ceram.* **2** (1981).
26. R. Falciai, G. Gironi, and A. M. Scheggi, Oxide glass hollow fiber for CO_2 laser radiation transmission. *Proc. SPIE—Int. Soc. Opt. Eng.* **494,** 84–87 (1984).

13

Integrated Optical Fiber Devices

Introduction

Conventional signal processing in optical fiber transmission systems often requires that the signal be converted into electrical format for easier processing. This makes the system less attractive because the added components can make it more complex, more expensive, limiting in useful bandwidth, and more unstable. Many advantages gained by using optical fibers are being consumed by this conversion. The purpose of integrated optics is to do as much signal processing as possible directly in the optical format. This could be done by building the optical fiber components on a single chip, as in the integrated microelectronic circuits, and performing the processing such as phase and polarization changes directly on the chip by external means (e.g., electrical or magnetic fields).

The basic concept of integrated optics was proposed by Anderson as early as 1965 [1]. Serious investigation began in the 1970s [2, 3]. Areas that could make use of the integrated technique most profitably include (1) the passive fiber optic communication components such as couplers, modulators, multiplexing couplers, switches, isolators, mode strippers, mode scramblers, and polarizers; (2) the active fiber optic components such as lasers and detectors; and (3) the sensors, particularly the pure fiber sensors. Integrated optics are easy to fabricate using the well-developed photolithographic technology. They can be made compatible with single-mode fibers (SMF) at a low cost. They are very compact, too. However, the density of integration can never be expected to reach a fraction of that achievable in the integrated microelectronic circuits. This is because fundamental integrated optical fiber devices require a minimum length of a few milli-meters to be operative and integrated optic devices need a substrate on which elements can be deposited and diffused in. Unfortunately, there are substrate materials that are suitable for passive optical de-vices but may not have the desired characteristics for active devices. For example, $LiNbO_3$ and $LiTaO_3$ crystals have excellent electroopti-cal properties for use as substrates in an optical waveguide, coupler, or phase shifter, and so on, but a light source, detector, and amplifier may need GaAs and other group III–V compound semiconductor ma-terials as substrates. Other limitations, such as couplings between ele-ments and impedance matching, will be pointed out as we proceed with the discussion.

Integrated optic technology is still in its infancy stage. Many new developments may be expected in the near future. In this chapter, we describe only those components that are the candidates for integra-tion, both passive—such as planar lightguides, directional couplers, switches, and modulators—and active components, such as lasers, detectors, and amplifiers. In no sense have we achieved true integra-tion yet, but this is a beginning.

Integrated Optic Fabrication Technology

In integrated optics, like integrated microelectronics circuitry, the component elements to be integrated are usually arranged in a planar geometry for easy processing. Active and passive elements are formed on top of a substrate by thin-film technology or more elaborate ion implantation and electron-beam microfabrication processes.

In microelectronic integration, the substrate serves to support the components and to dissipate heat. In integrated optics, an additional function of the substrate may have to be that of serving as the inter-action ground for the optical, electrical, or acoustic fields. For in-stance, if the optoelectric property is called for in the interaction, then $LiNbO_3$, GaAs, or GaAlAs crystals are used for the substrate. The orientation of the crystal plane relative to the functional component becomes important. Otherwise, a quartz or pure glass substrate can be used.

Optical integration differs from microelectronic integration in an-other respect—the ultimate density of integration. In conventional in-tegration, the density of integration is often limited by the linewidth that an electron beam can be written. In integrated optics, it is the functional dimension of the components and, in particular, the opti-cal waveguide dimensions that limit the density. For example, an op-tical directional coupler at optical frequencies may need at least a couple of millimeters of interaction length to function. If the optical waveguide involves bends, each bend needs a bend radius of at least 3 mm. Thus, the density is limited.

The typical method of fabricating an integrated optical device can be outlined as follows. After a proper substrate material has been selected, it is grown into a planar, polished flat surface. In the example cited here a single lightwave guide is to be built on a glass substrate. First, a boron silicate glass fiber approximately 0.65 μm thick is deposited on the substrate. The refractive indices of the substrate and the glass layer are 1.515 and 1.54, respectively, well suited for guiding a two-dimensional optical wave with confined guidance.

The procedure may be outlined in the following steps, which are shown diagramatically in Fig. 13.1: (a) a thick film of polymethylmethacrylate (PMM) photoresist of about 0.6 μm is spin-coated on the glass film; (b) the photoresist is exposed by an electron beam, which writes a desired pattern, say, a straight lightguide with a bent input end—the photoresist is developed; (c) manganese is selected as the mask material (a thin coating, ca. 2000 Å) and is deposited on top of the photoresist; manganese is used instead of the usual aluminum because it gives a better edge roughness after processing and can be

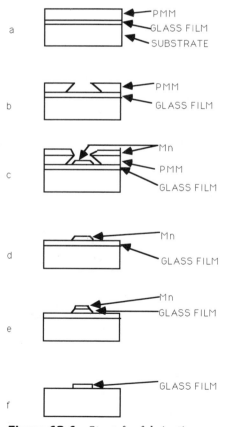

Figure 13.1 Steps for fabricating an integrated optical waveguide (a–f).

removed fairly easily by evaporation [4]; (d) the unwanted manganese and photoresist are removed with acetone; (e) the sample is then spatter-etched in argon; and (f) the remaining manganese is removed with hydrochloric acid. A rectangular optical waveguide is thus formed. The depth of the guide is about 0.6 μm.

When two such glass waveguides are deposited in parallel, each of a length of 3 mm and separated by about 1 μm, a directional coupler can be formed.

A 3-dB coupler can be formed using the following dimensions: guide width 3 μm, guide depth 0.6 μm, separation 1 μm, refractive-index difference Δ [= $(n_1 - n_2)/n_1$] = 0.01, and length (parallel) = 0.45 mm. If bends are used, the bend radii should be no less than 3 mm.

Lightguide stripes can be fabricated either by sputtering one type of glass onto another, by indiffusion of a deposited layer of titanium metal onto a substrate of $LiNbO_3$, or by liquid-phase epitaxy (GaAs and GaAlAs) and delineating the stripe pattern using electron beam lithography.

Active films are being developed for integrated optics. Improved electron-beam writing techniques and new film materials should make feasible the fabrication of numerous complex integrated optical circuits.

Passive Integrated Optical Devices

With growing interest in optical fiber communication and sensing systems, the demand for integrated optics is rapidly increasing. Although full-scale integration of optical fiber devices may still be years away, individual components such as optical waveguides, directional couplers, switches, modulators, and filters have been fabricated. These passive devices are the subject of discussion in this section.

Single-Mode Optical Waveguide and Couplers

The building block of integrated optic devices is a single-mode planar lightguide, exemplified in Fig. 13.1. Since a single-mode lightguide must employ two-dimensional guidance with proper confinement configuration, a planar, stripped lightguide seems to be a logical choice. Levi [5] developed an expression that serves as a guide to design of such a rectangular lightguide. Let n_1 and n_2 be the refractive indices of the guiding layer and the substrate, respectively, and assume that the top layer of the planar lightguide is air. Then for the lightguide to support a mode of order m, the depth of the guiding layer d must satisfy the expressions

$$d \geq \frac{\lambda(m + 1/2)}{2\sqrt{n_1^2 - n_2^2}} \tag{13-1}$$

If d is thinner than that described by Eq. (13-1), it becomes impossible to support any mode at all. If $n_1 = 1.5$, $n_2 = 1.49$, then to support a single-mode lightwave ($0 < m < 1$), according to Eq. (13-1), the guide thickness must be $1.4\lambda < d < 4.3\lambda$. Practical lightguide thickness may be approximately 2λ thick.

A single lightguide of this type is not useful in integrated optics. But a pair of these lightguides can be arranged to form a coupler, a switch, or a modulator. Let us start out with a coupler.

An optical directional coupler consists of a pair of parallel lightguides closely spaced to enable energy to be transferred from guide 1 to guide 2 if sufficient coupling length L is provided. This is shown in Fig. 13.2. It differs from the general structure of couplers discussed in Chapter 5 in two respects. First, it is small in size. Because of the extremely small separation (several micrometers) required for efficient coupling, the deposited lines must be bent to accommodate connections as shown in the figure. Second, the lines are deposited on a substrate to provide coupling between lines through the evanescent field penetrating into the substrate.

Let $A_1(z)$ and $A_2(z)$ be the amplitudes of the fields in guides 1 and 2, respectively. Then by using the coupled-mode theory [4] we can write

$$A_1(z) = \cos kz \, e^{i\beta z} \tag{13-2}$$

$$A_2(z) = i \sin kz \, e^{i\beta z} \tag{13-3}$$

where k is the coupling coefficient (assume $k_{12} = k_{21} = k$) and β is the propagation constant for a lossless guide. The power-flow equation becomes

$$P_1(z) = A_1(z)A_1^*(z) = \cos^2 kz \tag{13-4}$$

$$P_2(z) = A_2(z)A_2^*(z) = \sin^2 kz \tag{13-5}$$

where the starred quantities are the complex conjugate values of the amplitudes. For a short guide length, as is generally used for a cou-

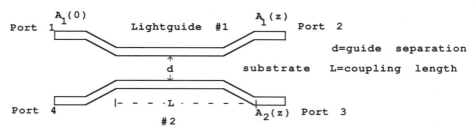

Figure 13.2 An integrated directional coupler.

pler, neglecting the line loss is justifiable. Equations (13-4) and (13-5) indicate the oscillatory nature of the power flow. Complete transfer of power from guide 1 to guide 2 as in a directional coupler is possible if $kL = \pi/2$, or $L = \pi/(2k)$. It is noted that for a longer length, power begins to transfer back to guide 1 again periodically. Thus, the coupling length is important for a good coupler. Also, once the coupling coefficient is known, L can be calculated.

Conversely, if the length required for complete power transfer can be measured, the effective coupling coefficient k can be determined. For example, if L is 200 mm, then $k = 0.79$ mm^{-1}.

Similarly, a 3-dB directional coupler where the output power is divided equally between two lightguides can be designed such that $kL = \pi/4$. This is a useful coupler, which is also known as a *power divider*. The inverse is known as a *combiner*. A pair of these form the important elements in an interferometric setup. For waveguides built on GaAs substrate with a substrate carrier density $n = 2 \times 10^{18}$/cm^3, if each lightguide dimension measures 3×3 μm, is separated by a 3-μm spacing, and has a $\Delta n = 0.005$, a 3-dB directional coupler can be built with $k = 0.79$ mm^{-1} [4].

For other couplers, the power transfer varies as $\sin^2 \pi l/4L$, where l is the coupling length and L is the guide length required for equal power division.

Besides the consideration of the dimension of the lightguides and their separation, other factors that influence the coupling are the refractive-index difference Δ, the edge roughness, and the bending radius. For significant coupling to be achieved in a short length (a few millimeters), the index difference Δ must be less than 0.02. To limit the bending losses to within 1 dB or less, bending radii must be larger than 0.5 mm. Furthermore, edge roughness of the sides of the guides must be held to within ± 500 Å for 3-μm-wide guides [6].

Switches and Switch Modulators

Deposition of optical waveguides on birefringent crystals such as LiNbO$_3$ and LiTaO$_3$ adds another dimension of flexibility. Electrodes are deposited over the guides. Voltages are applied to the electrodes to create electric fields, as shown in Fig. 13.3. By varying the applied voltage, the amount of light crossover can be controlled. If the voltage is just enough to stop the crossover completely, then by turning the voltage on and off, one can switch the lightwave from one guide to another, forming an electrically controlled switch of optical waves.

It is difficult to deposit the electrodes of the exact length to obtain the right coupling coefficient, however, because the birefringence of the crystal causes the two components (the *TE* and *TM* components) of the beam to have slightly different propagation velocities traveling down the waveguides. There is no single length that will give a com-

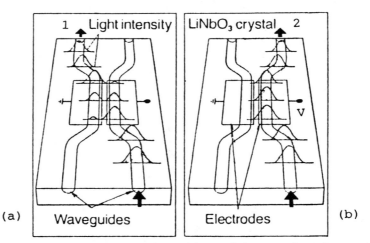

Figure 13.3 An integrated optical switch: (a) no voltage is applied, output appears at 1; (b) voltage is applied, output appears at 2. [After R. Haavind, *High Technol.*, v. 1 Nov./Dec.–August, p. 35, (1981).]

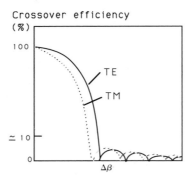

Figure 13.4 Crossover efficiency as a function of $\Delta\beta$

plete crossover for both waves. This means that when either the *TE* or *TM* mode attains a zero crossover, the other mode will still have a residual nonzero crossover. This is indicated clearly in Fig. 13.4, where the crossover efficiency of the component waves is plotted as a function of $\Delta\beta$ $[=(2\pi/\lambda)(n_1-n_2)]$. The side lobes indicate the non-clean crossover. Perfect crossover should exhibit no side lobes.

Splitting the electrodes and putting a voltage $+V$ on one and a voltage $-V$ on the other as shown in Fig. 13.5a has been attempted [7]. It is expected that a pair of voltages can be found that will allow desired operation between zero and full crossover. Again, as the differential indices are different for the *TE* and *TM* waves, this changed configuration is not sufficient to correct this difficulty.

Figure 13.5b shows another scheme. On the split electrodes, sepa-

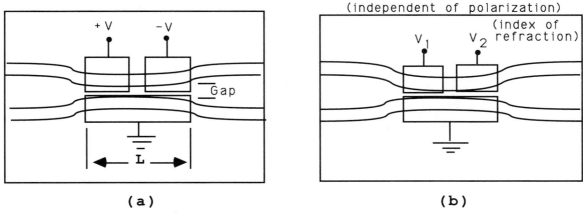

Figure 13.5 A method to ensure complete crossover: (a) a split-electrode configuration; (b) a weighted coupler.

rate voltages, V_1 and V_2, are independently applied to them as shown. By adjusting both voltages V_1 and V_2, complete elimination of the side lobes and thus full control of the crossover of power can be achieved [8]. With this scheme, known also as a *weighted coupler*, the crossover point, which occurs at $\Delta\beta\, L = \pi$, can be controlled solely by adjusting the voltages V_1 and V_2 without other critical dimensional changes. The switch works independent of the beam polarization. This is an advantage over other schemes that worked well only with polarization control.

Electrooptic directional couplers and switches have been reported with lightguides deposited on GaAs substrates. A pair of 6-μm guides separated by 7 μm with a coupling length of 8 mm have successively performed the switching action by passing an applied voltage (35 V) to the electrode. The driving power required is about 180 μW/MHz; a bandwidth of 100 MHz and an extinction ratio of 13 dB was reported [9].

The bandwidth of the device can be affected by the substrate material. A strongly birefringent material (e.g., LiNbO$_3$) gives a smaller bandwidth and vice versa. Strong birefringence also means that a shorter electrode will be sufficient to achieve power transfer of the modes.

Acoustooptic Transducer

Perhaps we should mention the acoustooptic transducer at this point because of its adaptability to miniaturized integrated devices. The basic principle of operation of this transducer is the same as that described in the bulk acoustooptic modulator in Chapter 8, except that this is a miniaturized version. Figure 13.6 shows the scheme of the design [10]. A thin layer of titanium is first diffused into a LiNbO$_3$

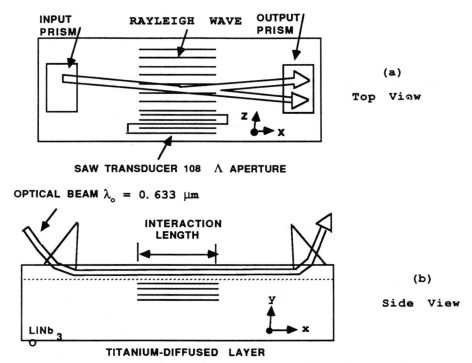

Figure 13.6 Acoustooptic transducer: (*a*) top view; (*b*) side view. [After C. S. Tsai, [10] © IEEE, 1979.]

substrate to form an optic waveguide. An array of interdigital electrodes are deposited directly onto the y-cut $LiNbO_3$ substrate. (See Fig. 13.6a, top view.) When an acoustic wave is fed into the interdigital structure, the acoustic wavefront presents a Bragg diffraction grating in the direction shown. Lightwaves of wavelength λ will then interact with the gratings. When the grating period Λ and the angle θ of the input beam satisfy the Bragg diffraction relation $\lambda = 2\Lambda \sin \theta$, beam modulation can be achieved.

Because the acoustic wave may be propagating, the diffracted light could be frequency-shifted by an amount equal to the acoustic frequency. The frequency shift can be downshifted or upshifted depending on the incident beam direction, or in other words, $\omega = \omega_o \pm \omega_m$, where ω_o and ω_m are the optical and acoustic angular frequencies, respectively.

Phase Shifter

In Chapter 8, where external modulation of optical power was introduced, we described a Pockel effect phase shifter that used an $LiNbO_3$ crystal with electrodes placed in proper orientation such that an applied voltage could affect the phase angle of the propagating lightwave through the crystal. There we commented that the required voltage was very high, on the order of a few thousand volts; thus, the device has only limited use in practical applications. Notice that in the phase-change equation

$$\Delta\phi = \frac{\pi}{\lambda}\, rn_1^3 V \frac{L}{D}$$

where r is the Pockel effect coefficient, n_1 the refractive index of the fiber, V the applied voltage, and L/D the length-to-separation ratio. In the bulk optical fiber modulator described earlier, the L/D ratio is about 10. Thus, π-radian phase-shifting requires a voltage of a few thousand volts. In integrated optics, this ratio can be made very large, say, 1000 or more. This fact makes the voltage required for a π-radian phase shift as low as a few volts, thus making a phase shifter easily realizable. A scheme of such a phase shifter is shown in Fig. 13.7a. Here the simple phase shifter has the dimensions of $L = 3$ mm and $D = 6$ μm; thus, $L/D = 500$. Since $r = 30.8 \cdot 10^{-12}$ m/V for this material and $n_1 = 2.203$, in order to achieve a π-radian phase shift, the voltage required is $V = 6.07$ V for $\lambda = 1.0$ μm.

When a pair of planar lightguides are deposited on an LiNbO$_3$ substrate with 3-dB directional couplers, with one arm equipped with a phase shifter and the other arm serving as a reference, an interferometric switch or modulator is formed as shown in Fig. 13.7b.

With no voltage applied on the phase shifter, no additional phase shift is introduced into the upper arm. A light signal injected at the input coupler divides itself into two branches equally and recombines at the output combiner with the output unaffected. If a π-radian of phase shift is achieved by applying a voltage on the phase shifter, the resulting output becomes zero. Thus, turning the applied voltage

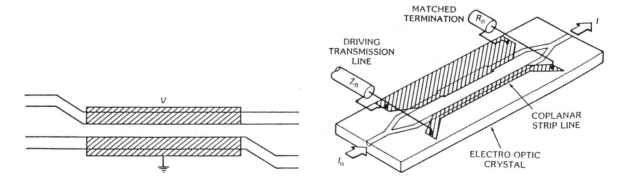

(a) A simple phase shifter   (b) An interferometric phase shifter

Figure 13.7 Integrated optical phase shifters: (a) a simple phase shifter; (b) an interferometric phase shifter. [After R. C. Alforness [11], © IEEE, 1982.]

on and off turns the light power on and off at the output; the device operates as an electronically controlled switch. The speed of switching is very fast because the rise time of the electrooptic effect is very fast, on the order of picoseconds. The actual rise time is then limited only by the electrode capacitance effect.

The same scheme can be used for a modulator. If the applied voltage on the electrodes is varied according to the modulation signal, the output light will be intensity-modulated accordingly [1, 3]. The bandwidth of the modulation is limited by the circuit parameter rather than by the rise time of the electrooptic effect. Since the electrode capacitance would be charged and discharged through a modulation cycle, the time constant would limit the bandwidth. This time constant could be in the nanosecond range. A figure of merit that has been used in electrooptic modulators is $V_\pi/\Delta f$, where V_π is the driving voltage required for a π-radian phase change and Δf is the bandwidth. For optimum modulation, this value should be $V_\pi/\Delta f = 0.5$ V/GHz at 0.633 μm and 1.5 V/GHz at 1.32-μm wavelength [11].

To relax the bandwidth-voltage limitation, a traveling-wave type of modulation is developed. This is shown in Fig. 13.7b. A radio-frequency (microwave) signal is applied to the electrodes over the lightguide to effect the phase shift. With a voltage of ± 3.5 V and a driving power of 120 mW, a 17-GHz bandwidth at 0.83-μm wavelength with an extinction ratio of 15 dB and $P/\Delta f = 7$ μW/MHz has been reported [12].

Filters

Filters are important components in many applications. In multiplex communication systems, good filters determine the total transmission capacity. The transmission characteristic of a good filter must satisfy two requirements, low transmission band attenuation and high interchannel isolation. Low transmission attenuation reduces the insertion loss and minimizes the system margin degradation. A passband loss of less than 0.1 dB is desirable. High isolation (at $\geq$ 30 dB) reduces interchannel cross-talk. In a wavelength-division multiplexing (WDM) system, the channel separation $\lambda_2 - \lambda_1$, $\lambda_3 - \lambda_2$ ... should be as small as permitted by the light source stability criterion. Another factor that affects the insertion loss is impedance matching. In bulk fiber optic communication systems, the two most commonly used filters are the interference filter and the diffraction grating filter [13, 14].

However, the added components make the multiplex system so complex and costly that serious considerations of other methods for increasing the system capacity, such as increasing the bit rate of digital communication, must be weighted.

In integrated optics, it is a different story. Efficient couplers and filters can be built inexpensively and in compact sizes. For illustration, we shall describe a polarization-independent filter that provides a narrowband, low passband attenuation, and high interchannel isolation with high compatibility with other couplers. A sketch of the filter is shown in Fig. 13.8. The basic principle of this filter is the $TE \leftrightarrow TM$ mode conversion. The device consists of a pair of planar titanium indiffused optical waveguides on an LiNbO$_3$ substrate. A pair of guided electrodes are deposited on top of the lightguide. The fingers on the electrodes play an important role in deciding the wavelength separation of the system. Since LiNbO$_3$ is a highly birefringent material, a lightwave usually has two polarizations, TE and TM components. When the TE and TM waves encounter a birefringent material, the waves see different refractive indices, depending on the crystal orientation. In a z-cut LiNbO$_3$ crystal, the differential index is about 0.09 between the TE and TM waves. The waves are thus traveling in the guide at different velocities. At a particular wavelength there is a periodic distance at which the two components will be in phase with each other. At this wavelength, the TE mode from one waveguide will be transferred to the other, and the TM mode wave, in the opposite direction. This enables only the TE mode (or the TM mode) at this particular wavelength to be transferred to an output waveguide. The electrodes with proper finger spacing and applied voltage act to enhance the switching. The selection or rejection properties of this filter for a particular wavelength is very sharp and adjustable with the applied voltage. Passband bandwidth as narrow as 5–50 Å has been demonstrated [11].

The filter shown in Fig. 13.8 can easily be converted into a tunable filter. If the two waveguides of the coupler are designed with slightly different sizes and different refractive indices, and when their amplitudes are independently plotted against the inverse of λ of the two waveguides, the curves will intersect at a given wavelength. This

Polarization independent filter
(using TE ◄──► TM conversion)

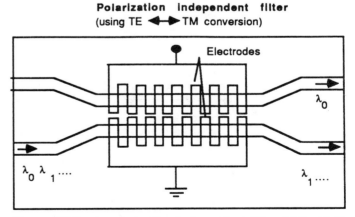

Figure 13.8 A polarization-independent filter. [After R. C. Alforness, *IEEE* [11] *Microwave Theory Tech.* (1982) © IEEE, 1982.]

means that at this particular wavelength, the waves in separate wave-guides are traveling at the same speed. Crossover will take place between these two lightguides. The wavelength at which this event takes place can be electrically shifted by the applied voltage (a few volts), thus making the selection of passband electronically tunable.

Multiplexing Couplers

Multiport optical couplers are the basic interconnection elements for power-distribution networks that employ optical fiber. The taper and star couplers were introduced in Chapter 5. The simplest taper coupler consists of two fibers (with their jacketing materials removed) that are spot-fusing together side-by-side to form a tapered structure. If more than two fibers are fused together, a multiport star coupler is formed. It can be used to distribute light power to many fiber networks [15]. The disadvantage of the star coupler is that the designer is unable to control the power distribution among the terminals. The power distribution is sensitive to mode excitation at the fused junction and often have excessive intermodal noise. This simple coupler is not suitable for a WDM coupler.

The wavelength-division multiplexer needs a channel separation of about 30 nm in wavelength. A low passband attenuation of 0.1 dB is required to minimize the insertion loss. A sharp cutoff slope of the channel transmission characteristic is also required to achieve sufficient interchannel isolation of 30 dB. Only interference-filter-type or the diffraction-grating-type couplers are suitable for this application [16].

Interference-Filter-Based WDM Couplers

A simple interference filter coupler uses the fiber end faces as an interference-filter. Several designs are possible as exemplified in Fig. 13.9. Figure 13.9a shows an all-fiber WDM coupler [17]. Figure 13.9b shows WDM couplers using graded-index rod (GRIN-rod) lenses and interference filters for multiplexing and demultiplexing [18]. All these couplers have one thing in common; they are used to combine or separate two wavelengths. Insertion loss of these designs could be very large if the airgaps were large. A seven-port polygonal prism WDM coupler is shown in Fig. 13.10 [19]. A seven-port polygonal prism with polished surfaces is the central structure of this coupler. To each port, a GRIN-rod lens, together with a spacer and an optical filter, is epoxied to the polished surface. The polygonal surfaces are cut such that the reflections from each surface are directed to the next surface in succession as indicated by the arrows. The structure has small interlens spacing to minimize insertion loss and possesses high mechanical stability and is easy to assemble.

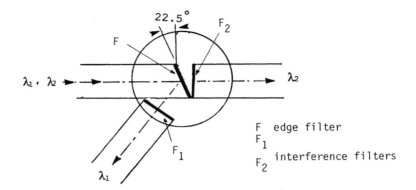

(a) A simple all fiber WDM coupler

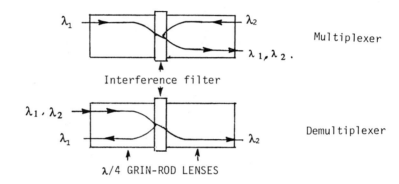

(b) Multiplexer and demultiplexer using λ/4 GRIN-ROD lenses and interference filter

Figure 13.9   Interference filter couplers (a) a simple all-fiber WDM coupler [17]; (b) multiplexer and demultiplexer using λ/4 GRIN-rod lenses and interference filter [18].

Diffraction-Grating-Based WDM Couplers

We have noticed that in the interference-filter-type coupler, each channel requires a separate selection element, making the device very bulky. The diffraction-grating-type coupler has the intrinsic ability to process all wavelength channels in parallel and is therefore very compact. Figure 13.11 shows such a coupler [20, 21]. Inputs and outputs are taken from the same side of a GRIN-rod lens, which is used as the focusing element. On the other end of the lens is the grating plate set at an angle from the end face. An airgap separates them. Insertion loss of 2.4 dB and −30 dB cross-talk at channel spacing of 27 nm have been reported. Channel numbers up to 12 have been reported with a large-diameter graded-index rod lens. The coupler maintains similar optical performance in all channels. The grating uses an anisotropic

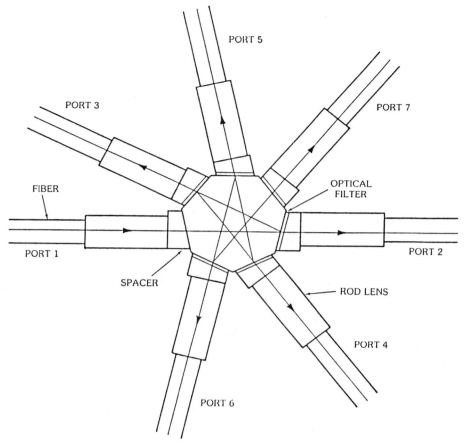

Figure 13.10 A seven-port polygonal prism coupler. [After S. Ishikawa *et al.*
[19] *Int. Soc. Opt. Eng.* **417,** 44 (1983).]

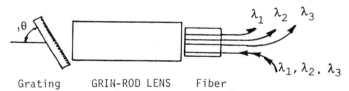

Figure 13.11 A difraction-grating coupler [20, 21].

etching of a silicon substrate instead of the commercial pressed gratings to achieve high optical efficiency and environmental stability.
Both are important for reliable operation of WDM couplers.

Hybrid and Monolithic Integration

The aim of monolithic integrated optics is to integrate all functions of
optical fiber signaling processes on a single chip as in microelectronic

integrated circuits. However, integrated optics have one more demand than in microelectronic integrated circuitry; the substrate would also provide lightwave guidance. Lightguides require the use of dielectric materials. Substrates suitable for laser and photodiode fabrication involve semiconductors. To integrate the passive functions, such as lightguide, coupler, and switch, with active functions such as light generation and detection, one needs to find a workable substrate material that satisfies both requirements. Unfortunately, passive lightwave components work best with insulating crystals. Crystals such as $LiNbO_3$, $LiTaO_3$, ADP–KDP, and $Eu_3Ga_5O_{12}$ are particularly useful because of their spectral properties with electrooptic, acoustooptic, or magnetooptic effect. On these substrates, no light generator or detector can be built successfully. On the other hand, the most useful substrates for producing lasers and photodetectors are GaAs and other group III–V semiconductor compounds. Although successful lightguides have been built with pure GaAs layers embedded in n-type GaAs substrate, the problem of optical guidance has not yet been adequately solved. In short, suitable substrate material has not yet been found, and total integrated optics may still be years away. All we can report in this section are partial integrations. Typical examples are (1) integration of a laser and a driving circuit, (2) integrated photodetector and field-effect transistor (FET), and (3) integrated preamplifiers.

Integration of a Laser and a Driving Circuit

Direct modulation of a laser requires a driver circuit to provide a fast but stable operating variable source with adequate power. With integrated lasers and driving circuits, great technical as well as economical advantages can be gained. Integration reduces the circuit connections, thus reducing the circuit time constant and enhancing the bandwidth. It also increases the system reliability and reduces cost.

In this section we report a monolithic integration of a multilayer quantum-well (MQW) laser with a FET driving circuit on a semiinsulating GaAs substrate [23]. Figure 13.12a is the schematic structural diagram of the integrated laser diode. MQW lasers have multiple active layers in the structure well separated by barrier layers. They are used because they give a better confinement factor, that is, a higher gain at a lower carrier density than do single quantum-well (SQW) lasers. The MQW active layers are formed in a groove on the substrate by using the metalloorganic chemical-vapor deposition (MOCV) method. After the formation of the laser structure, the multilayers grown on the outside of the groove are etched off to the semiinsulating GaAs surface. Then an integrated electronic circuit pattern is processed for the FET and resistors. The exposed n^+–GaAs layer in the groove is used as the n-side electrode of the laser and is connected to the electronic circuit with wires. The front- and back-side facets of

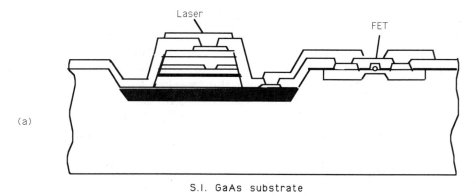

(a)

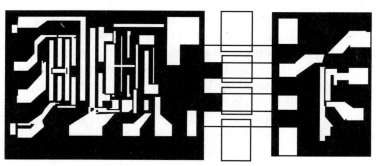

Figure 13.12   Integrated MQW laser and driving circuit: (*a*) structure; (*b*) chip.

the laser cavity are formed by ion-beam etching and cleaving. The final chip is shown in Fig. 13.12*b*.

The laser cavity length is 200 μm. A driving current of more than 15 mA at an input pulse amplitude of 0.8 V_{pp} (peak-to-peak) can be obtained. The driving current can also be adjusted by changing the gate voltage of one of the FETs. Altogether, 10 FETs and resistors are integrated in the circuit. All the FETs have a gate length of 1.5 μm. A transconductance as high as 60 mA/V_{mn} has been obtained. The chip size is 1.9 mm $\times$ 0.8 mm, and the threshold current of the laser is 40 mA at room temperature.

Monolithic Integration of Photodiode and a GaAs Preamplifier

Monolithic integration of a photodiode with a FET preamplifier has the advantage of increasing the sensitivity and speed primarily by reducing the parasitic reactance that plagues the performance of an ordinary discrete receiver. In this section we describe two types of devices. The first one is a monolithically integrated p–i–n/FET photoreceiver and the second, an MSM/FET integrated receiver. They differ only in fabrication technology. The former tries to integrate a

p–i–n photodiode (vertical) with a planar FET device. The processes used to separate these devices and then reconnect them together are rather tedious. The second structure is much more suitable for monolithic integration because both the MSM photodiode and FET are planar devices.

The schematic diagram of a p–i–n/FET structure is shown in Fig. 13.13a. The AlGaAs/GaAs multilayer structure is grown on the (100)-oriented semi-insulating GaAs structure. A p–i–n diode is processed vertically as shown on the left-hand side of the diagram. An n_+–GaAs layer contact (Au/AuGe) and a zinc diffused p-layer contact separated by an i-layer (n^-–GaAs with $n = 5 \times 10^{14}/cm^3$) forms the vertical p–i–n diode. On the left-hand side, a FET is formed horizontally whose source side is metallically bonded to the p–i–n diode by an aluminum interconnection. The substrate of the FET is isolated from the rest of the substrate by a high-resistivity layer (AlGaAs–HR layer). A photomicrograph of this device is shown in Fig. 13.13b. The metallic connection is clearly visible. The reported device has a photosen-

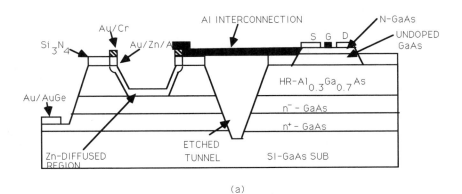

(a)

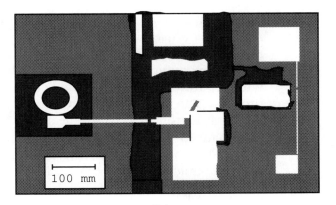

(b)

Figure 13.13 p–i–n/FET intergration: (a) structure; (b) chip. [After S. Miura et al. [23] *IEEE Device* (1983), © IEEE, 1983.]

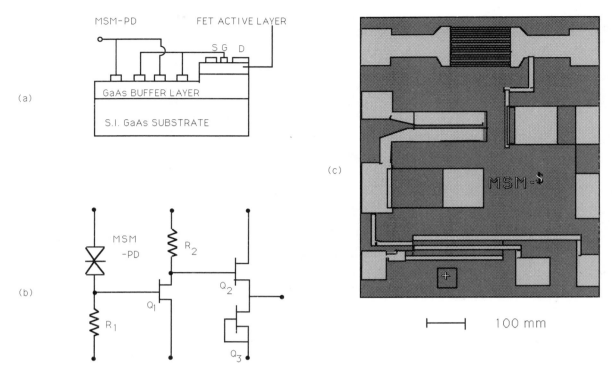

Figure 13.14   MSM/FET. (a) structure; (b) circuit; (c) chip. [After M. Ito *et al.* [24] *IEEE Device* (1984) © IEEE, 1984.]

sitive area with a diameter of 100 μm, and the FET channel width, channel length, and the gate length are 80.6 and 1.8 μm, respectively. With a reverse-bias voltage of 5 V, the dark current is 7×10^{-10} A, and the current amplification is approximately 10 [23].

The schematic cross-sectional view of an integrated MSM/FET is shown in Fig. 13.14a. The simplicity of this schematic diagram compared to the previous one is quite obvious. On a semi-insulator (100)-oriented GaAs substrate, a buffer layer of undoped GaAs (3 μm in thickness) is deposited. The density of this layer is 10^{17}/cm³. AuGa/Au films are used as n-type ohmic contacts for the source and drain. A mesa is etched to a depth of 0.5 μm for the FET active layer. Source and drain contacts are then deposited. The MSM diode is evaporated with aluminum film. The gate is also formed during the evaporation process. The carrier concentration of the buffer layer varies in the depth direction. It is 10^{15}/cm³ at the surface of the MSM diode and less than 10^{14}/cm³ at 1-μm depth. The circuit diagram is shown in Fig. 13.14b, and the photomicrograph is shown in Fig. 13.14c.

The characteristic of the MSM shows a leakage current of 5 μA when dark at a 15-V reverse-bias voltage. The photocurrent shows linear relations with the input power at 15 V. The external photosensitivity is about 2.2 A/W, indicating that a possible photocurrent gain is present. The mechanism responsible for this gain has not yet been explained. With a 50-Ω termination, a 3-dB cutoff frequency at 1 GHz was recorded.

When the FET pinch-off voltage is recorded as 3.5 V, a transconductance g_m of 10 ms (50 ms/mm) and a source drain saturation current of 30 mA are observed. The sensitivity of this receiver is about 26 mV/μW, which is much higher than that for the $p-i-n$/FET device previously reported [24].

In a recent paper [24] these authors proposed a hybrid GaAs integrated optical transimpedance amplifier. When combined with an external $p-i-n$ photodiode, this device operates at 140 Mbit/s and demonstrates a -38-dBm sensitivity for a 10^{-9} error rate.

A monolithic planar, linear 10-element GaAs detector–amplifier arrays using a GaAs/AlGaAs/GaAs epitaxial structure on semi-insulating GaAs has been reported recently by Anderson et al. [25]. Operating at 0.84-μm wavelength, it has a rise time and a fall time of 650 ps and 1.1 ns, respectively. The sensitivity of single, discrete detector–amplifier channels was better than -34 dBm.

Conclusion

Much research is in progress in integrated optics. The examples given in this chapter represent only a few ideas. There are two chief obstacles to large-scale integrated optics.

First, active components such as lasers and photodiodes are built primarily on heterostructures of GaAs and other group III–V semiconductor compounds, whereas passive components such as lightguides, couplers, switches, and filters can more effectively be fabricated on other materials that are less lossy and have special effects (electrooptic, acoustooptic or magnetooptic effects) with which to take advantage.

Second, the density of integrated optics can never approach even the small-scale microelectronic circuit integration because the electromagnetic field effect with which the lightwave operates could present a limit to the minimum length and a minimum separation between elements for a coupler to be effective.

Efforts to use GaAs and other group III–V semiconductor compounds as a common substrate for integration of both active and passive components are still continuing. However, the return is less encouraging. For a lightguide, the carrier density of GaAs should be low. But for laser operation, a much higher carrier density is required. These requirements are difficult to compromise.

In the foreseeable future, a logical solution could be a "hybrid integrated optics," that is, the integration of all passive components together on their favorable substrate and the connection of this to another unit containing the integration of all active components on their desirable substrate.

An alternative way is to search for a better way to build a new substrate that could contain various layers of materials for different functions. Here, separate components could be built on their most favorite substrates and then coupled together internally. But this is a big order for material and metallurgical engineers and it involves many interrelated problems. Only time will tell if there will be true success.

A recent research paper describing a new type of optoelectronic device appeared in the *IEEE Circuit and Devices Magazine*, May 1989 [28]. Thin films of crystalline organic semiconductor are deposited onto inorganic semiconductor substrate to form heterojunctions and devices that can function as photodetectors and transistors at optical frequencies. It may promise a new generation of versatile optoelectronic components to electronic engineering applications.

References

1. D. B. Anderson, *Optical and Electro-optical Information Processing*, pp. 221–234. M.I.T. Press, Cambridge, Massachusetts, 1965.
2. A. Miller, D. A. B. Miller, and S. D. Smith, Dynamic nonlinear optical processes in semiconductors. *Adv. Phys.* **30**, 697–800 (1981).
3. T. Tamir, ed., *Integrated Optics*, 2nd ed. Springer-Verlag, Berlin, 1979.
4. M. K. Barnoski, ed., *Introduction to Integrated Optics*. Plenum, New York, 1973.
5. L. Levi, *Applied Optics*, Vol. 2. Wiley, New York, 1980.
6. E. A. J. Marcatili, Dielectric rectangular waveguides and directional coupler for integrated optics. *Bell Syst. Tech. J.* **48**, 2071–2101 (1969).
7. M. Papuchon, Y. Combemele, X. Mehiew, D. B. Ostrowsky, L. Reiber, A. M. Roy, B. Sejourne, and M. Werner, Electrically switched optical directional coupler: COBAR. *Appl. Phys. Lett.* **27**, No. 289 (1975).
8. H. Kogalnik and R. V. Schmidt, Switched directional couplers with alternating $\Delta\beta$. *IEEE J. Quantum Electron.* **QE-12**, 396 (1976).
9. J. C. Campbell, F. A. Blum, D. W. Shaw, and K. L. Lawley, GaAs electro-optic directional-coupler switch. *Appl. Phys. Lett.* **27**, 202 (1975).
10. C. S. Tsai, Guided-wave acousto-optic Bragg modulator for wide-band integrated optic communication and signal processing. *IEEE Trans. Circuits Syst.* **CAS-26**(12), 1072–1098 (1979).
11. R. C. Alforness, Waveguide electro-optic modulators. *IEEE Trans. Microwave Theory Tech.* **MTT-30**(8), 1112–1137 (1982).
12. C. M. Gee, D. Thurmond, and H. W. Yen, 17 GHz bandwidth electron optic modulator. *Appl. Phys. Lett.* **43**(11), 998–1000 (1983).
13. S. P. Basdettini, Optical filters for wavelength division multiplexing. *Proc. SPIE—Int. Soc. Opt. Eng.* **417**, 67 (1983).
14. W. J. Tomlinson and C. Lin, Optical wavelength division multiplexer for 1-1.4 μm spectral region. *Electron. Lett.* **14**, 345 (1978).
15. J. C. Williams, S. E. Goodman, and R. L. Coon, Fiber optic subsystem considerations of multimode star coupler performance. *Proc. Conf. Opt. Fiber Commun.*, New Orleans, *1984*, Paper WC6 (1984).
16. E. Miyauchi, T. Iwama, H. Nakajima, N. Tokoyo, and K. Terai, Compact wavelength multiplexer using optical fiber pieces. *Opt. Lett.* **5**, 321 (1980).
17. M. Rode and E. Weidel, Compact and rugged all-fibre coupler for wavelength division multiplexing. *Electron. Lett.* **18**, 898 (1982).
18. K. Kobayashi, R. Ishikawa, K. Minemura, and S. Sugimoto, Microscoptic devices for fiber optic communications. *Fiber Integr. Opt.* **2**, 1 (1979).

19. S. Ishikawa, F. Matsumura, K. Takahashi, and K. Okuno, High stability wavelength division multi/demultiplexer with polygonal structure. *Proc. SPIE—Int. Soc. Opt. Eng.* **417,** *FiberOpt. Multiplex. Modul.,* p. 44 (1983).

20. W. J. Tomlinson and G. D. Aumiller, Optical multiplexer for multimode fiber transmission systems. *Appl. Phys. Lett.* **31,** 179 (1977).

21. W. J. Tomlinson and C. Lin, Optical wavelength-division-multiplexer for the $1-1.4$ μm spectral region. *Electron. Lett.* **14,** 345 (1978).

22. M. Hirao, S. Yawashita, T. P. Tanaka, and M. Nakano, Monolithic integration of a laser and driving circuit in GaAs/GaAlAs systems for high speed optical transmission, *Integr. Opt., Proc. Eur. Conf., ECIO '85, 3rd,* Berlin, *1985.*

23. S. Miura, O. Wada, H. Haniaguchi, M. Ito, M. Makinchi, K. Nakai, and T. Saturai, A monolithically integrated AlGaAs/GaAs p-i-n/FET photoreceiver by MOCVD. *IEEE Electron Device Lett.* **EDL-4**(10), 375−376 (1983).

24. M. Ito, O. Wada, K. Nakai, and T. Sakarai, Monolithic integration of a metal-semiconductor-metal photodiode and GaAs preamplifier, *IEEE Elec. Dev Lett.* **EDL-5,** 531−532 (1984).

25. Y. Archambault, D. Pavlidis, and J. P. Gult, GaAs monolithic integrated optical preamplifier. *IEEE J. Lightwave Technol.* **LT-5**(3), 355−366 (1987).

26. G. W. Anderson, N. A. Papanicolaou, D. I. Ma, I. A. G. Mack, J. A. Modolo, F. J. Kub, X. W. Young, Jr., P. E. Thompson, and J. B. Boos, Planar, linear GaAs detector-amplifier array with an insulating AlGaAs spacing layer between the detector and transistor layers. *IEEE Electron Device Lett.* **EDL-9**(10), 550−552 (1988).

27. S. R. Forrest, Organic-to-inorganic semiconductor heterojunctions: Building blocks for the next generation of optoelectronic devices? *The Circuits and Devices Magazine,* **5,** 33 (1989).

28. Forrest, S. R., Organic-to-Inorganic Semiconductor Heterojunctions: Building Blocks for the Next Generation of Optoelectronic Devices. *IEEE Circuit and Devices Magazine,* **5**(3), 33−37 (1989).

Index